Valentina Zharkova, Lakhmi C. Jain (Eds.)

Artificial Intelligence in Recognition and Classification
of Astrophysical and Medical Images

Studies in Computational Intelligence, Volume 46

Editor-in-chief
Prof. Janusz Kacprzyk
Systems Research Institute
Polish Academy of Sciences
ul. Newelska 6
01-447 Warsaw
Poland
E-mail: kacprzyk@ibspan.waw.pl

Further volumes of this series
can be found on our homepage:
springer.com

Vol. 28. Brahim Chaib-draa, Jörg P. Müller (Eds.)
*Multiagent based Supply Chain
Management,* 2006
ISBN 978-3-540-33875-8

Vol. 29. Sai Sumathi, S.N. Sivanandam
*Introduction to Data Mining and its
Application,* 2006
ISBN 978-3-540-34350-9

Vol. 30. Yukio Ohsawa, Shusaku Tsumoto (Eds.)
Chance Discoveries in Real World Decision Making,
2006
ISBN 978-3-540-34352-3

Vol. 31. Ajith Abraham, Crina Grosan, Vitorino
Ramos (Eds.)
Stigmergic Optimization, 2006
ISBN 978-3-540-34689-0

Vol. 32. Akira Hirose
Complex-Valued Neural Networks, 2006
ISBN 978-3-540-33456-9

Vol. 33. Martin Pelikan, Kumara Sastry, Erick
Cantú-Paz (Eds.)
*Scalable Optimization via Probabilistic
Modeling,* 2006
ISBN 978-3-540-34953-2

Vol. 34. Ajith Abraham, Crina Grosan, Vitorino
Ramos (Eds.)
Swarm Intelligence in Data Mining, 2006
ISBN 978-3-540-34955-6

Vol. 35. Ke Chen, Lipo Wang (Eds.)
Trends in Neural Computation, 2007
ISBN 978-3-540-36121-3

Vol. 36. Ildar Batyrshin, Janusz Kacprzyk, Leonid
Sheremetor, Lotfi A. Zadeh (Eds.)
*Preception-based Data Mining and Decision Making
in Economics and Finance,* 2006
ISBN 978-3-540-36244-9

Vol. 37. Jie Lu, Da Ruan, Guangquan Zhang (Eds.)
E-Service Intelligence, 2007
ISBN 978-3-540-37015-4

Vol. 38. Art Lew, Holger Mauch
Dynamic Programming, 2007
ISBN 978-3-540-37013-0

Vol. 39. Gregory Levitin (Ed.)
*Computational Intelligence in Reliability
Engineering,* 2007
ISBN 978-3-540-37367-4

Vol. 40. Gregory Levitin (Ed.)
*Computational Intelligence in Reliability
Engineering,* 2007
ISBN 978-3-540-37371-1

Vol. 41. Mukesh Khare, S.M. Shiva Nagendra (Eds.)
*Artificial Neural Networks in Vehicular Pollution
Modelling,* 2007
ISBN 978-3-540-37417-6

Vol. 42. Bernd J. Krämer, Wolfgang A. Halang (Eds.)
Contributions to Ubiquitous Computing, 2007
ISBN 978-3-540-44909-6

Vol. 43. Fabrice Guillet, Howard J. Hamilton (Eds.)
Quality Measures in Data Mining, 2007
ISBN 978-3-540-44911-9

Vol. 44. Nadia Nedjah, Luiza de Macedo
Mourelle, Mario Neto Borges, Nival Nunes
de Almeida (Eds.)
Intelligent Educational Machines, 2007
ISBN 978-3-540-44920-1

Vol. 45. Vladimir G. Ivancevic, Tijana T. Ivancevic
*Neuro-Fuzzy Associative Machinery for
Comprehensive Brain and Cognition Modelling,*
2007
ISBN 978-3-540-47463-0

Vol. 46. Valentina Zharkova, Lakhmi C. Jain (Eds.)
*Artificial Intelligence in Recognition and
Classification of Astrophysical and Medical
Images,* 2007
ISBN 978-3-540-47511-8

Valentina Zharkova
Lakhmi C. Jain (Eds.)

Artificial Intelligence
in Recognition
and Classification
of Astrophysical
and Medical Images

With 137 Figures and 11 Tables

 Springer

Prof. Valentina Zharkova
Department of Computing and Department of Cybernetics
Bradford University
Bradford BD7 1DP
United Kingdom
E-mail: v.v.zharkova@brad.ac.uk

Prof. Lakhmi C. Jain
Knowledge-Based Intelligent Engineering Systems Centre
School of Electrical and Information Engineering
University of South Australia
Mawson Lake Campus
South Australia, SA 5095
Australia
E-mail: Lakhmi.jain@unisa.edu.au

Library of Congress Control Number: 2006934859

ISSN print edition: 1860-949X
ISSN electronic edition: 1860-9503
ISBN-10   3-540-47511-7 Springer Berlin Heidelberg New York
ISBN-13   978-3-540-47511-8 Springer Berlin Heidelberg New York

Springer is a part of Springer Science+Business Media
springer.com
© Springer-Verlag Berlin Heidelberg 2007

Cover design: deblik, Berlin
Typesetting by the editors and SPi using a Springer LaTeX macro package
Printed on acid-free paper SPIN: 11671244   89/SPi   5 4 3 2 1 0

Preface

During the past decade digital imaging has significantly progressed in all imaging areas ranging from medicine, pharmacy, chemistry, biology to astrophysics, meteorology and geophysics. The avalanche of digitized images produced a need for special techniques of processing and knowledge extraction from many digital images with minimal or even without human interaction. This has resulted in a new area in the digital processing called pattern recognition that becomes increasingly necessary owing to a growing number of images to be processed. The first applications of pattern recognition techniques were for the analysis of medical X-rays and MMR images that enabled the extraction of quantified information in terms of texture, intensity and shape and allowed to significantly improve a diagnosis of human organs. These techniques were significantly developed over the last few years and combined feature detection and classification by using region based and artificial intelligence methods. By using growing databases of medical images processed with pattern recognition and classification techniques, one can produce fast and consistent diagnosis of diseases based on the accumulated knowledge obtained from many other similar cases from the stored databases.

The use of CCD cameras for astrophysical instruments on the ground and space produce digitized images in various fields of astrophysics. In the past decade, many space and ground-based instruments provide large numbers of digitized images of the night skies and of the Sun, our closest star. These images provide more and more valuable knowledge about the evolution of celestial bodies and the physical processes occurring in them. This ample information can be processed with relatively new methods of feature recognition and classification developed in other imaging fields. With every new instrument and space mission, the archives of digital images are growing enormously in size. This imposes increasing requirements for the development of automated pattern recognition methods in applications to these archives.

The progress in digital imaging led to the application of pattern recognition techniques developed for medical and biomedical image to astrophysical images. They have proven to be the revolutionary way for the data processing in astrophysics and solar physics. In spite of difference between the images in medicine, astrophysics and solar physics, there are many common approaches and techniques that are applicable to them all while some alterations are required to accommodate differences in the specific data.

Unlike features in medical images that are, in a general way, understood by wider range of readers as related to a human body, astrophysical and solar images contain the information about physical processes in the stars and the Sun that often affect the Earth and many aspects of human lives. These processes can be also uncovered with the pattern recognition techniques similar to those applicable to medical images but modified to accommodate the differences in recognized patterns. This book makes use of domain knowledge in astrophysical and medical image processing areas by employing the techniques used for general object recognition for an automated recognition of the features on astrophysical and medical images.

The book is intended for astrophysicists, medical researches, engineers, research students and technically aware managers in the Universities, Astrophysical Observatories, Medical Research Centres working on the processing of large archives of astrophysical or medical digital images. This book can be used as a text book for students of Computing, Cybernetics, Applied Mathematics and Astrophysics.

We are indebted to the authors and the reviewers for their wonderful contribution. Special thanks go to Berend Jan van der Zwaag, Rajesh Kannah, and Nandini Loganathan for their excellent help in the preparation of the camera ready copy. The editorial assistance provided by Springer is acknowledged.

Editors

Foreword

The interest in classification of objects into categories is as old as civilisation itself. It doesn't matter whether we are classifying types of animal species, rocks, weather patterns, or more modern applications such as automated character recognition, segmentation of retail customers, or identifying which stocks are likely to increase in value over a certain period of time - the task is still the same: classification is the search for common features in the data to enable us to group the data into distinct categories or classes. Once these similar features have been found the class of similar data can be labelled with a common label.

The techniques used in this book fall into two classes themselves. The first are the set of artificial intelligence methods used for pattern recognition and classification across a broad set of application domains. These include neural networks, decision trees, and more sophisticated spectral methods such as wavelet transforms and fractal analysis. The second class of techniques used relate to the specific application domain that is the focus of this book: images arising in medical and astrophysical applications. As soon as we start to apply classification techniques to images, we require a whole suite of specific techniques to assist with the unique data preparation and processing issues that images present. Thus, this book also discusses techniques for image standardization, enhancement and segmentation, as well as source separation. The role of all of these techniques in computer-aided recognition and diagnosis (in the case of the medical images) is explored. The collection of questions and solutions is a useful addition that makes this book ideal as a textbook.

Of all the application domains that these techniques are applicable to, none have more significance for the human race than recognising and classifying images belonging to our bodies and our universe. While there are plenty of volumes tackling pattern recognition problems in finance, marketing, and the like, I commend the editors and the authors for their efforts to tackle the big questions in life, and their excellent contributions to this book.

Professor Kate Smith-Miles
Head, School of Engineering and Information Technology
Deakin University
Australia

Contents

1 Introduction to Pattern Recognition and Classification in Medical and Astrophysical Images....................................1
V.V. Zharkova and L.C. Jain

    1.1   Introduction to Pattern Recognition in Medical Images....1
    1.2   Introduction to Pattern Recognition in
            Astrophysical and Solar Images 5
            1.2.1   Astrophysical Features 5
            1.2.2   Solar Features...6

2 Image Standardization and Enhancement 19
S.S. Ipson, V.V. Zharkova and S.I. Zharkov

    2.1   Digital Image Distortions and Standardization
            of Shape and Intensity 19
            2.1.1   Geometrical Standardization 19
            2.1.2   Intensity Standardization....................................36
    2.2   Digital Image Enhancement and Morphological
            Operations...42
            2.2.1   Image Deblurring 42
            2.2.2   Morphological Operations...................................45

3 Intensity and Region-Based Feature Recognition in Solar Images 59
*V.V. Zharkova, S.S. Ipson, S.I. Zharkov and
Ilias Maglogiannis*

    3.1   Basic Operations in Recognition Techniques.................59
            3.1.1   Histograms 59
            3.1.2   Intensity Thresholds 63
    3.2   Intensity-Based Methods for the Solar Feature
            Detection...66
            3.2.1   Threshold Method 66
            3.2.2   Histogram Methods...67
            3.2.3   Simulated Annealing..69

3.3   Edge-Based Methods for Solar Feature
        Detection.. 72
        3.3.1   Lagrangian of Gaussian (LgOG) Method........... 72
        3.3.2   Canny Edge Detector 74
        3.3.3   Automatic Sunspot Recognition in Full-Disk
                Solar Images with Edge-Detection Techniques.. 76
3.4   Region-Based Methods for Solar Feature Detection 91
        3.4.1   Introduction 91
        3.4.2   Segmentation and Seed Selection 91
        3.4.3   Automated Region Growing Procedure
                for Active Regions.. 94
        3.4.4   Region Growing Procedure
                for Filaments.. 108
        3.4.5   Region Growing Methods for Different Classes
                of Features 110
3.5   Other Methods for Solar Feature Detection................... 112
        3.5.1   Bayesian Inference Method for Active
                Region Detection 112
        3.5.2   Detection of Flares on H_α Full-Disk
                Images (Veronig Method)................................. 114
        3.5.3   Detection of Coronal Mass Ejections............... 116
        3.5.4   Magnetic Inversion Line Detection.................. 118
3.6   Skin Lesion Recognition with Region Growing
        Methods 123
        3.6.1   Introduction 123
        3.6.2   Building a Computer Vision System
                for the Characterization of Pigmented
                Skin Lesions 125
        3.6.3   Reported Experimental Results
                from Existing Systems...................................... 138
        3.6.4   Case Study Application..................................... 140
        3.6.5   Conclusions 143

4   **Advanced Feature Recognition and Classification Using
        Artificial Intelligence Paradigms** **151**
        *V. Schetinin, V.V. Zharkova, A. Brazhnikov, S. Zharkov,
        E. Salerno, L. Bedini, E.E. Kuruoglu, A. Tonazzini,
        D. Zazula, B. Cigale and H. Yoshida*

4.1   Neural-Network for Recognizing Patterns
        in Solar Images 151
        4.1.1   Introduction 151

4.1.2	Problem Description	152
4.1.3	The Neural-Network Technique for Filament Recognition	153
4.1.4	Training Algorithm	156
4.1.5	Experimental Results and Discussion	159
4.2	Machine Learning Methods for Pattern Recognition in Solar Images	160
4.2.1	Introduction	161
4.2.2	Neural-Network-Based Techniques for Classification	161
4.2.3	Neural-Network Decision Trees	165
4.2.4	Conclusion	168
4.3	The Methodology of Bayesian Decision Tree Averaging for Solar Data Classification	168
4.3.1	Introduction	169
4.3.2	The Methodology of Bayesian Averaging	171
4.3.3	Reversible Jumps Extension	173
4.3.4	The Difficulties of Sampling Decision Trees	175
4.3.5	The Bayesian Averaging with a Sweeping Strategy	176
4.3.6	Performance of the Bayesian Decision Tree Technique	179
4.3.7	The Use of the Bayesian Decision Tree Techniques for Classifying the Solar Flares	180
4.3.8	Confident Interpretation of Bayesian Decision Tree Ensembles	183
4.3.9	Conclusions	188
4.3.10	Questions and Exercises	189
4.4	The Problem of Source Separation in Astrophysical Images	200
4.4.1	Introduction	200
4.4.2	A Linear Instantaneous Mixture Model for Astrophysical Data	202
4.4.3	The Source Separation Problem	204
4.4.4	Source Models Parametrizing the Mixing Matrix	205
4.4.5	Noise Distribution	208
4.4.6	Conclusion	209
4.5	Blind and Semi-Blind Source Separation	209
4.5.1	Introduction	210

| | 4.5.2 | Independent Component Analysis Concepts | 212 |

4.5.2   Independent Component Analysis
        Concepts 212
4.5.3   Application-Specific Issues............................ 213
4.5.4   Totally Blind Approaches 218
4.5.5   Independent Factor Analysis.......................... 221
4.5.6   Bayesian Source Separation Using
        Markov Random Fields 227
4.5.7   A Semi-Blind Second-Order Approach 231
4.5.8   Particle Filtering... 240
4.5.9   Future Trends 245
4.5.10 Conclusion.. 246

4.6   Intelligent Segmentation of Ultrasound Images
      Using Cellular Neural Networks 247
      4.6.1   Introduction 248
      4.6.2   Basics of Cellular Neural Networks................ 249
      4.6.3   CNN Simulation... 256
      4.6.4   Training of CNNs....................................... 262
      4.6.5   Segmentation of Overian Ultrasound Images .. 278
      4.6.6   Conclusion.. 283

4.7   Computer-Aided Diagnosis for Virtual
      Colonoscopy 302
      4.7.1   Introduction 303
      4.7.2   CAD Scheme for the Detection
              of Polyps 305
      4.7.3   Detection of Polyp Candidates...................... 310
      4.7.4   Characterization of False Positives 316
      4.7.5   Discrimination from False Positives
              and Polyps.. 319
      4.7.6   Performance of CAD in the Detection
              of Polyps 322
      4.7.7   Improvement of Radiologists' Detection
              Performance by use of CAD.......................... 323
      4.7.8   CAD Pitfalls.. 325
      4.7.9   Conclusion.. 326

**5   Feature Recognition and Classification Using
     Spectral Methods**..**339**
     K. Revathy

     5.1   Feature Recognition and Classification
           with Wavelet Transform 339
           5.1.1   Introduction 339

	5.1.2	Introduction to Wavelets	340
	5.1.3	Multiresolution Analysis	343
	5.1.4	Wavelets in Astronomical Image Processing	344
5.2		Feature Recognition and Classification with Fractal Analysis	354
	5.2.1	Introduction	354
	5.2.2	Fractal Dimension	356
	5.2.3	Fractal Signature	359
	5.2.4	Concept of Multifractals	359
	5.2.5	Fractals in Astronomical Image Processing	359
	5.2.6	Conclusion	369

List of Contributors

L. Bedini
Istituto di Scienza e Tecnologie
dell'Informazione – CNR
Via Moruzzi
1, I- 6124 Pisa, Italy

A. Brazhnikov
Anvaser Consulting
Toronto, Canada

B. Cigale
Faculty of Electrical Engineering and
Computer Science
Smetanova ulica 17
SI-2000 Maribor, Slovenia

Ilias Maglogiannis
Department of Information and
Communication Systems Engineering
University of Aegean
83200 Karlovasi, Samos
Greece
E-mail: imaglo@aegean.gr

S.S. Ipson
EIMC Department,
University of Bradford
Bradford BD7 1DP, UK

L.C. Jain
Knowledge-Based Intelligent
Engineering Systems Centre
School of Electrical and Information Engineering
University of South Australia
Mawson Lake Campus
South Australia, SA 5095
Australia
E-mail: Lakhmi.jain@unisa.edu.au

E.E. Kuruoglu
Istituto di Scienza e Tecnologie dell'
Informazione – CNR
Via Moruzzi
1, I- 6124 Pisa, Italy

K. Revathy
Department of Computer Science
University of Kerala
Kariavattom,
Trivandrum 695581,
India

E. Salerno
Istituto di Scienza e Tecnologie
dell'Informazione – CNR
Via Moruzzi
1, I- 6124 Pisa, Italy

V. Schetinin
Department of Computing and
Information Systems
University of Luton
Luton, LU1 3JU, UK

A. Tonazzini
Istituto di Scienza e Tecnologie dell'
Informazione – CNR
Via Moruzzi
1, I- 6124 Pisa, Italy

H. Yoshida
Department of Radiology
Massachusetts General Hospital and
Harvard Medical School
75 Blossom Court
Suite 220
Boston, MA 02114, USA

D. Zazula
Faculty of Electrical Engineering and
Computer Science
Smetanova ulica 17
SI-2000 Maribor, Slovenia

S.I. Zharkov
Department of Applied Mathematics
University of Sheffield
Sheffield SH3, UK

V.V. Zharkova
Department of Computing and
Department of Cybernetics
University of Bradford
Bradford BD7 1DP, UK
E-mail: v.v.zharkova@brad.ac.uk

1 Introduction to Pattern Recognition and Classification in Medical and Astrophysical Images

V.V. Zharkova and L.C. Jain

1.1 Introduction to Pattern Recognition in Medical Images

The field of medical image processing is continually evolving. During the past ten years, there has been a significant increase in the level of interest in medical image morphology, full-color image processing, image data compression, image recognition, and knowledge-based medical image analysis systems. Medical images are increasingly used by doctors and healthcare providers for diagnosing diseases and recommending therapy. There are various types of images such as mammograms, ultrasound images, and fine needle aspiration slides (see, for example, Jain et al. 2000). Generally, the amount of diagnostic information contained in most patterns is small compared to a great volume of information represented by the entire image. Thus, the information to process represents only a small set of the different features of the original human organ.

The importance of images and feature recognition in medical diagnosis, prognosis, and therapy is undeniable. In many countries there are specific programs developed for medical imaging and pattern recognition. An example is the Biomedical Imaging Program at the National Cancer Institute at the National Institute of Health, Australia (Clark 2000) that initiated several new ways to improve the imaging techniques by making them increasingly accurate and automated. Some of the initiatives include highly innovative image acquisition and enhancement methods. This section presents a brief introduction to the various medical imaging techniques.

A steady growth of computer performance and a decrease in cost have influenced all computer-assisted applications and services. One of the most rapidly developing fields encompasses medical imaging with a variety of imaging facilities and technologies (Bronzino 1995). Ultrasound, for example, is relatively inexpensive, so its use has spread widely. Although his interpretation of an ultrasound image is not a straightforward or an easy task but requires advanced pattern recognition methods to be used.

V.V. Zharkova and L.C. Jain: *Introduction to Pattern Recognition and Classification in Medical and Astrophysical Images,* Studies in Computational Intelligence (SCI) **46**, 1–18 (2007)
www.springerlink.com   © Springer-Verlag Berlin Heidelberg 2007

Researchers have to extract, process, and recognize a vast variety of visual structures from the captured images. The first step is to reduce an image complexity, i.e., the image must be segmented into smaller constituent components, such as edges and regions. The task is not simple even when the environment conditions in the whole image are stable. When the conditions change or the observed scene contains a dynamic factor, the processing methods need to use adaptive solutions (Perry et al. 2002). The human brain has attained a high degree of adaptive processing ability. It also possesses an efficient ability of learning. Both the adaptability and learning abilities can be used as an example for the developers of computer recognition and classification algorithms.

For example, in dentistry the X-ray images are widely used for a diagnostics and treatment of periodontal disease (Burdea et al. 1991). The small changes in the bony structure supporting a tooth can be detected by using difference images taken over time rather than a single one. Computer-aided interpretation of quantification of bone defects on dental radiographs is also reported in Van der Stelt and Geraets (1991). The computer-aided procedure was able to rank a series of artificial bone lesions as accurately as experienced clinicians. Images are also used for detecting skin diseases. A radial search technique is used for detecting skin tumor borders in clinical dermatology images (Zhang 2000).

Another example is chest images, which provide the information on several organs systems including cardiovascular, respiratory, and skeletal. There are some automated techniques developed for the detection of posterior rib borders by isolating the area of search of the ribs by extracting the thoracic cage boundary by using a knowledge-based Hough transform.

There is also much research world wide in the diagnosis and prognosis of breast cancer. Some studies show that a woman has a "1 in 8" chances of developing breast cancer in her lifetime (Jain et al. 2000). Automatic feature extraction from the breast images may play a significant role in the early and accurate detection of breast cancer. A number of studies are focused on the mammographic images, ultrasound images, and images of fine needle aspiration. The artificial intelligence-based paradigms are investigated with the hope for accurate diagnosis, in order to make it possible to detect this disease at an early stage.

Numerous attempts have been made in the past decade to develop computerized methods to process, analyze, and display multidimensional medical images in radiology. A typical example is the 3-dimensional (3-D) visualization of semiautomatically segmented organs (for example, segmentation of the liver, endoluminal visualization of the colon and bronchus), or image processing of a part of an organ for the generation of an image that is more easily interpreted by readers (e.g., peripheral equalization of the breast in mammograms, digital subtraction bowel cleansing in virtual colono-scopy). These computerized methods often still automate only one of the image-processing tasks and depend on user interaction for the remaining tasks.

Computer-aided diagnosis (CAD) goes beyond these semiautomated image-processing applications, and moves into the area of automated medical image understanding or interpretation. In its most general form, CAD can be defined as a diagnosis made by a radiologist who uses the output of a computerized scheme for automated image analysis as a diag-nostic aid. Conventionally, CAD acts as a "second reader," hopefully, pointing out to abnormalities to the radiologist that might have otherwise been missed. This definition emphasizes that the intent of CAD is to sup-port rather than to substitute for a human reader in the detection of polyps, and as ever the final diagnosis is made by the radiologist.

The concept of CAD is universal across the different modalities and dis-ease types. Historically, CAD has been most popular in the diagnosis of breast cancers, for example, to detect of microcalcifications and classifi-cation of masses in mammograms (Astley and Gilbert 2004). CAD is even more important and beneficial for those examinations that have became feasible only recently due to the advanced digital imaging technologies. The latter examples are the detection of lung nodules in Computed-Tomographic (CT) scans of the lung and the detection of polyps in *Virtual Colonoscopy* (VC), also known as *CT colonography* (Levin et al. 2003; Morrin and LaMont 2003), in which a large number of images need to be rapidly interpreted in order to find a lesion with low incidence.

VC uses CT scanning to obtain a series of cross-sectional images of the colon for the detection of polyps and masses. The CT images are reformat-ted into a simulated 3-D "endoluminal view" of the entire colon that is comparable to that seen using optical colonoscopy. Radiologists can "fly through" the virtual colon, from the rectum to the cecum and back, search-ing for polyps and masses. Therefore, by using virtual colonoscopy, one can noninvasively examine the interior of the colon without physically invading it that is a safer procedure compared to optical colonoscopy.

Dermatologists and general physicians base the conventional diagnosis of skin lesions on visual assessment of pathological skin and the evaluation of skin macroscopic features. Therefore the correct assessment is highly dependent on the physician's experience and on his or her visual acuity. Moreover, the human vision lacks accuracy, reproducibility, and quantifi-cation in gathering visual information. These elements are of substantial importance during the follow-up studies in diseases monitoring. During the last years however, computer vision-based diagnostic systems for the assessment of digital skin lesion images have made significant progress.

Such systems have been used in several hospitals and dermatology clinics, aiming mostly at the early detection of malignant melanoma tumor, which is among the most frequent types of skin cancer in comparison to other types of nonmalignant coetaneous diseases.

The significant interest in melanoma is due to the fact that its incidence has increased faster than that of almost all other cancers and the annual incidence rates have increased on the order of 3–7% in fair-skinned populations in recent decades (Marks 2000). The advanced coetaneous melanoma is still incurable, but when diagnosed at early stages it can be cured without complications. However, the differentiation of early melanoma from other nonmalignant pigmented skin lesions is not trivial even for experienced dermatologists. In some cases primarycare physicians can underestimate melanoma in its early stage (Pariser and Pariser 1987).

Visual information conveyed by the observed images to the viewer depends on the inhomogeneity in images. Variations among different images are large but yet only a small number of image feature types are needed to characterize them. These can be attributed to smooth regions, textures and edges (Perry et al. 2002) and, thus, have to be recognized by any live or artificial image processing systems.

Often the structure and functions of the human central neural system has to be mimicked, in order to design flexible and intelligent machine vision systems capable of performing the demanding tasks of image processing and pattern recognition (Bow 2002). Human brains master much more than this task by providing the abilities of perceptual interpretation, reasoning, and learning. The processing power needed originates in the billions of interconnected biological neurons. They create neural networks, which act as a very complex nonlinear system.

In the last three decades, the idea of copying this structure matured in a variety of classes of artificial neural networks (ANNs). Technically speaking, ANNs belong to the category of distributed systems, consisting of many identical processing elements, i.e., the neurons, which perform a rather simple but nonlinear function. Only the ensemble of all neurons, the network, shows the recorded behaviour (Hänggi and Moschzty 2000). Hence, ANNs must be trained first to adopt the desired behavior.

A special network structure was introduced later called cellular neural networks (CNNs). Its structure resembles the continuous-time ANN, but implements strictly local connections with the basic building elements incorporating the characteristics of essentially simple dynamic systems. Structurally, they represent the equivalents of ANN's neurons named in CNNs as cells (Chua and Yang 1988).

Current developments in medical imaging present a wide variety of pattern recognition and classification methods. The most recent and advanced of them are described in the following chapters of this book.

1.2 Introduction to Pattern Recognition in Astrophysical and Solar Images

1.2.1 Astrophysical Features

An important problem in astrophysics and cosmology is to determine various radiation components that form the maps obtained from radioastronomical surveys. As an example, the microwave signal coming from the sky is the sum of radiation with different origins. Thus, the sky maps reconstructed from this signal measured at any frequency are superpositions of individual component maps, each representing the angular distribution of emission in a particular astrophysical source. Among these components, the cosmic microwave background (CMB) is of utmost importance. This is what remains of the first photons propagated in the Universe, as early as some 300,000 years after the big bang.

The CMB anisotropies were first detected by the NASA's COBE-DMR experiment (Smoot et al. 1992) and contain information about cosmological parameters, which help us to assess the present competing theories for the evolution of the Universe. The total signal also contains other components, called *foregrounds*, coming from less remote locations in space and time. Some of them come from our galaxy, such as the galactic dust radiation, the synchrotron radiation, and the free–free emission. Other radiation originates from outside the Milkyway, such as the signals from individual galaxies or clusters of galaxies.

Each of the foreground radiation sources contains its own interest for Astrophysics. As an example, maps of the galactic synchrotron emission give us information about the structure of the magnetic field, the spatial and energy distribution of relativistic electrons, and the variations of electron density, electron energy, and magnetic field induced by supernova shocks. In addition, the simultaneous study of dust and free–free emission is useful to investigate the relationships between different phases of the interstellar medium. Thus, rather than just looking for a clean CMB map, one need to extract each of the sources from the mixture maps by using a *source separation* strategy instead of mere noise cancellation.

Separating the individual radiation sources from the measured signals is a common problem in astrophysical data analysis for the case of CMB studies. In particular, a wealth of data is expected from several satellite or balloon-borne missions, for example, from NASA's *Wilkinson Microwave Anisotropy Probe* (*WMAP*: http://cmbdata.gsfc.nasa.gov/product/map/, see Bennet et al. 2003), and the forthcoming ESA's *Planck Surveyor Satellite* mission   (http://www.rssd.esa.int/index.php?project=PLANCK),   whose launch is expected for August 2007. The present or expected availability of such a volume of data gives strong motivation for the search of valid separation methods.

In synthesis, the astrophysical source separation problem can be formulated as follows: a number of radiometric sky maps reproducing the total sky emission in different frequency bands are available. Each observed map is the sum of a number of component maps reproducing the angular distribution of the individual astrophysical radiation sources. If we assume that any single source map is the product between a frequency-independent angular template and a characteristic frequency spectrum, then the observed maps result from a linear combination of the individual angular templates.

The mixture coefficients thus depend on the frequency responses of the measuring instruments and, through the emission spectra, on the radiative properties of the source processes taking emission spectra into account. The instrumental noise affecting the data is normally very strong. Sometimes, owing to the different observation time allowed for different points in the sky, the noise power also depends on the particular pixel. Given this information, the problem is to reconstruct the maps of the individual source of this radiation. The source emission spectra are normally known more or less approximately. Assuming that they are perfectly known leads to the so-called *nonblind* separation procedures. Allowing them to be estimated by the algorithm as additional unknowns leads to the *blind* or *semiblind* separation procedures and they depend on the possible introduction of additional knowledge.

1.2.2 Solar Features

There are a number of features appearing in solar images obtained in various wavelengths that are related to different solar phenomena occurring at various temporal and spatial scales (Gibson 1973). These features include *sunspots,* dark compact features on the quiet Sun background which are the oldest activity measures: *plages, faculae, or active regions,* related to the centers of solar activity associated with *flares* and *jets*; *filaments,* seen as elongated dark features that have a lifetime of a few solar rotations; and *coronal mass ejections*, large-scale eruptions from the upper solar atmosphere into the interplanetary space. This list can be complemented by the catalogs of solar flares, extremely dynamic events occurring over periods of minutes up to about an hour, which are extensively investigated and well cataloged manually. The appearance of these features also varies with a 11- or 22-year solar cycle that implies the suggestion that they are affected by the global changes of the solar magnetic field during this period of time (Priest 1984; Priest and Forbes 2000). These data can be used for a short term activity forecast in conjunction with other recognized features.

Sunspot identification and characterization including location, lifetime, contrast, etc. are required for a quantitative study of the solar cycle. Sunspot studies also play an essential part in the modeling of the total solar irradiance during the solar cycle. As a component of solar active regions, sunspots and their behavior are also used in the study of active region evolution and in the forecast of solar flare activity (see Steinegger et al. 1997).

Manual sunspot catalogs in different formats are produced at various locations all over the world such as the Meudon Observatory (France), the Locarno Solar Observatory (Switzerland), the Mount Wilson Observatory (USA), and many others. The Zurich relative sunspot numbers (or since 1981 Sunspot Index Data (SIDC)), compiled from these manual catalogs, are used as a primary indicator of solar activity (Hoyt and Schatten 1998a, b and Temmer, Veronig, and Hanslmeier 2002).

Sunspots are one of the types of features, whose behavior reveals some secrets about the global solar activity. The sunspots and their groups have been investigated for hundreds of years and now are assembled from all the observatories into manually created databases kept by the Solar Influence Data Centre http://sidc.oma.be/ at the Royal Observatory of Belgium, Brussels. Only recently, with the launch of the space missions SMM, YOHKOH, and SOHO in 80–90s, other features like UV and hard X-ray bright points, X-ray flares and loops, have been discovered in solar images. In the present book we describe detection techniques for the features, which are related to long-term (years) and medium-term (days) of the solar activity and solar cycle. These are the part of the active feature database known as Solar Feature Catalogues (http://solar.inf.brad.ac.uk) (Zharkova et al. 2005), which can be used for feature classification and the solar activity forecast.

Sunspots and sunspot groups are the most conspicuous features of solar activity in white light and Ca II K radiation that appear with sizes and numbers varying during the eleven-year cycle. Over centuries, *sunspots* observed in while light images (see Fig. 1.1), generally, consist of dark inner umbras surrounded by lighter penumbral regions that are the evidences of a lower

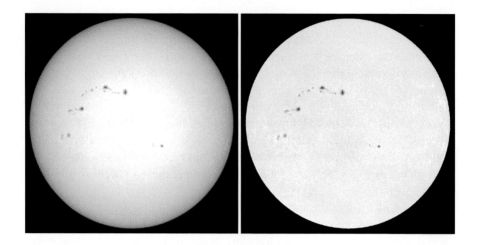

Fig. 1.1. A full disk solar white light image taken by SOHO/MDI instrument with darkening toward the limb (left) and after correction without darkening (right). Below the segment of a high-resolution SOHO/MDI image shows close-up of sunspots with dark umbras and lighter penumbras on a gray quiet Sun background

surface temperature comparing to a background. The change in intensity from the quiet photosphere near a sunspot to the sunspot intensity occurs over a distance no more than 2–4 arcsec (or 1–2 pixels for the data). The area covered by sunspots on a given day is the important parameter used to investigate the solar activity. The determination of area becomes more complicated when attempting to separate the individual areas of umbra and penumbra since the ratio of umbra to penumbra areas depends on a spot size.

The first thresholding methods for the extraction of sunspot areas used an a priori estimated intensity threshold on white-light full-disk solar images (Chapman and Groisman 1984; Chapman et al. 1994a, b). Sunspots were defined as features with intensity 15% (CG84) or 8.5% (C94) below the quiet Sun background and simply counted all pixels below these values.

Similar methods were applied to high-resolution nonfull disk images containing a sunspot or a group of sunspots using constant intensity boundaries for the umbra–penumbra and the penumbra–photosphere transitions at 59% and 85% of the photospheric intensity, respectively, (Steinegger et al. 1990; Brandt et al. 1990).

The thresholding methods were significantly improved by using image histograms that help to determine the threshold levels in sunspots. Steinegger et al. (1996) used the difference histogram method to determine the intensity boundary between a penumbra and the photosphere that was defined for each individual spot. Another method based on the cumulative sum of sunspot area contained in successive brightness bins of the histogram (Steinegger et al. 1997) that was applied to determine the umbral areas of sunspots observed with high resolution nonfull disk observations (Steinegger et al. 1997). Another method using sunspot contrast and contiguity based on region growing technique was also developed (Preminger et al. 2001) that has produced more accurate results compared to thresholding and histogram methods but was very time-consuming.

More accurate approach to sunspot area measurements was developed by utilizing edge detection and boundary gradient intensity was suggested for high-resolution observations of individual sunspot groups, and/or nonfull disk segments (Győri 1998). The method is very accurate when applied to the data with sufficiently high resolution. However, this method was not suitable for the automated sunspot detection on full disk images of the low and moderate resolutions that are present in most archives and needed some adjustments.

There is also a Bayesian technique developed for the detection and labeling of active region and sunspot (Turmon et al. 2002) that is rather computationally expensive. Moreover, the method is more oriented toward the faculae detection and does not detect sunspot umbras. Although the algorithm performs well when trained appropriately, the training process itself can be rather difficult to arrange on images with different background variations corresponding to varying observing conditions.

Therefore, in their original forms, all the existing techniques described earlier are not suitable for the automatic detection and identification of sunspots on full disk images, since their performance depends on the images with high resolution (Győri 1998) and/or quality (Turmon et al. 2002) that cannot be guaranteed for full disk images. This motivated the development of automated edge-based procedures for sunspot detection described in Sect. 3.3.

The most clearly distinguished in the Hα-line images at the chromosphere are active regions (ARs) (Fig. 1.2, bottom plot), which are physically associated with sunspots seen in the white light images and Plages

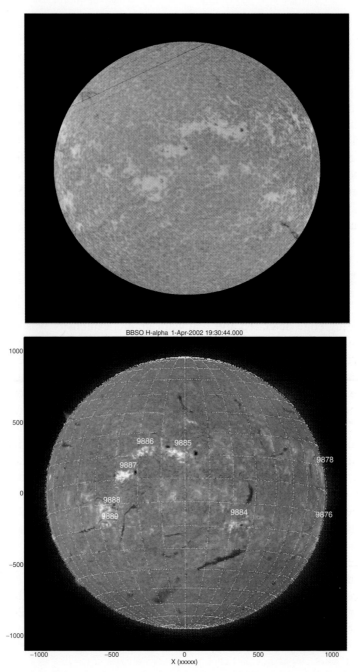

Fig. 1.2. A full disk solar image obtained in Ca II K3 line (02/04/02, Paris-Meudon Observatory, France, upper plot) and in Hα-line (01/04/02, Big Bear Astronomical Observatory, BBSO, California, USA, lower plot)

seen in Ca II K3 spectroheliogram observations of the photosphere (Fig. 1.2, upper plot). The plages can be large in extent and normally are associated with a strong magnetic field and increased brightness. Bipolar ARs are widely accepted to be the manifestations of emerging buoyant magnetic flux tubes (van Driel-Gesztelyi 2002; Harvey and Zwaan 1993 and the references therein). These flux tubes are seen on the photosphere in white light images as sunspots or in the corona as very diffusive bright areas in extreme ultraviolet (EUV) emission. The same ARs are often found to produce numerous flares during their life time, in a process that is related to the changes in their magnetic configurations (Priest 1984; Priest and Forbes 2000). These magnetic field changes can occur on even smaller scales related to the emergence of mini and nano-flux tubes causing smaller flaring events (Aschwanden et al. 2002; Schmieder et al. 2002). This produces different responses at different levels of the solar atmosphere, which can be used in the testing of theoretical models describing these responses. Therefore, by determining the structure of ARs at various levels of the solar atmosphere one can find a key to the understanding and proper forecast of solar activity manifestations such as: solar flares, coronal mass ejections (CMEs), eruptive filaments etc.

There are three different approaches identified in the literature for the automatic identification of bright ARs (plages). The first is based on the selection of a threshold to separate an object from the background and is straightforward if the intensity histogram is bimodal, but otherwise can be difficult (Preminger et al. 2001; Steinegger and Brandt 1998; Worden et al. 1996). The second approach is based on region growing techniques for segmenting images into bright and dark regions (Hill et al. 2001; Veronig et al. 2001) that has been applied to solar images in various wavelengths, including Hα. Finally, the third approach uses the Bayesian inference method for automatically identifying various surface structures on the Sun (Turmon et al. 1998).

All these approaches can give what is considered on manual inspection to be a reasonable accuracy of detection with suitable images, but the Bayesian-based methods are the most computationally expensive. The intensity threshold-based methods are simple and fast, but are relatively sensitive to noise and background variations, which affect the reliability of the segmentation results obtained. In order to replace the existing manual detection methods, the techniques were required to combine the elements of the first two approaches earlier, threshold and region growing (TRG), for the automated detection of ARs (plages) at different heights in the solar atmosphere, which are described in Sect. 3.4.

Another features associated with many aspects of the solar activity are solar filaments, or prominences if seen on the solar limb, present in Hα-line images at the chromosphere (see Fig. 1.2, bottom plot). *Filaments* appear on a solar disc as elongated or compact features with darker than the surrounding background representing clouds of a cooler gas supported by magnetic forces. Filaments often appear over magnetic inversion lines and are used as the indicators of a large scale solar magnetic field. They can be stable for a long period of time or sometimes can erupt or show dynamics associated with change of magnetic configuration they reside on.

Prominence is an object in the chromosphere or corona that is denser and cooler than its surroundings (Tandberg-Hanssen 1995). When seen against the intense Há light of the disk, the prominences appear in absorption as dark features, i.e., filaments. The extraction process used for filament detection is therefore based on the segmentation of the darkest regions in the image. Even if filaments, especially quiescent ones, are commonly described as elongated features, their shape is not taken into account in this work. Indeed, the variety in filament forms due to the wide range of possible magnetic field structures and sources of condensed material (Gibson 1973) prohibits any definitive postulate about their shapes, as long as completeness is a priority. Many blob-like filaments, or filament parts, observed in Há could be missed if such a condition was applied. Also a small number of filaments are often falsely detected as sunspots and have to be removed using the physical criteria.

Semiautomated detection methods have already been developed in order to follow the evolution of a filament in time by detecting the moments of their disappearance (Gao et al. 2002) or to investigate solar surface rotation (Collin and Nesme-Ribes 1992). To extract the filaments they use either thresholding or region-based techniques, particularly, a region growing one. In these studies the simple matching of a filament from one observation to another can be sufficient and the exact shape of the filament is not required. More recently, Wagstaff et al. (2003) presented the results of a region-based method to detect both filaments and sigmoids, and Shih and Kowalski (2003) described a method based on directional filtering to identify thin and elongated objects (the features obtained are then considered as seeds for a region growing operation) and Zharkova and Schetinin (2005) detected filaments by using artificial neural networks.

However, all these methods use some manual interaction either in detection process itself or background reduction. For processing large data sets a fully automated method for filament detection is required that is present in the Sect. 3.4, Chap. 2.

Flares and coronal mass ejections are transient events accompanied by a rapid release of energy and the transport of energy material as downward to the solar surface far away from the surface of the Sun, contributing to interplanetary particles, emission and shocks. The recent techniques are presented for the detection of these features in solar images using combinations of edge-based, region-growing, and Hough transform techniques.

An autonomous procedure for detecting coronal mass ejections (CMEs) in image sequences from LASCO C2 and C3 data was reported by Bergman et al. (2003). This procedure uses a Hough transform approach, which, in general, is used to detect the presence of edge pixels in an image which fall on a specified family of curves and uses an accumulator array whose dimension equals the number of parameters in the curves to be detected. For example, in the case of a straight line expressed in the normal form $\rho = x\cos\theta + y\sin\theta$, the pl arameters are ρ and θ and the two indices of the accumulator array correspond to quantized values of ρ and θ over the range of values of these parameters which are of interest.

Measurements of *solar magnetic field* using either line-of-sight (LOS) magnetographs (Babcock and Babcock 1955; Scherrer et al. 1995), or recent full vector magnetographs (Wang et al. 1998) provide the important data resource for understanding and predicting solar activity, in addition, to spectro-heliograms of the solar atmosphere (see Fig. 1.3). In every model of the solar activity solar magnetic fields play the most essential role defining the major source transforming the fusion reaction energy released in the solar core into plasma motions beneath and on the solar surface and into the radiation in many wavelengths emitted from the Sun to the solar system and Earth. Therefore, the automated detection will not be complete if magnetic field is not automatically detected alongside the other solar features mentioned earlier.

The importance of the magnetic field is emphasized in many theoretical works on topics about heating of the solar atmosphere; the formation and support of multilevel magnetic loops and filaments resting on them, located along magnetic inversion lines, the activation of energy releases in solar flares and coronal mass ejections; and many other small and large scale events occurring in the solar atmosphere and interplanetary space (Priest 1984; Somov 2000). A difference of magnetic field magnitudes is often assumed to account for the solar events of different scales. The resulting scenario of these events is defined not only by the magnitudes themselves but also by configurations of magnetic loops with the opposite magnetic polarities (Priest 1984; Vainshtein and Parker 1986; Priest and Forbes 2000). Therefore, in addition to the major features used for the classification of solar

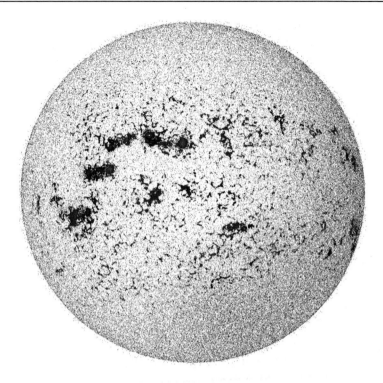

Fig. 1.3. Line-of-sight magnetogram obtained by the MDI instrument aboard SOHO satellite on 01/04/02. The red and blue false colors denote negative and positive magnetic polarities, the magnetic field B is measured by the relevant pixel values

activity which include sunspots, active regions, filaments, flares and CMEs, one needs to consider simultaneous measurements of the magnetic configurations associated with each event and their location in relative to a magnetic neutral line (NML, thereafter).

Different scales of magnetic field magnitudes are assumed to account for solar events of different scales and these event scenarios result mostly from magnetic configurations of loops with opposite magnetic polarities. For example, solar filaments often appear on the boundaries of coronal holes (Kuperus and Raadu 1974; Lerche and Low 1980) or above a middle line between two-ribbon flares (Fox et al. 1998; McIntosh 1979; Sturrock and Jardin 1994; Grigoriev and Ermakova 1999) that usually is very close to the location of magnetic inversion lines, or apparent magnetic neutral lines (AMNLs or MNLs thereafter). Filaments are well established to be flux ropes with a longitudinal magnetic field (twisted or not) occurring above the bipolar magnetic regions in the vicinity of magnetic inversion lines (Kippenhahn and Schluter 1957; Hood and Priest 1979; Priest et al. 1989, 1996; Demoulin and Priest 1993; Rust and Kumar 1994).

Therefore, line-of-sight (LOS) magnetic observations were often used to mark the filament locations in many wavelengths (see, for example, Gaizauskas et al. (1997), Grigoriev and Ermakova (1999)). However, it is apparent that filaments occurring at higher atmospheric levels in the chromosphere or corona should not necessarily match the photospheric inversion lines because of the possible different magnetic structure at these levels. In this case, magnetic observations at the photosphere can be brought up to higher atmospheric layers by smoothing, or averaging, the magnetic data within a particular window (Duvall et al. 1977; Durrant et al. 2001). This smoothing results in a cancellation of smaller-scale magnetic fields, or smaller loops, and approximates higher magnetic structures. By repeating this smoothing process several times it is possible to find the degree of smoothing that allows matching the inversion line maps with filament skeletons (Durrant 2002). This technique was tested on many magnetic data from MWO and NSOKP for filament elongation not only in the solar center but in polar plumes and polar crown gaps and revealed a very good agreement between the MNLs and filaments at various levels of smoothing (Durrant 2002) and during different phases of the solar cycle (McCloughan and Durrant 2002). Hence, in order to extend this research to the data obtained in the solar feature catalogs and to produce sufficient statistical analysis of the filaments of their locations with respect to MNLs in the relationship between the level of smoothing an automated technique is required. These techniques are described in Sect. 3.5.

The automated detection and online classification of the solar activity features using the digital images obtained by ongoing space missions and ground-based observations will allow the online analysis of a solar activity and its short-term and long-term forecast.

References

Aschwanden, M.J., De Pontieu, B., & Schrijver, C.J.A. (2002). *Sol. Phys.*, *206*, 99

Astley, S.M., & Gilbert, F.J. (2004). Computer-aided detection in mammography. *Clin. Radiol.*, *59*(5), 390–399

Babcock, H.V., & Babcock, H.D. (1955). *Astrophys. J.*, *121*, 349

Bennett, C., Hill, R.S., Hinshaw, G., Nolte, M.R., Odegard, N., Page, L., Spergel, D.N., Weiland, J.L., Wright, E.L., Halpern, M., Jarosik, N., Kogut, A., Limon, M., Meyer, S.S., Tucker, G.S., & Wollack E. (2003). First-year Wilkinson Microwave Anisotropy Probe (WMAP) observations: foreground emissiion. *Astrophys. J. Suppl. Ser., 148*, 97–117

Bergmans, D. (2003). Automated technique for detection of coronal mass ejections in LASCO images. In: *Proceedings of the 1st SIPW*, 7–8 Nov'03, Brussels, Belgium

Bow, S.-T. (2002). *Pattern recognition and image processing*. New York, USA: Dekker

Brandt, P.N., Schmidt, W., & Steinegger, M. (1990). On the umbra–penumbra area ratio of sunspots. *Solar Phys.*, *129*, 191

Bronzino, J.D. (1995). *The biomedical engineering handbook*. Boca Raton, USA: CRC

Burdea, G.C., et al. (1991). Real-time sensing of tooth position for dental digital subtraction radiography. *IEEE Trans. Biomed. Eng.*, *38*(4)

Chapman, G.A., & Groisman, G.A. (1984). Digital analysis of sunspot areas. *Solar Phys.*, *91*, 45

Chapman, G.A., Cookson, A.M., & Dobias, J.J. (1994a). Observations of changes in the bolometric contrast of sunspots. *Astrophys. J.*, *432*, 403–408

Chapman, G.A., Cookson, A.M., & Hoyt, D.V. (1994b). Solar irradiance from Nimbus-7 compared with ground-based photometry. *Solar Phys.*, *149*, 249

Chua L.O., & Yang L. (1988). Cellular neural networks: Theory. *IEEE Trans. Circuits Syst.*, *35*(10), 1257–1272

Clarke, L.P. (2000). NCI Initiative: Development of novel imaging technologies. *IEEE Trans. Med. Imaging*, *19*(11)

Collin, B., Nesme-Ribes, E. (1992). Pattern recognition applied to H_{alpha} spectroheliograms solar rotation and meridional circulation, ESA. *Solar Phys. Astrophys. Interferometric Res.* 145–148

Demoulin, P., & Priest, E.R. (1993). *Solar Phys.*, *144*, 283

van Driel-Gesztelyi, L. (2002). In *Proceedings of the SOLMAG, the magnetic coupling of the solar atmosphere*, Greece: Euroconference & IAU Colloquium, Vol. 188, p.113

Durrant, C.J. (2002). *Solar Phys.*, *211*, 83

Durrant, C.J., Kress, J.M., & Wilson, P.R. (2001). *Solar Phys.*, *201*, 57

Duvall, T.L., Wilcox, J.M., Svalgaard, L., Scherrer, P.H., & McIntosh, P.S. (1977). *Solar Phys.*, *55*, 63

Fox, P., McIntosh, P.S., & Wilson, P.R. (1998). *Solar Phys.*, *177*, 375

Gaizauskas, V., Zirker, J.B., Sweetland, C., & Kovacs, A. (1997). *Astrophys. J.*, *479*, 448

Gao, J., Wang, H., & Zhou, M. (2002). Development of an automatic filament disappearance detection system. *Solar Phys.*, *205*(1), 93–103

Gibson, E.G. (1973). *The quiet Sun, NASA SP*. Washington: National Aeronautics and Space Administration (NASA), 663 pp.

Grigoriev, V.M., & Ermakova, L.V. (1999). *Astron. Astrophys. Trans.*, *17*, 355

Győri, L. (1998). *Solar Phys.*, *180*, 109–130

Hänggi M., & Moschzty G.S. (2000). *Cellular neural network: Analysis, design and optimisation*. Boston, USA: Kluwer

Harvey, K.L., & Zwaan, C. (1993). *Solar Phys.* 148(1), 85

Hill, M., Castelli, V., Chung-Sheng, L., Yuan-Chi, C., Bergman, L., Smith J.R., & Thompson, B. (2001). *International conference on image processing*, Thessaloniki, Greece, 7–10 October 2001, Vol. 1, p. 834

Hood, A.W., & Priest, E.R. (1979). *Solar Phys.*, *64*, 303

Hoyt, D.V., & Schatten, K.H. (1998a). Group sunspot numbers: A new solar activity reconstruction, *Solar Phys.*, *179*, 189–219

Hoyt, D.V., & Schatten, K.H. (1998b). Group sunspot numbers: A new solar activity reconstruction, *Solar Phys.*, *181*, 491–512

Jain, A. et al. (Eds.) (2000). *Artificial intelligence techniques in breast cancer diagnosis and prognosis*. Singapore: World Scientific

Kippenhahn, R., & Schluter, A. (1957). *Astrophysik*, *43*, 36

Kuperus, M., & Raadu, M.A. (1974). *Astron. Astrophys.*, *31*, 189

Lerche, I., & Low, B.C. (1980). *Solar Phys.*, *66*, 303

Levin, B., Brooks, D., Smith, R.A., & Stone, A. (2003). Emerging technologies in screening for colorectal cancer: CT colonography, immunochemical fecal occult blood tests, and stool screening using molecular markers. *CA Cancer J. Clin.*, *53*(1), 44–55

Marks, R. (2000). Epidemiology of melanoma. *Clin. Exp. Dermatol.*, *25*, 459–463

McCloughan, J., & Durrant, C.J. (2002). *Solar Phys.*, *211*, 53

McIntosh, P.S. (1979). *UAG Report 70*. Boulder, CO: World Data Centre for SIP

Morrin, M.M., & LaMont, J.T. (2003). Screening virtual colonoscopy—ready for prime time? *N. Engl. J. Med.*, *349*(23), 2261–2264

Pariser R.J., & Pariser D.M. (1987). Primary care physicians errors in handling cutaneous disorders, *J. Am. Acad. Dermatol.*, *17*, 239–245

Perry S.W., Wong H.-S., & Guan L. (2002). *Adaptive image processing: A computational intelligence perspective*. Boca Raton, USA: CRC

Preminger, D.G., Walton, S.R., & Chapman, G.A. (2001). Solar feature identification using contrast and contiguity. *Solar Phys.*, *202*, 53

Priest, E.R. (1984). Solar magneto-hydrodynamics. *Geophys. Astrophys. Monographs*, Dordrecht: Reidel

Priest, E.R., & Forbes, T. (2000). *Book review: Magnetic reconnection*. Cambridge: Cambridge University Press

Priest, E.R., van Ballegooijen, A.A., & Mackay, D.H. (1996). *Astrophys. J.*, *460*, 530

Priest, E.R., Hood, A.W., & Anzer, U. (1989). *Astrophys. J.*, *344*, 1010

Rust, D.M., & Kumar, A. (1994). *Solar Phys.*, *155*, 69

Schmieder, B., Engvold, O., Lin, Y., Deng, Y.Y., & Mein, N. (2002). In *Proceedings of the SOLMAG 2002, the magnetic coupling of the solar atmosphere*. Santorini, Greece: Euroconference and IAU Colloquium 188, June 2002, p. 223

Scherrer, P.H., Bogart, R.S., Bush, R.I., Hoeksema, J.T., Kosovichev, A.G., Schou, J., Rosenberg, W., Springer, L., Tarbell, T.D., Title, A., Wolfson, C.J., Zayer, I. (1995). MDI engineering, the solar oscillations investigation – Michelson Doppler Imager. *Solar Phys.*, *162*, 129–188

Shih, F.Y., Kowalski, A.J. (2003). Automatic extraction of filaments in Hα solar images. *Solar Phys.*, *218*(1), 99–122

Smoot, G.F., Bennett, C.L., Kogut, A., Wright, E.L., Aymon, J., Boggess, N.W., Cheng, E.S., Deamici, G., Gulkis, S., Hauser, M.G., Hinshaw, G., Jackson, P.D., Janssen, M., Kaita, E., Kelsall, T., Keegstra, P., Lineweaver, C.,

Loewenstein, K., Lubin, P., Mather, J., Meyer, S.S., Moseley, S.H., Murdock, T., Rokke, L., Silverberg, R.F., Tenorio, L., Weiss, R., & Wilkinson, D.T. (1992). Structure in the COBE differential microwave radiometer 1st-year maps. *Astrophys. J., 396*(1), L1

Somov, B.V. (2000). *Cosmic plasma physics.* Boston, MA: Kluwer, Astrophysics and space science library, p. 251

Steinegger, M., & Brandt, P.N. (1998). *Solar Phys., 177*(1), 287

Steinegger, M., Brandt, P.N., Pap, J., & Schmidt, W. (1990). Sunspot photometry and the total solar irradiance deficit measured in 1980 by ACRIM. *Astrophys. Space Sci., 170*, 127–133

Steinegger, M., Vazquez, M., Bonet, J.A., & Brandt, P.N. (1996). On the energy balance of Solar Active Region. *Astrophys. J., 461*, 478

Steinegger, M., Bonet J., A., & Vazquez, M. (1997). Simulation of seeing influences on the photometric determination of sunspot areas. *Solar Phys., 171*, 303

Sturrock, P.A., & Jardin, M. (1994). *Book review: Plasma physics.* Cambridge: Cambridge University Press

Tandberg-Hanssen, E. (1995). *The nature of solar prominences*, Dordrecht: Kluwer, p. 3

Temmer, M., Veronig, A., & Hanslmeier, A. (2002). Hemispheric sunspot numbers R_n and R_s: Catalogue and N--S asymmetry analysis. *Astron. Astrophys., 390*, 707–715

Turmon, M., Pap, J.M., & Mukhtar, S. (1998). *Structure and dynamics of the interior of the sun and sun-like stars.* Boston: Proceedings of the SOHO 6/GONG 98 Workshop

Turmon, M., Pap, J.M., & Mukhtar, S. (2002). *Astrophys. J., 568*, 396–407

Vainshtein, S.I., Parker, E.N. (1986). Magnetic nonequilibrium and current sheet formation. *Astrophys. J., Part 1, 304*, 821–827

Van der Stelt, P., & Geraets, W.G.M. (1991). Computer-aided interpretation and quantification of angular periodontal bone defects on dental radiographs. *IEEE Trans. Biomed. Eng., 38*(4)

Veronig, A., Steinegger, M., Otruba, W., Hanslmeier, A., Messerotti, M., & Temmer, M. (2001). *HOBUD7, 24*(1), 195–2001

Wagstaff, K., Rust, D.M., LaBonte, B.J., & Bernasconi, P.N. (2003). Brussels: Solar image recognition workshop, 23–24 October 2003, p. 331

Wang, H., Denker, C., Spirock, T., et al. (1998). *Solar Phys., 183*, 1

Worden, J.R., White O.R., & Woods, T.N. (1996). *Solar Phys., 177*(2), 255

Zhang, Z. (2000). Detection of digitized skin tumor images. *IEEE Trans. Med. Imaging, 19*, 11

Zharkova, V.V., Ipson, S.S., Zharkov, S.I., Aboudarham, J., Benkhalil, A.K., & Fuller, N. (2005). Solar feature catalogues in EGSO. *Solar Phys., Solar Imaging, 228*(1), 139–150

Zharkova, V.V., & Schetinin, V. (2005). ANN techniques for filament recognition in solar images. *Solar Phys., Solar Imaging, 228*(1), 363–377

2 Image Standardization and Enhancement

S.S. Ipson, V.V. Zharkova and S.I. Zharkov

2.1 Digital Image Distortions and Standardization of Shape and Intensity

Abstract

This section describes a number of geometrical and intensity procedures which are frequently used in the processing of solar images. Geometrical operations have a range of applications from compensating for different sizes and viewpoints of images, to displaying an image in a different co-ordinate system. Intensity operations ranging from removing radial and non-radial background illumination variations to removing dust lines; image de-blurring can be useful preliminary steps before the application of feature detection algorithms.

2.1.1 Geometrical Standardization

This section describes procedures which are used to change the shape or remap a digital image by applying geometrical transformations such as the one shown in Fig. 2.1.

A common approach, used when the transformation is invertible, is to first apply the forward transformation to define the space occupied by the new transformed image. This space is then scanned pixel by pixel, and for each pixel the inverse transformation is applied to determine where it came from in the original image. This position generally falls between the integer pixel locations in the original position and an interpolation procedure is applied to determine the appropriate image value at that point. "Image Resampling and Interpolation" describes in detail this commonly used approach to applying a geometrical transformation to an image and "Removal of Nonradial Background Illumination and Removal of Dust Lines" describe specific examples in which different geometrical transformations are applied to solar images.

S.S. Ipson et al.: *Image Standardization and Enhancement*, Studies in Computational Intelligence
(SCI) **46**, 19–58 (2007)
www.springerlink.com   © Springer-Verlag Berlin Heidelberg 2007

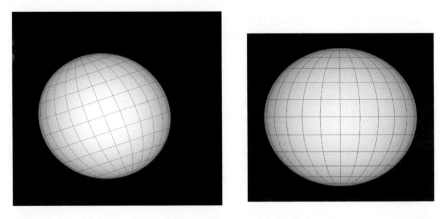

Fig. 2.1. An affine transformation comprising translation, rotation, and rescaling applied to an artificial elliptical image on the left to produce a centered, circular image of standardized size on the right

Image Resampling and Interpolation

The general mathematical form of a continuous geometrical transformation applied to an image is

$$x' = f(x, y) \qquad y' = g(x, y), \qquad (2.1)$$

where the image intensity at the point (x, y) in the original image has been moved to the point (x', y') in the transformed image under the functions f and g. A simple case is the affine transformation (illustrated in Fig. 2.1) which can be used to standardize the position and size of a solar image and is sometimes used to correct the shape of the solar disc from an instrumentally distorted elliptical shape back to the true circular shape. The affine transformation is expressed in matrix form with homogeneous coordinates as follows:

$$\begin{bmatrix} x' \\ y' \\ 1 \end{bmatrix} = \begin{bmatrix} S_x & \alpha & T_x \\ \beta & S_y & T_y \\ 0 & 0 & 1 \end{bmatrix} \begin{bmatrix} x \\ y \\ 1 \end{bmatrix}, \qquad (2.2)$$

where S_x and S_y magnify (or shrink) the image independently in the x and y directions, respectively, T_x and T_y translate the image in the x and y directions, respectively. The coefficients α and β can be used to rotate the image through an angle θ about the origin, when $\alpha = -\beta = \sin\theta$, or, for other values, to impart a skew deformation to the image.

Digital images are sampled, usually on a square grid, and the samples quantized to some specified number of bits. Applying a geometrical transformation to an individual sample (pixel) in the original image will, in general, move it to a noninteger position in the transformed image. Placing it at the nearest integer grid position will, when applied to all the pixels in the original image, generally leave grid positions without a value. Rather than applying a procedure to fill in these holes in the transformed image, it is more common, as mentioned earlier, to apply the inverse geometrical transform (assuming it exists) to each pixel in the transformed image to determine where it came from in the original image. In general this position is a noninteger grid position and the image value at this position has to be computed by interpolation to determine the appropriate value to place in the transformed image pixel. Common interpolation procedures are: zero hold where the image value is taken from the transformed pixel position obtained by setting the fractional part to zero; nearest neighbor, where the image value is taken from the pixel which is closest to the transformed pixel position; bilinear, where the image value is computed from the bilinear function fitted to the four pixels closest to the noninteger position; bicubic, where the image value is computed from a bicubic function fitted to the 16 pixels closest to the noninteger position (Press et al. 1992). Figure 2.2 shows a small region of a Meudon H_α solar image which has been magnified by a factor of four in the vertical and horizontal directions using nearest neighbor, bilinear and bicubic interpolation.

On close inspection, the nearest neighbor version appears blocky because the individual pixels are large enough visible. By comparison, the bilinear and bicubic versions appear much smoother, with individual pixels not visible. The bicubic version appears only marginally better than the bilinear version in this example.

In general, bilinear and bicubic interpolation give smoother transformed images than zero-hold and nearest neighbor interpolation at the cost of increased computational effort and introduce image values in the transformed image which are not present in the original image. By contrast, zero-hold and nearest-neighbor interpolation do not introduce pixel values into the transformed image which are not in the original image and some authors, for example Turmon et al. (2002) use these interpolation procedures so that the statistical distributions of the pixel values are not altered. When creating a transformed digital image the consequences of the sampling theorem should be considered. Aliasing artefacts may be introduced

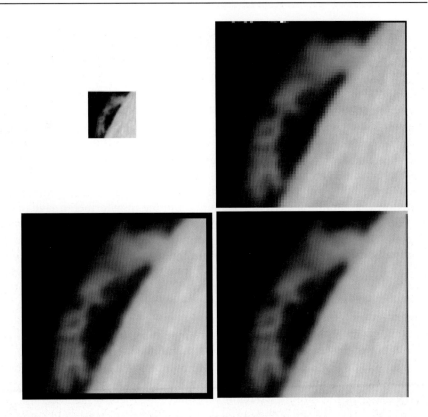

Fig. 2.2. A fragment of a solar image (H_α 26/07/2001) near the limb showing a solar prominence (top left), magnified four times using nearest neighbor (top right), bilinear (bottom right) and bicubic (bottom left) interpolation

in regions of the transformed image where the effect of the transformation is to subsample the image as illustrated by the example in Fig. 2.3. To avoid such effects a smoothing transformation to limit the spatial frequencies present below the Nyquist limit of the transformed sampling rate should be applied to the original image before the geometrical transformation is applied as illustrated in Fig. 2.3.

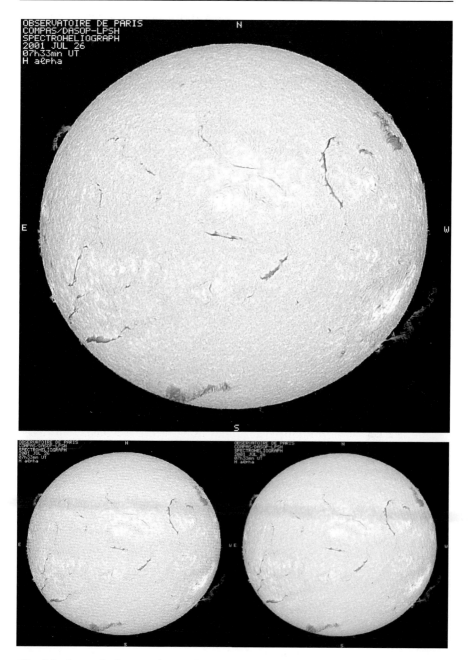

Fig. 2.3. At top is the Meudon H_α image from Fig. 2.1a after sharpening. At bottom left is the image after reduction in size by factor two with visible fringe-like artifacts. At bottom right is the image with smoothing before size reduction with no visible artifacts

Example 2.1.1.

The following code in the IDL language reads an image, determines its size, performs a rectangular to polar coordinate transformation on the image and displays the result using nearest neighbor interpolation.

```
Pro Rect_to_Polar
    filename = dialog_pickfile(path='c:\Images')
    fitsread_win, header, image, FILENAME = filename
    info = size(image)
    imagex = info[1]
    imagey = info[2]
    xc = (imagex-1)/2.0
    yc = (imagey-1)/2.0
    R = 420
    Rmax = 420
    Rmax2 = Rmax*Rmax
    Inorm = 200
    window,1, xsize=imagex, ysize=imagey,title='Original Image'
    TVSCL, image
    R2PI = LONG (2 * !DPI * Rmax +0.5)
    bim = MAKE_ARRAY(1024,1024, /INT, VALUE = 0)
    Polar_array = MAKE_ARRAY(1024,Rmax, /INT, VALUE = 0)
    FOR i=0, Rmax-1 DO BEGIN
        FOR j=0, imagex-1 DO BEGIN
            theta = j * ((2*!DPI)/1024)
            x = i * COS(theta)
            y = i * SIN(theta)
            X= LONG(x+xc+0.5)
            Y= LONG(y+yc+0.5)
            Polar_array(j,i) = image(X,Y)
        ENDFOR
    ENDFOR
    window,2, xsize=imagex, ysize=Rmax,title='Polar_Array'
    TVSCL, Polar_array
END
```

Change of Viewing Position and Time

When comparing features in a pair of solar images it is necessary to compensate for different locations and viewing times of observations. The Sun is about 150 million km (152.6 million km in July to a minimum of 147.5 million km in January) from Earth so simultaneous observations from the Earth and satellites in near earth orbits (earth diameter about 13,000 km) observe essentially the same view of the Sun. However, not all satellites based solar instruments are in near earth orbits. The Solar Heliospheric Observatory (SOHO) satellite (SOHO 2005), is positioned at the inner Lagrange point (L1) between Earth and the Sun and from this viewpoint the Sun appears approximately 1% bigger than it does from Earth. Geometrical adjustments are required whenever SOHO images (of size $1,024 \times 1,024$ pixel) are compared with images from ground-based observatories, or satellites in low Earth orbit. A project currently in the planning stage called STEREO will involve a pair of spacecraft carrying solar imaging instruments operating at large distances from the Earth Sun line to achieve stereo imaging of the solar atmosphere. The geometrical differences between STEREO images and near-earth images will be much greater than for SOHO and these will have to be taken into account when making comparisons with images from near earth instruments.

As the earth moves in orbit about the sun, the viewpoint changes by about 1° per day. With a full disc image of the Sun of radius 400 pixel, this corresponds to a movement of about 7 pixel at the center of the disk and this effect should be taken into account when comparing images taken several days apart. In addition the sun rotates on its axis, but it is not a solid body so different parts of the sun rotate at different rates. The equatorial regions of the sun rotate with a sidereal period (relative to the fixed stars) of 25.38 days but this becomes about 30 days near the poles. The period seen from the Earth is about two-days longer because the Earth is moving in its orbit in the same direction as the solar rotation and the synodic period of rotation of the Sun seen from the Earth varies from 27.21 to 27.34 days according to the position of the Earth in its orbit. The positions of features on the solar disc are often given in units of Carrington rotations (27.2753 days) with rotation number one starting on 9 November 1853. A time difference of one hour results in a movement of about four pixels at the center of a solar disc of about 400 pixel radius.

Solar software routines are available, which take account of all the factors mentioned earlier, to transform an image from one viewpoint and time to another. Figure 2.4 illustrates the synchronization of two observations of sunspots taken about three days apart at the Meudon observatory in Paris in Ca K1 radiation and the L1 point by the SOHO satellite in white light continuum radiation. The SOHO Mdi image at the top right has been rotated to move features to the right and bring them into coincidence positions with the corresponding features on the Ca K1 image. At the same time, the two images at the bottom have been adjusted to the same size and solar disk position and center. The rotated Sun shown at the bottom right in Fig. 2.4 is not circular because part of the invisible rear surface of the Sun has been rotated into view and displayed in gray.

The following IDL code was used to synchronize the views of the two source images shown in Fig. 2.4.

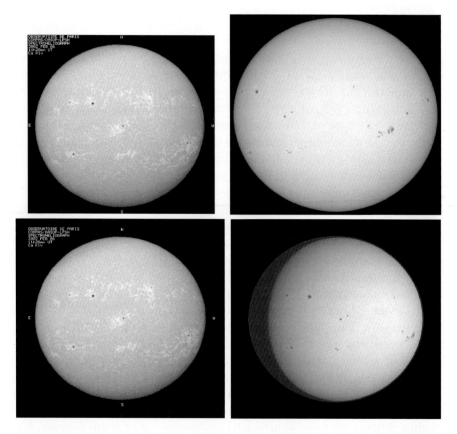

Fig. 2.4. Sunspots seen in Ca K1 radiation at the Meudon observatory (06/02/2002) at left and in white light continuum radiation from SOHO mdi (03/02/2002) at top right. The rotation of the solar disk in the interval between observations is clearly visible. The images at the bottom have been transformed to a standard size and centered and the bottom right image has been transformed through differential rotation to correspond to the view on 06/02/2002

Example 2.1.2.

```
PRO fits2map_earth_view_stand1
   pa1='c:\Images\MDI\'
   pa2='c:\Images\meudon\k1\'
   fn1=dialog_pickfile(path=pa1, FILTER = '*.FITS')
   data1=readfits(fn1, h1)
   window, 0, TITLE='Input data', xs=1024, ys=1024
   tvscl, data1
   index2map, h1, data1, map1
   emap1=map2earth(map1)
   gmap1=grid_map(emap1, 1024, 1024, Dx=NewDx, Dy=NewDx)
   window, 1
   plot_map,gmap1
   dmap1=drot_map(gmap1, 59.475)
   window, 2
   plot_map,dmap1
   map2index,dmap1, h1, newdata1
   fn2=dialog_pickfile(path=pa2, /multiple_files, FILTER = '*.FITS')
   mreadfits, fn2, h2, data2
   window, 4, TITLE='Input data', xs=1024, ys=1024
   tvscl, data1
   index2map, h2, data2, map2
   window, 4
   plot_map,map2
END
```

Correction of Solar Shape

Solar images are often elliptical in shape due to different spectrograph slit sizes at the top and bottom, or a slight inclination of the slit from the vertical line, or asynchronous motion of the slit with respect to the solar motion during the image capture (Kohl et al. 1994). Several authors (Kohl et al. 1994; Bornmann et al. 1996; Walton et al. 1998; Denker et al. 1999; Zharkova et al. 2003) have described techniques for fitting a circle or ellipse to the solar disc to determine its center and shape, generally prior to the removal of limb darkening or transformation to a standard size for comparison with other images. Determination of the solar disc radius and center may also be necessary because the values provided in the FITS file header are incorrect. With knowledge of the shape and true location of the solar disc in the image and the nature of the distortion, if any, a geometrical transformation may be applied to correct the shape (if necessary) and transform the solar disc to a standard size, centered in a standard sized image. Figure 2.5 shows four

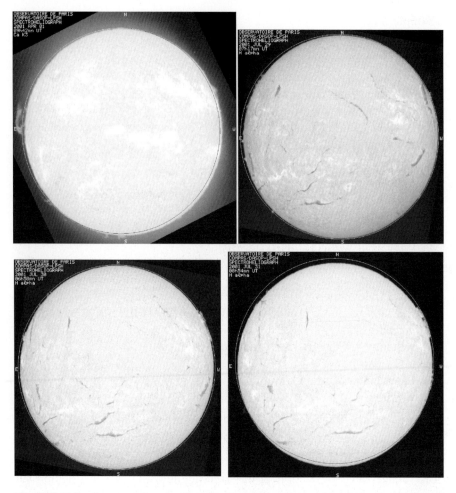

Fig. 2.5. Solar images taken by the Meudon observatory with circle superimposed in reverse gray value showing the position of the circular solar limb according to the information in the file header

examples of H_α and Ca K line full disk images taken at the Meudon Observatory in Paris and available from the website Bas2000 (2005). Superimposed on the images are circles, drawn in the reverse background intensity, representing the position of the solar limb according to the information contained in the associated file header. It can be seen that the images are not at the specified locations and they are all slightly elliptical in shape.

Limb Fitting

The automatic procedures for determining the solar disc described in Walton et al. (1998) and Zharkova et al. (2003) both make a preliminary estimate of the solar disc position. The reason for this is illustrated in Fig. 2.6 which shows an example image after the application of a Canny edge detector (Canny 1986). If all the edge points shown in this figure are supplied to an ellipse fitting routine then the result will be very unreliable so a preliminary estimate of the limb location is used to reject the majority of the spurious limb points.

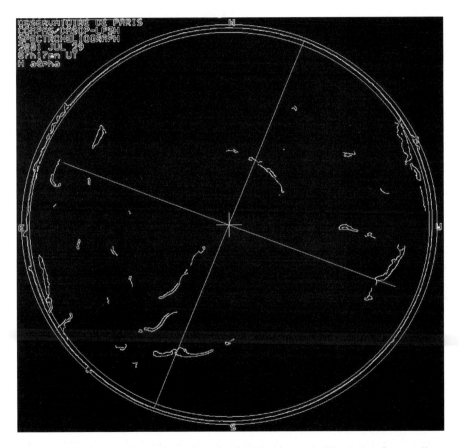

Fig. 2.6. The Meudon Ca K image (29/07/2001), shown in Fig. 2.1a after application of the Canny edge detector, see text. The *small cross* in the center marks the initial estimate of the disk center. The two concentric *circles* mark the initial region within which edge points are fitted to an ellipse, which is shown in gray beneath the edge points. The *large cross* shows the major (longer line) and minor axes of the fitted ellipse

The procedure for making the initial estimate of limb location in Zharkova et al. (2003) includes the following steps:

1. Compute the image histogram which has two or more large peaks, as illustrated by the example in Fig. 2.7, corresponding to the solar disc and outside disc background. Analyze the histogram to determine an intensity threshold between the solar disc peak and the next lower intensity peak corresponding to off-limb background.
2. Threshold the image at the intensity value found in (1) to form a binary image in which the largest object is the solar disc, possibly containing some holes due to dark features in the original image. Fill the region within the outer boundary of the disc object and determine its area (number of pixels) and centroid. This centroid is the initial estimate of the solar center and the square root of the area divided by pi is the initial estimate of the solar radius.

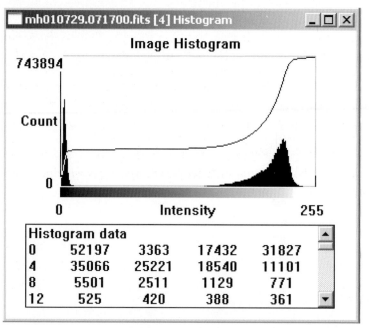

Fig. 2.7. The intensity histogram of the Meudon Ca K image (29/072001) shown in Fig. 2.6 which is the source of the edge points in Fig. 2.1a. The intensities have been rescaled from 12 to 8 bit. The histogram has two major peaks associated with the low-intensity region outside the solar disk and the high-intensity region inside the disk. The curve is the cumulative distribution function for the histogram. The numbers shown below the graph are the lowest 16 values of the histogram which include the lower peak

The ellipse fitting procedure in Zharkova et al. (2003) includes the following steps:

1. A Canny edge detector is applied to the image (see Sect. 3.3.2). This includes a Gaussian smoothing kernel of size 5×5 pixel, gradient computation, non-maximum gradient suppression and hysteresis tracking to produce a binary edge map of the original image. The two thresholds used by the hysteresis tracking are 25 and 5 for 8 bit images and 200 and 40 for 12 bit images.

2. The edge map produced by (1) and illustrated in Fig. 2.6 is analyzed to determine an annulus enclosing the initial circular estimate of the limb within which edge pixels are selected for input to the ellipse fitting procedure. The analysis is done by computing a histogram of pixel distances from the initial estimate of center and, starting at the distance corresponding to the initial estimate of the radius R, accumulating the histogram counts working alternately inward and outward away from R until the count reaches the value $2.5\,\pi\,R$. If this total is not reached because the edge gradient is low, due to high off-limb background intensity for example, step (1) is repeated with reduced upper threshold value. If a sufficient number of candidate points cannot be obtained, maybe because of the shape of the limb edge, then the process is terminated. The candidate edge candidates are checked for the presence of continuous angular gaps and if gaps greater than $30°$ are found then step (1) is repeated with reduced upper threshold value.

3. An ellipse fit is performed using the candidate edge pixels from (2) by minimizing the algebraic distance by using Singular Value Decomposition as follows:

The equation of a conic:

$$ax^2 + bxy + cy^2 + dx + ey + f = 0 \qquad (2.3)$$

In the matrix form becomes:

$$x^T A = 0, \qquad (2.4)$$

where $x^T = \begin{bmatrix} x^2 & xy & y^2 & x & y & 1 \end{bmatrix}$ and $A^T = \begin{bmatrix} a & b & c & d & e & f \end{bmatrix}$.

The sum of the squared algebraic distances

$$\sum_{i=1}^{N} \left| x_i^T A \right|^2 = A^T X^T X A, \tag{2.5}$$

where

$$X = \begin{bmatrix} x_1^2 & x_1 y_1 & y_1^2 & x_1 & y_1 & 1 \\ x_2^2 & x_2 y_2 & y_2^2 & x_2 & y_2 & 1 \\ & & \vdots & & & \\ & & \vdots & & & \\ x_N^2 & x_N y_N & y_N^2 & x_N & y_N & 1 \end{bmatrix}$$

is minimized, ignoring the trivial solution $A = 0$, by computing the singular value decomposition $X^T X = U D V^T$. The required least square deviation solution $\tilde{A}$ is the vector in V corresponding to the smallest (ideally zero) singular value in D (Press et al. 1992). Although this approach does not constrain the solution to an ellipse in general, in this application with the edge points on all sides of the limb, a constrained solution is unnecessary. However, while the limb edge is strongly represented in the candidate edge points there are usually extra points associated with structures such as prominences and interior features that will bias the ellipse fit. To deal with them the fit is performed iteratively, removing the outliers with the biggest radial errors after the previous fit, until the standard deviation of radial distance between the points and the fitted ellipse is less than 0.75 pixel. The result of this process applied to the source image for Fig. 2.6 is shown in Fig. 2.8.

Geometrical Correction

Having determined the elliptical geometry of a solar limb, in order to correct the shape back to a circle before applying the limb darkening corrections, a single transformation combining all the geometrical corrections should be applied. Applying the individual transformations using a sequence of the rotation and resizing functions will result in a build up of the interpolation errors. The transformation applied should reverse the process that caused the distortion. In the correction of Ca K line images from the San Fernando Observatory by Walton et al. (1998), a shearing transformation was applied. In the case of the H_α and Ca k line images from the Meudon observatory by Zharkova et al. (2003), a transformation rescaling the ellipse in the directions of the principal axes to make it circular was applied.

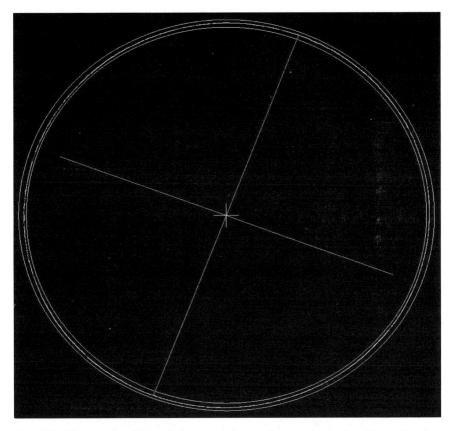

Fig. 2.8. The result of iteratively removing outlying edge points from the data shown in Fig. 2.6 from the ellipse until the standard deviation between the radial distance between points and ellipse is less than 0.75 pixel. The edge points used in the final fit are shown in white and are superimposed on the fitted ellipse which is shown in gray. The fitted ellipse is only visible in the gaps in the edge points where the original limb boundary is obscured by filaments or dark features

To rescale in a particular direction, the transformation combines the following steps: a translation to move the co-ordinates origin from the corner to the center of the ellipse; a rotation to make the principal axes of the ellipse parallel to the image edges; a rescaling in the horizontal and vertical directions to make the major and minor axes of the Sun the same required size; the inverse rotation; a translation to move the center of the Sun to the center of new image. The geometrical transformation from the new image to original image determines the point in the original image from which a particular pixel in the new image came, and its value can be obtained by a nearest neighbor, bilinear or bicubic interpolation.

The overall transformation combining rotation and translation in homogeneous co-ordinates is

$$
\begin{pmatrix}
a & b & ad_x + bd_y \\
b & c & bd_x + cd_y \\
0 & 0 & 1
\end{pmatrix},
\tag{2.6}
$$

where d_x and d_y are the x and y components of the displacements from the new position of the solar center to its initial position. The matrix combining the rotation and rescaling transformations from the new image to the original one is

$$
\begin{pmatrix}
a & b \\
b & c
\end{pmatrix}
=
\begin{pmatrix}
\dfrac{1}{S_x}\cos^2\theta + \dfrac{1}{S_y}\sin^2\theta & -\left(\dfrac{1}{S_y} - \dfrac{1}{S_x}\right)\sin\theta\cos\theta \\[3ex]
-\left(\dfrac{1}{S_y} - \dfrac{1}{S_x}\right)\sin\theta\cos\theta & \dfrac{1}{S_x}\sin^2\theta + \dfrac{1}{S_y}\cos^2\theta
\end{pmatrix}
\tag{2.7}
$$

where θ is the orientation of the principal axes of the ellipse to the image axes and S_x and S_y are the ratios of the semimajor axes to the required radius. The transformation from the final image to the original one is applied with resampling done using either nearest neighbor, bilinear, or bicubic interpolation depending on whether speed or quality is most important. Figure 2.9 shows the source image for Figs. 2.6, 2.7, and 2.8 after limb fitting and correction to circular shape and standardized size.

Change of Coordinate Systems

Transformation of coordinate systems has a number of applications in the development of the Solar Feature Catalogue (Zharkova et al. 2005).

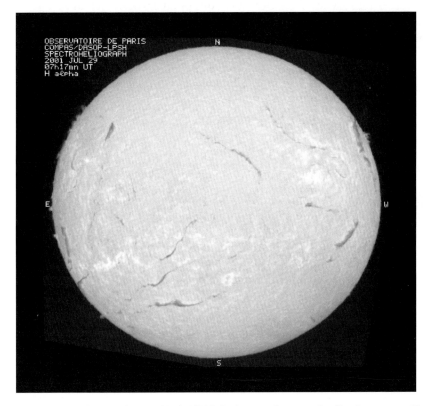

Fig. 2.9. A Meudon H$_\alpha$ image (29/07/2001) shown after standardization to a radius of 420 pixel centered in an image of size 1,024 × 1,024 pixel. The ellipse fitting procedure gave values of 411.6 and 406.7 pixel for the major and minor axes, respectively, and center at 427.6, 432 pixel in the original image of size 866 × 859 pixel

One is the transformation of the solar disc from the original rectangular (x, y) coordinate system to the polar (r, θ) coordinate system, illustrated in Fig. 2.10, to facilitate the computation of the median intensity at different radial distances from the center of the solar disc described in Removal of Radial Background Illumination and the detection of prominences on the edge of the disc. Another application is the transformation from the solar disc from the original rectangular (x, y) coordinate system to the heliographic latitude longitudinal system or Carrington coordinate system. This is needed to generate synoptic charts for the display of solar activity like filaments, sunspots, and active regions over several solar rotations.

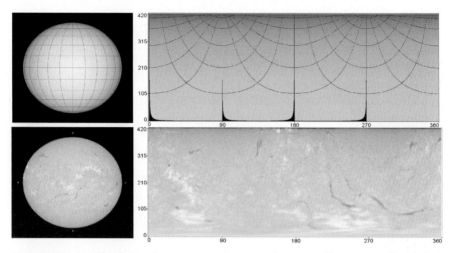

Fig. 2.10. Application of a rectangular to polar co-ordinate transformation (horizontal axis: angle starting at right most point on disk, vertical axis: distance from disk center) to an artificial image at top and to an H_α solar image at bottom

2.1.2 Intensity Standardization

The radiation-sensing elements (films or CCDs) used to capture the solar image may have a position dependent linear or nonlinear response to the brightness of the solar disc. For example, photographic negatives have nonlinear transmission characteristics and CCD sensing elements have a linear but nonuniform spatial brightness response (Denker et al. 1999; Preminger et al. 2001). These effects should be removed by the instrument calibration before the images are archived and are not discussed in this chapter. However there are other noninstrumental causes of variation of the background brightness of the solar disc with position.

The solar atmosphere is not completely transparent and due to the scattering of radiation only the part within an optical depth of unity of the observer are visible and deeper layers are invisible. This means that at the center of the solar disc an observer can see deeper into the Sun than toward the limb. At visible wavelengths the temperature of the sun decreases with increasing radius and limb darkening is seen. This effect is illustrated in Fig. 2.11. At ultraviolet wavelengths the temperature of the Sun increases with increasing wavelength and limb brightening is observed. In order to improve the accuracy of feature recognition near the limb, these variations may be removed prior to detection by the methods described in "Removal of Radial Background Illumination."

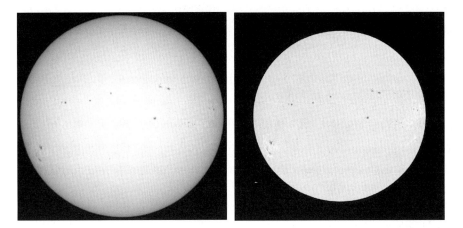

Fig. 2.11. At left a SOHO Mdi white light continuum image (01/01/2002) and at right the same image after limb darkening removal and standardization of size

Solar images obtained by ground based observatories are also affected by atmospheric transparency, turbulence etc., which vary from day to day or even during the time of observations and may be nonuniform over the solar disc. This type of nonradial solar background variation may be removed using the methods described in "Removal of Nonradial Background Illumination." Solar images produced by spectroheliograms sometimes exhibit dark straight lines caused by the combination of a grain of dust lying on the entrance slit and the scanning mechanism. These lines are particularly difficult to differentiate from the filaments seen in H_α images, especially if they overlap. They can be removed by the method described in "Removal of Dust Lines."

Removal of Radial Background Illumination

The first step in removing the varying part of the solar disc background illumination to leave the local features is to estimate that variation. The radial profile of the background intensity may be estimated by using a standard radial intensity profile (Allen 1973) or directly from the solar image (Walton et al. 1998; Zharkova et al. 2003). Both Walton et al. and Zharkova et al. use the median intensity of the solar disc at distance R pixels from the center as the estimate of the background intensity at R.

Whatever method is used to estimate the background, the corrected solar intensity is obtained by dividing the observed solar intensity by the estimated background intensity at the same point. The result is an intensity variation over the solar disc which deviates from unity only where there are local darker features like sunspots and filaments or brighter features like active regions and flares. The resulting intensity range may be renormalized to a desired standard range or to standardize the mean intensity and mean intensity variation, as required, by multiplying by an appropriate linear scaling function.

The formula in Allen (1973) for the radial variation of the solar background intensity is as follows:

$$I(\theta) = I(0)\left[1 - u(1 - \cos(\theta)) - v(1 - \cos^2(\theta))\right] \qquad (2.8)$$

The intensity at a point on the solar disk is a function of the angle θ between the vectors from the Sun's center to the point and toward the observer. θ varies between 0 and $\pi/2$. The values of u and v are calculated as polynomials in the observation wavelength λ in Angstroms using the following equation taken from the SolarSoft library (Slibrary 2005) darklimb_u and darklimb_v functions, valid in the range 4,000–15,000 Angstroms.

$$u = \sum_{i=0}^{5} A_i \times \lambda^i \qquad\qquad v = \sum_{i=0}^{5} B_i \times \lambda^i$$

The values of the coefficients are: $A_0 = -8.9829751$, $A_1 = 0.0069093916$, $A_2 = -1.8144591e-6$, $A_3 = 2.2540875e-10$, $A_4 = -1.3389747e-14$, $A_5 = 3.0453572e-19$, $B_0 = 9.2891180$, $B_1 = -0.0062212632$, $B_2 = 1.5788029e-6$, $B_3 = -1.9359644e-10$, $B_4 = 1.1444469e-14$, $B_5 = -2.599494e-19$.

The procedure for estimating the radial background intensity variation of the solar disk presented in Zharkova, et al. (2003) includes the following steps:

1. The solar disc standardized to circular shape is mapped onto a rectangular grid using a Cartesian-to-Polar co-ordinate transformation where the rows contain samples at fixed radius and the columns at fixed angle, as illustrated in Fig. 2.10.
2. The median value of each row is computed.
3. A radial intensity distribution is obtained by fitting a low-order polynomial function to the median values and this is used to renormalize the intensity as illustrated in Fig. 2.11.

Removal of Nonradial Background Illumination

Manual inspection of images from the Meudon observatory by displaying the intensity variation in false color reveals that some of them have a discernable linear spatial variation of intensity. The amplitude of this linear variation can be measured (Zharkova et al. 2003) by applying a Fourier transform to rows of the rectangular image obtained by applying the

Cartesian-to-Polar co-ordinate transformation. The zero-order Fourier coefficient is the average intensity of the row and the first-order coefficient is the amplitude of the linear spatial variation in the original image. Any detected systematic variation of first order coefficients in different rows can be removed to eliminate the associated linear intensity variation.

Variations in solar background illumination which do not have a simple known mathematical structure are difficult to estimate accurately by automatic procedures because this requires prior identification of the regions of the disc which do not include local features. If a wide spread of candidate background points are available over the image then a suitable background function may be fitted to them (Veronig et al. 2001). Alternatively the paper by Fuller et al. (2005) presents a method for correcting nonradial intensity variations.

To identify this kind of large-scale variations of the intensity over the disk Fuller et al. (2005) first use median filtering with a large window. The median filter eliminates the highest and lowest values (corresponding to bright plages and dark filaments) to give a first approximation of the background fluctuations which can then be subtracted from the original image. This first normalized image can be used to more efficiently locate pixels corresponding to bright and dark regions by defining two suitable thresholds. These thresholds are obtained from the histogram of the new image and the values of pixels outside this range are replaced by corresponding values in the background map. Applying median filtering again (with a smaller window) then gives a more reliable estimate of the large scale variations. The following pseudocode contains the details of the algorithm and Fig. 2.12 shows an example with the resulting estimated background.

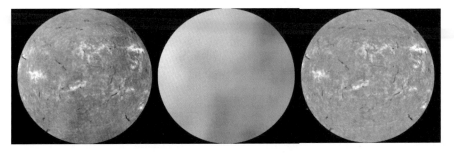

Fig. 2.12. Intensity normalization using median filtering. On the left is the original image, in the center the estimated background intensity on the right the image after background removal

Example 2.1.3.

> **begin**
> Rescale I to a smaller size: I_{small} // *To save computer time*
> $B_{small} = median(I_{small}, W_{size})$ // W_{size} *is the filter size*
> $I'_{small} = I_{small} - B_{small} + mean(B_{small})$ // *Subtract and get back to original intensity Level*
> $Hist = histogram(I'_{small})$
> $H_M = mode(Hist)$ // intensity with highest count
> Let V_M be the intensity value corresponding to H_M
> Let V_1 be the intensity value corresponding to H_M/a_1 ($V_1 < V_M$)
> Let V_2 be the intensity value corresponding to H_M/a_2 ($V_2 > V_M$) // a_1, a_2 *are constants*
> Let S be the set of pixels in I'_{small} lower than V_1 and greater than V_2
> $I'_{small}[S] = B_{small}[S]$
> $B'_{small} = median(I'_{small}, W_{size}/2)$
> Rescale B'_{small} to original size: B'
> $I_n = I - B' + mean(B')$ // I_n: *image with background removed*
> **end**

Removal of Dust Lines

Fuller and Abourdarham (2004) present the following method which they applied to the removal of dark lines from Meudon H_α spectroheliograms. A binary image is first computed from an intensity normalized image obtained as described in "Removal of Radial Background Illumination" or "Removal of Nonradial Background Illumination," using the cumulative distribution function computed from the image histogram to set the threshold controlling the ratio of nonzero feature pixels to zero background pixels. A morphological thinning operator (see "Use of skeletons") is then used to reduce the number of feature pixels by thinning connected features to a thickness of one pixel and speeding up the next step, a Hough transform line detector (Hough 1962).

The Hough transform approach, in general, is used to detect the presence of edge pixels in an image which fall on a specified family of curves and uses an accumulator array whose dimension equals the number of parameters in the curves to be detected. For example, in the case of a straight line expressed in the normal form $\rho = x \cos \theta + y \sin \theta$, the parameters are ρ and θ and the two indices of the accumulator array correspond to quantized values of ρ and θ over the range of values of these parameters which are of interest. The accumulator array is used as follows. Each edge pixel in the image with co-ordinates x, y is considered in turn and those elements in the accumulator array which satisfy the straight line $\rho = x \cos \theta + y \sin \theta$, with quantized values of ρ and θ are incremented. After all the edge pixels in the image have been processed in this way, the accumulator array is examined to locate those elements with largest values. The largest elements indicate the parameters of the straight lines within the image which have the most supporting evidence. Using this information, the image can be searched to find the points lists of the line segments with corresponding parameter values. The implementation of the Hough transform is similar for curves with more parameters but the execution time increases very rapidly with the number of model parameters. Fast variants of the Hough transform exist, which take account of the gradient of edge about each edge point (O'Gorman and Clowes 1976; Burns et al. 1986).

The Hough transform is also used in the detection of coronal mass ejections by Bergman et al. (2002), to find bright ridges in height-time maps generated from an image sequence. In the line cleaning application by Fuller and Abourdarham the Hough transform is applied to the thinned thresholded image containing line and other edge points. The data in the transform space is then thresholded at a value corresponding to the minimum number of points needed to identify straight lines. The straight lines obtained are then used to identify lines of dark pixels whose values are corrected by interpolating from neighboring pixels (see Fig. 2.13).

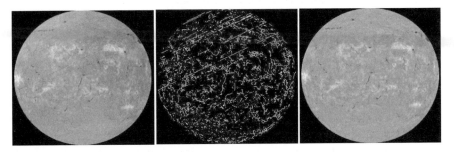

Fig. 2.13. Steps in the dust line removal procedure. On the left original image with dust lines, in the center thinned binary image and on the right

2.2 Digital Image Enhancement and Morphological Operations

S.S. Ipson and S.I. Zharkov

Abstract

This section describes a number of basic digital image processing operations which are frequently used either to preprocess an image to improve the effectiveness of a feature recognition procedure or as intermediate steps within automated solar feature recognition procedures. Included here are sharpening and deblurring operations and a number of morphological operations.

2.2.1 Image Deblurring

Solar images may be blurred by the imaging instrument or by atmospheric conditions. Sharpened images may be obtained by applying simple spatial domain convolution or frequency domain filters (Gonzalez and Woods 2001). Alternatively, images may be corrected by first estimating the point-spread-function (PSF) of the blurring process and then applying a deblurring transformation such as a Wiener filter (Walton and Preminger 1999).

Sharpening Filters

Fuller and Abourdarham (2004) apply a sharpening operation to H_α solar images in order to enhance the filament edges and increase the detection efficiency for the thinnest parts of the filaments. The sharpening filter they apply is a convolution with the following 3×3 kernel.

$$\begin{vmatrix} -1 & -1 & -1 \\ -1 & 9 & -1 \\ -1 & -1 & -1 \end{vmatrix} \tag{2.9}$$

This convolution has the effect of adding to the original image, those high frequency components resulting from the convolution of the image with the following high-pass filter kernel.

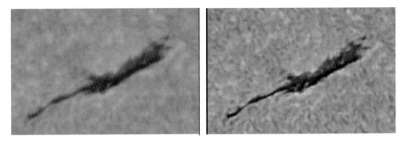

Fig. 2.14. At left is the original image of the local region about a filament seen in an H_α solar image with sharpened version shown at right

$$\begin{vmatrix} -1 & -1 & -1 \\ -1 & 8 & -1 \\ -1 & -1 & -1 \end{vmatrix} \tag{2.10}$$

The example from Fuller and Aboudarham shown in Fig. 2.14 illustrates the effect of the sharpening process on an image which contains blurred detail similar in scale to the size of the convolution kernel. The result is a visible sharpening effect with increased local contrast.

The amount of sharpening obtained can be reduced by decreasing the proportion of the high-frequency component, for example by varying the parameter W in the following sharpening convolution kernel, and this may be preferred if the original image has a significant amount of noise.

$$\begin{vmatrix} -1 & -1 & -1 \\ -1 & W & -1 \\ -1 & -1 & -1 \end{vmatrix} \times \frac{1}{(W-8)} \quad where \ W \rangle 8 \tag{2.11}$$

Example 2.2.1.

The following IDL code was used to sharpen the image in Fig. 2.14.

```
PRO sharp
     a = dialog_read_image (PATH = 'c:\ Test images', FILE = fa,
       IMAGE = im1)
   info=size(im1)
   image1x = info[1]
   image1y = info[2]
```

```
      window,0, xsize=image1x, ysize=image1y,title='Source image'
      TVSCL, im1
      kernel = [[-1, -1, -1], [-1, 9, -1], [-1, -1, -1]]
      im2 =  convol(im1, kernel)
      window,1, xsize=image1x, ysize=image1y,title='Sharp'
      TVSCL, im2
   END
```

Image Deblurring Using the System PSF

A commonly used model for the image formation process is the convolution of the true image by the PSF of the image capturing system followed by the addition of noise. The PSF is the image captured by the system of a point object and can have a nonzero extent for several reasons including aberrations in the optical system, incorrect focusing, motion of the object or camera system, and light scattering. The technique described in "Sharpening Filters," although straightforward and fast, is only an enhancement technique and does not give a reconstruction of the ideal image from the blurred image taking account of the system PSF and noise. In the frequency space the convolution operation becomes multiplication and, in the absence of noise, reconstruction can be performed by dividing the Fourier transform of the blurred image by the Fourier transform of the PSF (the optical transfer function or OTF) and inverse Fourier transforming the result. When noise is present, which is the practical situation, this approach gives very poor results unless the division is performed only over a limited range of frequencies about the origin. The range is adjusted interactively to give the best results. An alternative method of deblurring which takes account of the noise and is optimal in the least-squares sense is the Weiner filter, adopted for images by Helstrom (1967). This filter G, applied in the frequency domain to the Fourier transform of the blurred image, has the following form:

$$G(u,v) = \frac{H^*(u,v)P_s(u,v)}{|H(u,v)|^2 P_s(u,v) + P_n(u,v)} \tag{2.12}$$

where u and v are the spatial frequencies in the x and y directions, H is the OTF and P_s and P_n are the power spectra of the signal and the noise, respectively. In the limit where the noise is zero, the Weiner filter reduces to the inverse filter, $1/H(u, v)$, which divides the spectrum of the blurred image by the OTF to produce the spectrum of the true image. In the presence of noise the Wiener filter introduces smoothing linked to the amount of noise present which prevents excessive amplification of the noise but reduces the accuracy of the reconstruction.

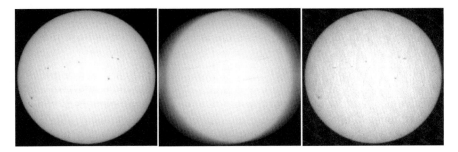

Fig. 2.15. Deblurring an image using the Wiener filter. Source image at left, image blurred by uniform motion at middle and deblurred image at right. Artifacts associated with the deblurring process are visible in the deblurred image

Example 2.2.2.

Figure 2.15 illustrates the application of the Wiener filter to image deblurring using the following MATLABT^{M} code.

```
I = imread('source.bmp);
figure;imshow(I);title('Original Image');
LEN = 31;
THETA = 11;
PSF = fspecial('motion',LEN,THETA);
Blurred = imfilter(I,PSF,'circular','conv');
figure; imshow(Blurred);title('Blurred Image');
wnr1 = deconvwnr(Blurred,PSF);
```

figure;imshow(wnr1); title('Restored image');

2.2.2 Morphological Operations

Morphological operations, see for example Gonzalez and Woods (2001), are frequently used as pre- and postprocessing steps in algorithms for the enhancement or analysis of binary and grayscale images. Example applications include noise reduction and smoothing, boundary extraction, object recognition, thinning, skeletonization, and pruning. The principal morphological operations used in the development of the Solar Feature Catalogue for the EGSO project are based on erosion and dilation operations described in "Erosion and Dilation," the hit-miss transform described in "The Hit-or-Miss Transform," skeletonization and pruning operations described in "Use of Skeletons," and the Watershed transform described in "Watershed Transform". The distance transform operation is described in "Distance Transform."

Erosion and Dilations

The erosion and dilation operations are the simplest morphological operations and many other morphological operations can be expressed in terms of
these two primitives. The application of an erosion operation to a binary
image consisting of background pixels and foreground object pixels is illustrated in Fig. 2.16 and involves the use of an object, small compared with the
whole image, called the structuring element. The structuring element is
commonly square, rectangular, or circular, but may be any shape, depending
on the result required. It contains a reference point which is often, but not
always, at its center. When the erosion operation is applied to a binary
image, the chosen structuring element is placed at all possible positions
within the image. The set of reference positions for which the structuring
element does not overlap any background pixels and is entirely contained
within foreground objects define the foreground pixels in the eroded
image. The extent of the erosion depends on the size of the structuring element and objects smaller than the structuring element, which may be due
to noise, disappear. Offsetting the reference point from the center of the
structuring element causes the amount of erosion to vary over the boundary of the object. In a similar manner to erosion, the dilation operation
applied to a binary image is defined as the set of all reference positions for
which the structuring element overlaps with at least one pixel with the
foreground set of pixels. The object on the left in Fig. 2.16 is the dilation
of the object on the right using the indicated structuring element.

Erosion Θ and dilation $\oplus$ may be defined using set terminology as
follows:

$$A \Theta B = \left\{ z \mid (\hat{B})_z \subseteq A \right\} \tag{2.13a}$$

and

$$A \oplus B = \left\{ z \mid \left[(\hat{B})_z \cap A \right] \subseteq A \right\}, \tag{2.13b}$$

where A and B are the image foreground and structuring element pixels,
respectively, $\hat{B}$ indicates the reflection of B about its reference point and z
indicates the translations of the structuring element. Applied to grayscale
images, the erosion and dilation operations replace each pixel in the image
to which they are applied by the minimum or maximum intensities, respectively, of the $n \times n$ region about that pixel (after summing the image pixels
with corresponding structuring element values if the latter are nonzero).

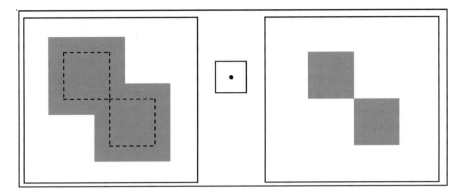

Fig. 2.16. An illustration of the morphological erosion process: at left is a binary image containing a foreground object. The structuring element, with reference point marked, is shown in the center and the resulting eroded object is shown on the right. A region of width d/2, where d is the dimension of the square structuring element, has been removed from the object as indicated by the *dashed lines*. Conversely, the object on the left is the result of dilating the object on the right using the same structuring element

The result of these operations is to erode or dilate, respectively, foreground features which are brighter than their surroundings and dilate and erode respectively background features which are darker than their surroundings.

Example 2.2.3.

Figure 2.17 (middle) shows the results of applying the following IDL code to a binary image.

```
PRO erode
    a = dialog_read_image(PATH = 'c:\, FILE = fa, IMAGE = im1)
    info=size(im1)
    image1x = info[1]
    image1y = info[2]
    window,0, xsize=image1x, ysize=image1y,title='Source image'
    TVSCL, im1
    im2 = erode (im1, REPLICATE(1,8,8))
    window,1, xsize=image1x, ysize=image1y,title='MORPH_ERODE'
    TVSCL, im2
    a = dialog_write_image (im2, PATH = 'c:\', FILENAME = 'erode.tif')
END
```

Fig. 2.17. Original binary image of leaves on left; eroded using a 5 × 5 structuring element in center and after dilation of the eroded image with the same structuring element on the right. The combination of erosion followed by dilation is morphological opening

Feature edges can be detected by subtracting an eroded image from a dilated image and then thresholding as illustrated in the upper part of Fig. 2.18. Feature boundary smoothing can be achieved using morphological opening and closing operations which are defined as erosion followed by dilation and as dilation followed by erosion respectively. Both generally smooth the boundaries of features at the scale of the structuring element, but opening also breaks narrow isthmuses and eliminates thin protrusions while closing also fuses narrow isthmuses and long thin gulfs and eliminates small holes. The lower part of Fig. 2.18 shows examples of opening and closing on a binary image. These operations are used in sunspot and active region detection (Zharkov et al. 2005; Benkhalil et al. 2005).

The Hit-or-Miss Transform

The morphological hit-or-miss transform is a basic tool for selecting objects on the basis of their sizes and shapes. It uses two structuring elements. The first identifies the positions where the hit structuring element lies entirely within binary objects and the second identifies positions where the miss structuring element lies entirely outside binary objects. The resulting image is the intersection of these two sets of pixels. The IDL code in Example 2.2.4. was used to generate the image shown at the right in Fig. 2.19, selecting the blobs of radius 6 pixel from the image on the left containing blobs of radii 2, 6, and 12 pixel.

Fig. 2.18. The upper images from left to right are the original image, original dilated by 5 × 5 structuring element, original eroded by 5 × 5 structuring element and difference of dilated and eroded image. The width of the resulting boundary map is the width of the structuring element. The lower images from left to right are the original image, original image after opening and original image after closing, both using a 5 × 5 structuring element

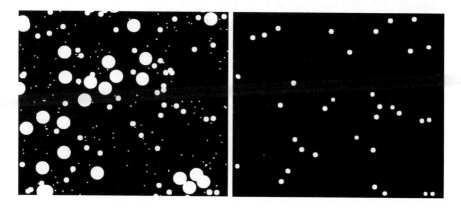

Fig. 2.19. An application of the Hit-or-Miss Transform on the image on the left containing blobs of three distinct sizes to produce the image on the right containing only the middle size blobs in the image on the left

Example 2.2.4.

```
PRO HitMiss2
   a = dialog_read_image(PATH = 'c:\ Images', FILE = fa, IMAGE =
   im1)
   info=size(im1)
   image1x = info[1]
   image1y = info[2]
   ; Display the original image
   window,0, xsize=image1x, ysize=image1y,title='Source'
   TVSCL, im1
   rh = 5 ;Radius of hit disc
   rm = 10 ;Radius of miss disc
   ;Create a binary disc of given radius.
   hit = SHIFT(DIST(2*rh+1), rh, rh) LE rh
   ;Complement of disc for miss
   miss = SHIFT(DIST(2*rm+1), rm, rm) GT rm
   ;Compute matches
   matches = MORPH_HITORMISS(im1, hit, miss)
   ;Expand matches to size of hit disc
   im2 = DILATE(matches, hit)*255
   ;Show matches
   window,2, xsize=image1x, ysize=image1y,title='Matches'
   TVSCL, im2
   a = dialog_write_image (im2, PATH = 'c:\Images', FILENAME =
      'match.bmp')
END
```

Use of Skeletons

The skeleton of a region may be defined as the medial axis which is the set of interior points which are each equidistant from two or more region boundary points (Blum 1967). Many algorithms (Arcelli et al. 1981; Pavlidis 1982; Zhang and Suen 1984) have been proposed for computing a medial axis representation of region, typically by iteratively removing edge pixels, which are computationally efficient, do not remove end points, do not break connectivity, and do not cause excessive erosion. Figure 2.20 shows the effect of the IDL thin operator on two images, one of a simple shape and the other of a complex shape.

Example 2.2.5.

The following IDL code was used to produce the images in Fig. 2.20.

```
PRO thin
   a = dialog_read_image(PATH = 'c:\ Images', FILE = fa, IMAGE =
   im1)
```

```
info=size(im1)
image1x = info[1]
image1y = info[2]
window,0, xsize=image1x, ysize=image1y,title='Source'
TVSCL, im1
im2 =  thin(im1)
im2 = im2
window,1, xsize=image1x, ysize=image1y,title='Thinned'
TVSCL, im2
a = dialog_write_image (im2, PATH = 'c:\Images', FILENAME =
'thinned.bmp')
END
```

Fig. 2.20. Two examples are shown of image thinning using the IDL thin function, source images on left and resulting thinned image on the right

Fig. 2.21. Original shape (1), full skeleton (2), pruned skeleton (3)

An example related to solar physics is shown in Fig. 2.21. For many years the positions of solar filaments have been recorded manually at the Meudon observatory in Paris by choosing a few points along the path of the filament and linking them together. To compare the results of automatic filament detection with the manually detected results it is therefore necessary to first compute the skeleton of the detected filament region. The skeleton, shown in the middle of Fig. 2.21 has several short spurs and these are removed by a pruning procedure (Gonzalez and Woods 2001) before the skeleton is suitable for comparison with the manually recorded data.

Watershed Transform

Interpreting a grayscale image as a surface, the Watershed transform (Digabel and Lantuejoul 1978; Beucher and Meyer 1992) detects the ridges or watershed boundaries between adjacent local minima. It does this by conditionally dilating the regions of local minima until they meet at watershed boundaries. These are the points, where viewed as a surface, a drop of water would be equally likely to fall to more than one minimum. Within each watershed region, a drop of water would be certain to fall to a single minimum. The regions formed all have unique labels which are given to the seed pixels located at the initially selected local minima. In practice, the Watershed transform is often used for the detection of nearly uniform objects from a uniform background and the transform applied to the gradient image rather than the original image. Figure 2.22 shows the application of the Watershed transform to both grayscale and binary images. In the latter case it identifies those regions which are completely enclosed by ridges.

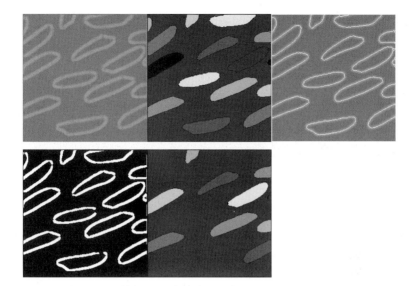

Fig. 2.22. Applications of the Watershed transform. From left to right upper row grayscale gradient source image, result of applying Watershed transform, merged watershed boundaries with source image. The bottom left binary image contains examples of complete and incomplete object boundaries and the bottom right image is the resulting Watershed transformed image with closed regions filled and labeled

Example 2.2.6.

The following IDL code was used to generate the images shown in Fig. 2.22.

```
PRO watershed
a = dialog_read_image(PATH = 'c:\Images', FILE = fa, IMAGE =
im0)
window,0, xsize=image1x, ysize=image1y,title='Source'
TVSCL, im0
info=size(im0)
image1x = info[1]
image1y = info[2]
;Create watershed image
im2 = WATERSHED(im1)
;Display it, showing the watershed regions
window,2, xsize=image1x, ysize=image1y,title='Watershed'
TVSCL, Im2
;Merge original image with boundaries of watershed regions
```

```
im3 = im1 > (MAX(im1) * (im2 EQ 0b))
window,3, xsize=image1x, ysize=image1y,title='Merged'
TVSCL, im3
END
```

In the solar physics application in Fig. 2.23, the Watershed transform is used in the sunspot detection procedures of the Solar Feature Catalogue after edge detection to fill in the sunspot areas completely enclosed by edges.

Distance Transform

The morphological distance transform produces a distance map in which points within a region of a binary image (which may be foreground or background) are labeled with their distance from the nearest boundary point. The distance measure used may be the Euclidean distance, which is exact, or chessboard or city-block distances which are less accurate but faster to compute. A fast implementation of the Euclidean distance transform can be found in Cuisenaire and Macq (1999). The distance transform may be used for a variety of morphological operations including erosion, dilation, and thinning by circular structure elements. Figure 2.24 shows the result of applying the Euclidean distance transform to the binary image of a crescent shaped object. As can be seen by inspecting the images in Fig. 2.24, the contours of constant distance from the object correspond to the dilation of the object by a circular structuring element and approximate circles at larger distances. The right-hand image in Fig. 2.24 shows the Euclidean distance transform of the crescent object and the points of local highest intensity correspond to the medial axial transform of the object.

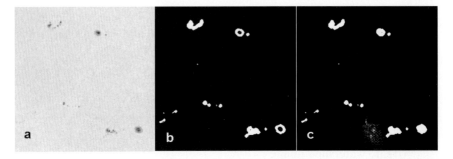

Fig. 2.23. Solar physics application of Watershed transform. At left is original solar image containing sunspots, at center is detected sunspot edges after application of morphological closing and at right is sunspot regions filled using the Watershed transform

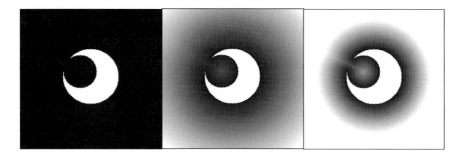

Fig. 2.24. Application of the distance transform to a binary image. From left to right, the source image, the result of applying the Euclidean distance transform to the source image with distance from object displayed as intensity value, the result of contrast stretching the middle image to make the distance variation in the region partially enclosed by the crescent more visible

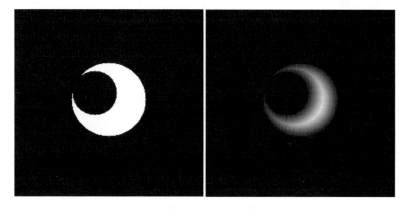

Fig. 2.25. Application of the distance transform to the interior of the binary image shown on the left. The medial axial transform of the left hand object correspond to the line of locally brightest pixels in the right hand image

Thresholding the right hand image in Fig. 2.25 at different intensities would be equivalent to the erosion of the object with circular structuring elements of different radii.

Example 2.2.7.

The following IDL code applies the morphological distance transform to an image. The valid values of the keyword NEIGHBOR_SAMPLING are 0, 1, 2, and 3 corresponding to a neighbor distances (no diagonal) equal to 1 pixel, chessboard distance, city block distance, and actual distance, respectively.

```
PRO distance
   a = dialog_read_image(PATH = 'c:\Images', FILE = fa, IMAGE = im1)
   info=size(im1)
   image1x = info[1]
   image1y = info[2]
   window,0, xsize=image1x, ysize=image1y,title='Source'
   TVSCL, im1
   ; compute the distance transform of the background
      im2=morph_distance(im1,/BACKGROUND,
   NEIGHBOR_SAMPLING = 3)
   im2 = im2*255
   window,1, xsize=image1x, ysize=image1y,title='distance transform'
   TVSCL, im2
END
```

References

Allen, C.W. (1973). *Astrophysical quantities*. London: Athlone

Arcelli, C., Cordella, L.P., & Levialdi, S. (1981). From local maxima to connected skeletons. *IEEE Trans. Pattern Anal. Mach. Intell.*, *PAMI 3*(2), 134–144

Benkhalil, A., Zharkova, V., Ipson, S., & Zharkov, S. (2005). An automated recognition of active regions on the full disk solar spectroheliograms using Ha, Ca II K3 and Fe XII 195 Å Lines. *Int. J. Comput. Appl., 12*(1), 21–29

Beucher, S., & Meyer, F. (1992). The morphological approach of segmentation: The watershed transformation. In Dougherty, E. (Ed.), *Mathematical morphology in image processing*. New York: Dekker

Blum, H. (1967). A transformation for extracting new descriptors of shape. In Walthen-Dunn, W. (Ed.) *Models for the perception of speech and visual form*. Cambridge, MA: MIT

Bornmann, P.L., Winkelman, D., & Kohl, T. (1996). *Automated solar image processing for flare forecasting*. In Proceedings of the solar terrestrial predictions workshop, 23–27. Japan: Hitachi

Burns, J.R., Hanson, A.R., & Riseman, E.M. (1986). Extracting Straight Lines. *IEEE Trans. Pattern Anal. Mach. Intell., 8*, 425–455

Canny, J. (1986). A computational approach to edge detection. *IEEE Trans. Pattern Anal. Mach. Intell. 8*, 679–698

Cuisenaire, O., & Macq, B. (1999). Fast and exact signed Euclidean distance transformation with linear complexity. In *Proceedings of the IEEE international conference on acoustics, Speech and Signal Processing (ICCASSP99), 6*, 3293–3296, Phoenix, AZ: ICCASSP99

Denker, C., Johannesson, A., Marquette, W., Goode, P.R., Wang, H., & Zirin, H. (1999). Synoptic H full-disk observations of the sun from big bear solar observatory. *Solar Phys., 184*, 87–102

Digabel, H., & Lantuejoul, C. (1978). Iterative algorithms. In Chermant, J.-L. (Ed.), Actes du second symposium european d'analyse quantitative des microstructures en sciences des materiaux, biologie et medecine, Caen, 4–7 October 1977 (1978), Stuttgart: Riederer Verlag, pp. 85–99

Fuller, N., & Aboudarham, J. (2004). Automatic detection of solar filaments versus manual digitization, the 8th international conference on knowledge-based intelligent information & engineering systems (KES2004). *Springer Lect. Not. Comput. Sci., LNAI 3215, 3,* 467–475

Fuller, N.J., Aboudarham, J., & Bentley, R.D. (2005). Filament recognition and image cleaning on meudon ha spectroheliograms *Solar Phys., 227*(1), 61–73

Gonzalez, R.C., & Woods, R.E. (2001). *Digital image processing.* Upper Saddle River, NJ: Prentice Hall

Helstrom, C.W. (1967). Image restoration by the method of least squares. *J. Opt. Soc. Am., 57*(3), 297–303

Hough, P.V.C. (1962). *Methods and means for recognizing complex patterns.* US Patent 3069654

Kohl, C.A., Lerner, R.A., Hough, A.A., & Loiselle, C.L. (1994). A stellar application of the IUE: Solar feature extraction. In *Proceedings of the DARPA image understanding workshop, Section IV image understanding environment (IUE),* 13–16, Monterey, CA: IUE

O'Gorman, F., & Clowes, M.B. (1976). Finding picture edges through collinearity of feature points. *IEEE Trans. Comput. C-25,* 449–454

Pavlidis, T. (1982). *Algorithms for graphics and image processing.* Rockville, MD: Computer Science Press

Preminger, D.G., Walton, S.R., & Chapman, G. (2001). A solar feature identification using contrasts and contiguity. *Solar Phys. 202*(1), 53–62

Press, W.H., Teukolsky, S.A., Vetterling, W.T., & Flannery, B.P. (1992). *Numerical recipes in C, 2nd edn.* Cambridge: Cambridge University Press

Slibrary (2005). http://www.astro.washington.edu/deutsch/idl/htmlhelp/slibrary32.html

SOHO (2005). http://sohowww.nascom.nasa.gov/

Turmon, M., Pap, J.M., & Mukhtar, S. (2002). Statistical pattern recognition for labelling solar active regions: Applications to SOHO/MDI imagery. *Astrophys. J. 568,* 396–407

Veronig, A., Steinegger, M., Otruba, A., Hanslmeier, A., Messerotti, M., Temmer, M., Gonzi, S., & Brunner, G. (2001). Automatic image processing in the frame of a solar flare alerting system. *Hvar. Obs. Bull., 24*(1), 195–205

Walton, S.R., & Preminger, D.G. (1999). Restoration and photometry of full-disk solar images. *Astrophys. J., 514,* 959–971

Walton, S.R., Chapman, G.A., Cookson, A.M., Dobias, J.J., & Preminger, D.G. (1998). Processing photometric full-disk solar images. *Solar Phys., 179,* 31–42

Zhang, T.Y., & Suen, C.Y. (1984). A fast digital algorithm for thinning digital patterns, *Commun. ACM, 27*(3), 236–239

Zharkov, S.I., Zharkova, V.V., Ipson, S.S., Benkhalil, A.K. (2005). Technique for automated recognition of sunspots on full disk solar images, *EURASIP J. Appl. Sign. Proc., 15,* 2573–2584.

Zharkova, V.V., Aboudarham, J., Zharkov, S., Ipson, S.S., Benkhalil, A.K., & Fuller, N. (2005). Solar feature catalogues in EGSO. *Solar Phys. Topical Issue, 228*(1–2), 139–160

Zharkova, V.V., Ipson, S.S., Zharkov, S.I., Benkhalil, A.K., Aboudarham, J., & Bentley, R.D. (2003). A full disk image standardisation of the synoptic solar observations at the Meudon observatory. *Solar Phys., 214*(1), 89–105

3 Intensity and Region-Based Feature Recognition in Solar Images

V.V. Zharkova, S.S. Ipson, S.I. Zharkov, and Ilias Maglogiannis

3.1 Basic Operations in Recognition Techniques

V.V. Zharkova and S.S. Ipson

3.1.1 Histograms

Histograms

A Histogram $H(I)$ is used to display in bar graph format measurement data distributed by categories. A histogram is used for:

1. Making decisions about a threshold after examining the intensity variations.
2. Displaying easily the variations in the process, i.e., as a result of faulty scanning.

Steps for constructing histograms:

1. Gather and tabulate data on a digital image.
2. Calculate the range of the pixel values by subtracting the smallest number in the data set from the largest. Call this value R.
3. Decide about how many bars (or classes) you want to display in your eventual histogram. Call this number K. With 1,024 pixels of data, $K = 11$ works well.
4. Determine the fixed width of each class by dividing the range, R, by the number of classes K. This value should be rounded to a "nice" number, generally a number ending in a zero. For example 11.3 would not be a "nice" number but 10 would be considered a "nice" number. Call this number i, for interval width. It is important to use "nice" numbers else the histogram created will have awkward scales on the X-axis.
5. Create a table of upper and lower class limits. Add the interval width i to the first "nice" number less than the lowest value in the data set to determine the upper limit of the first class. This first "nice" number becomes the lowest lower limit of the first class. The upper limit of the first class

V.V. Zharkova et al.: *Intensity and Region-Based Feature Recognition in Solar Images,* Studies in Computational Intelligence (SCI) **46**, 59–149 (2007)
www.springerlink.com

becomes the lower limit of the second class. Adding the internal width (i) to the lower limit of the second class determines the upper limit for the second class. Repeat this process until the largest upper limit exceeds the biggest piece of data. You should have approximately K classes or categories in total.

6. Sort, organize, or categorize the data in such a way that you can count or tabulate how many pixels in image fall into each of the classes or categories in your table above. These are the frequency counts and will be plotted on the Y-axis of the histogram.

7. Create the framework for the horizontal and vertical axes of the histogram. On the horizontal axis plot the lower and upper limits of each class determined above. The scale on the vertical axis should run from zero to the first "nice" number greater than the largest frequency count determined above.

8. Plot the frequency data on the histogram framework by drawing vertical bars for each class. The height of each bar represents the number or frequency of values occurring between the lower and upper limits of that class.

9. Interpret the histogram for skew and clustering problems using the formulae:

Interpreting Skew Problems

The data of some processed images are skewed. This situation appears in their intensity variable over the image (for example, the instrument slit is not vertical) or in the service processes (for example, the scanning process is not at a constant speed).

Data may be skewed to the left or right. If the histogram shows a long tail of data on the left side of the histogram, the data is termed left or negatively skewed. If a tail appears on the right side, the data is termed right or positively skewed. Most processed data should not typically appear skewed. If data is seriously skewed either to the left or right this may indicate that there are inconsistencies in the process or procedures, etc. This can require the determination of the appropriateness of the direction of the skew and its correction, either in the instrument (preferable) or in the digital images in the first is not possible.

Interpreting Clustering Problems

Data may also be clustered on the opposite ends of the scale or display two or more peaks indicating serious inconsistencies in the process or procedure or the measurement of a mixture of two or more distinct groups or processes that behave very differently. These data are likely to be disregarded from analysis and investigated for inconsistencies.

Example 3.1.1.

The data used are the spelling test scores for 20 students on a 50 word spelling test. The scores (numbers correct) are: 48, 49, 50, 46, 47, 47, 35, 38, 40, 42, 45, 47, 48, 44, 43, 46, 45, 42, 43, 47.

The largest number is 50 and the smallest is 35. Thus, the range, $R = 15$. We will use class classes, so $K = 5$. The interval width $i = R/K = 15/5 = 3$.

Then we will make our lowest lower limit, the lower limit for the first class 35. Thus the first upper limit is 35+3 or 38. The second class will have a lower limit of 38 and an upper limit of 41. The completed table (with frequencies tabulated) is as follows:

Class	Lower Limit	Upper Limit	Frequency
1	35	38	1
2	38	41	2
3	41	44	4
4	44	47	5
5	47	50	8

The corresponding data histogram is shown in Fig. 3.1.

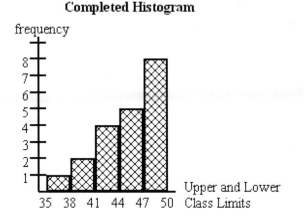

Fig. 3.1. Completed histogram for the presented example

Example 3.1.2.

Histograms of digital solar images taken in white light (SOHO/MDI) and CaII K1 (Meudon Observatory, France).

For full-disk solar images free of limb-darkening, the quiet Sun intensity value I_{QSun}, is established from the histograms of the on disk regions (Fig. 3.2.) as the intensity with the highest pixel count.

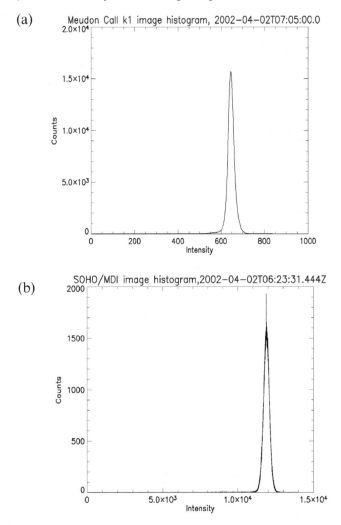

Fig. 3.2. Flat image histograms for the Meudon Observatory CaII K1 line (**a**) and SOHO/MDI white-light images (**b**) (from Zharkov et al. 2005a)

Thus, in a manner similar to the above, by analyzing the histogram of the "flat" image (free of a disk-to-limb darkening), an average quiet Sun intensity, I_{QSun}, can be determined.

3.1.2   Intensity Thresholds

Thresholding is a labeling operation, which marks pixels with the intensities above and below a threshold intensity T with binary values. It is the simplest procedure for segmenting an image into the object and background regions that can work well if the assumption of two well separated classes of object and background intensities is satisfied. When the image histogram has two nonoverlapping dominant modes, the choice of threshold intensity is easily chosen as a point in the valley between them.

The choice becomes more difficult as the two modes overlap and many procedures have been proposed for automatically choosing the threshold. These methods often give similar results when applied to images with two well separated dominant modes but rather different results otherwise. An example is shown in Fig. 3.3. Three most used threshold methods are briefly described below.

Iterative Procedure

Let us use the following procedure to calculate intensity threshold T automatically as described by Gonzalez and Woods (2002).

1. Make an initial estimate of T, for example the middle intensity.
   Segment the image histogram into a group G_1 of intensities $<T$ and a group G_2 of intensities $\geq T$ and compute the mean intensities μ_1 and μ_2 of the pixels p(I) in the two groups as follows:

$$\mu = \sum_{I=0}^{N-1} I\, p(I) \qquad (3.1)$$

$$p(I) = n_I / N , \qquad (3.2)$$

where N is the total number of pixels and n_I is the number of pixels with intensity I.

2. Compute a new threshold $T = 0.5(\mu_1 + \mu_2)$.
3. Repeat steps 2 and 3 until the difference in success values of T is less than a predefined limit.

Minimizing Within-Group Variance

Otsu (1979) suggested that the best criterion for choosing a threshold T is to choose the value which minimizes the weighted sum of the variance of the two groups of pixel intensities which are less than and greater than T, respectively. The weights q_1 and q_2 are the probabilities of the two groups. This criterion leads to the same result as choosing the threshold, which maximizes the between group variance because the sum of the within-group variance and between-group variance is a constant. The between-group variance $\sigma_B^2(T)$ is defined by

$$\sigma_B^2 = q_1(T)(\mu_1(T) - \mu)^2 + q_2(T)(\mu_2(T) - \mu)^2 \tag{3.3}$$

Where

$$q_1(T) = \sum_{I=0}^{T-1} p(I) \tag{3.4a}$$

$$q_2(T) = \sum_{I=T}^{N-1} p(I) \tag{3.4b}$$

$$\mu_1(T) = \sum_{I=0}^{T-1} I\, p(I)/q_1(T) \tag{3.5a}$$

$$\mu_2(T) = \sum_{I=T}^{N-1} I\, p(I)/q_2(T) \tag{3.5b}$$

and N is the total number of pixels, n_I is the number of pixels with a given intensity I with μ and $p(I)$ being calculated from formulae (3.19) and (3.3). A sequential search through the possible values of T is used to determine the value that maximizes $\sigma_B^2(T)$. The MATLAB™ function graythresh (image) uses the Otsu method to automatically threshold matrix images.

Minimizing Kullback Information Distance

A different criterion to that of Otsu for finding the best threshold was suggested by Kittler and Illingworth (1985). The image histogram is assumed to be a mixture $f(I)$ of two Gaussian distributions h_1 and h_2 having means μ_1 and μ_2, variances σ_1^2 and σ_2^2 and weights q_1 and q_2, respectively. They determine the threshold T and Gaussian distribution parameters which minimize the Kullback directed divergence J (Kullback 1959). This is interpreted as a measure of the distance from the image histogram $p(I)$ to the unknown mixture distribution $f(I)$ defined as follows.

$$J = \sum_{I=0}^{N-1} p(I) \log \left[\frac{p(I)}{f(I)} \right] \tag{3.6}$$

$$f(I) = q_1 h_1(I) + q_2 h_2(I) \tag{3.7}$$

Minimizing J is equivalent to minimizing H defined as follows:

$$H = -\sum_{I=0}^{N-1} p(I) \log f(i). \tag{3.8}$$

Assuming the distributions are well separated, H can be simplified to

$$H = \frac{1 + \log 2\pi}{2} - q_1 \log q_1 - q_2 \log q_2 + \frac{1}{2}(q_1 \log \sigma_1^2 + q_2 \log \sigma_2^2) \tag{3.9}$$

Where

$$\sigma_1^2 = \sum_{I=0}^{T-1} (I - \mu_1(T))^2 \, p(I) / q_1(T) \tag{3.10a}$$

$$\sigma_2^2 = \sum_{I=T}^{N-1} (I - \mu_2(T))^2 \, p(I) / q_2(T) \tag{3.10b}$$

The parameters in H are estimated for the two parts of the histogram separated by T, as for the Otsu method and H evaluated for each threshold to find the one, which minimizes H.

Example 3.1.3.

The Blobs image shown in Fig. 3.3 was thresholded by the iterative and Otsu methods using the following MATLAB™ code.

```
imname = input ('Type an image file name:','s') ;
a = imread (imname) ;
imshow(a) ;
T = 0.5*(min(a(:))+max(a(:))) ;
for i = 1:20
b = a >= T ;
T1 = 0.5*(mean(a(b))+mean(a(~b))) ;
if abs(T-T1)< 1, break, end
T = T1
end
figure
imshow(b) ;
T2 = graythresh(a)*255 ;
C = a >= T2 ;
figure
imshow(c) ;
```

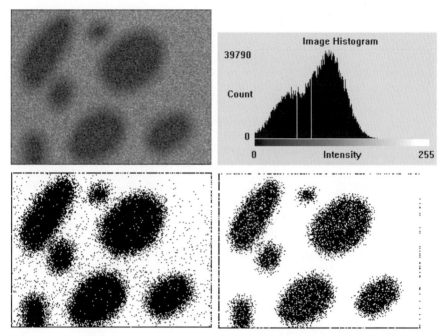

Fig. 3.3. *Top-left* noisy image of blobs with thresholds computed at 83, 81, and 61 by the iterative, Otsu and Kittler methods described in the text. The iterative method took six iterations. At *top-right* the blobs image histogram is shown with the thresholds at 61 and 81 marked with vertical white lines. At *bottom-left* and *right* are shown the blobs image thresholded at 81 and 61 respectively

3.2 Intensity-Based Methods for the Solar Feature Detection

V.V. Zharkova, S.S. Ipson and S.I. Zharkov

3.2.1   Threshold Method

Any solar features which are either darker than the background (like sunspots and filaments) or brighter (like plages and active regions) can be easily detected by a straightforward *thresholding approach*. Although the accuracy of this detection is dependent on the selected threshold that, it turn depends on the image quality. In good quality solar images with a constant variance and small skewness (see Sect. 2.1) this method can produce good results if they are checked manually by the observers.

This technique for solar features can be refined using the notion of contrast, i.e., the measurements are often used the intensity of sunspot in relation to a "quiet Sun" intensity. Segmentation of images solely on the basis of the pixel values is conceptually the simplest approach; choosing the best global intensity thresholds to use from an analysis of the histogram of the intensity levels in the image (Lee et al. 1990). The size of the segmented object varies with a threshold value and in order to avoid bias the knowledge of a type of edge between the object and background is often required. This approach gives very good results on a uniformly illuminated background allowing the objects to be segmented within distinct ranges of intensity.

For example, since sunspots are defined as features with intensity 15% (Chapman et al. 1994a) or 8.5% below the quiet Sun background (Chapman et al. 1994a,b; Steinegger et al. 1990) then sunspot areas can be estimated by simply counting all pixels below these values. Similar methods were applied to high-resolution images of solar disk regions containing a sunspot or a group of sunspots using constant intensity boundaries for the umbra–penumbra and the penumbra–photosphere transitions at 59% and 85% of the photospheric intensity, respectively (Brandt et al. 1990; Steinegger et al. 1997).

3.2.2 Histogram Methods

The two histogram methods presented below are not autonomous but are fast interactive methods for determining sunspot areas that are especially suited for the measurement of complex sunspot groups and large data sets. The first method is claimed to be especially well suited for the measurement of umbral areas while the second method yields results that are less influenced by stray light.

The use of photographic negatives with nonlinear characteristics will have affected the position of the maximum gradient but the effect on the calculated areas should be small if the gradients are large. Nevertheless, the method should work better with intensity calibrated images. The methods as described have difficulty when applied to sunspots near the solar limb and corrections for center-to-limb intensity variation should be made before the histogram methods are used. Correction for the foreshortening of sunspots near the limb should also be made.

The Difference Histogram Method

The difference histogram method (DHM) for measuring total sunspot area calculates the difference between correctly selected and normalized intensity histograms of the quiet sun and the sunspot (Steinegger et al. 1997). The sunspot area is taken to be equal to the sum of all bins of the difference histogram. The steps involved in the implementation of the algorithm are as follows:

(a) Two rectangular regions, one containing only quiet photosphere, the other a sunspot under study, are selected at the same radial position on the solar disk.
(b) From the photosphere region a mean intensity is obtained that is used to normalize the intensities of both regions as well as the intensity of histogram of the quiet photosphere.
(c) The photosphere histogram is then scaled so that its maximum count equals the local maximum count corresponding to the photosphere in the sunspot histogram.
(d) The normalized and scaled quiet photosphere histogram is subtracted from the sunspot histogram, eliminating the contribution of the quiet photosphere from the sunspot histogram.
(e) The total sunspot area is found by adding up all the bins in the difference histogram.

The Cumulative Histogram Method

The cumulative histogram method (CHM) for measuring umbral areas (Pettauer and Brandt 1997) uses an intensity histogram that is computed from a photographic negative of a sunspot region. The intensities at which large changes of gradient in the cumulative histogram occur are then used to define a first approximation to the boundaries between the umbra and penumbra and the penumbra and photosphere. These boundaries are refined by visual inspection to fit the observed boundary. The areas of the umbral and penumbral regions are obtained by adding up the pixels between the appropriate boundaries.

The steps involved in the implementation of the algorithm are as follows:

(a) Calculate the intensity histogram $H(I)$ of the sunspot region. Then
compute the cumulative distribution function as

$$C(I_b) = \sum_{I=255}^{I_b} H(I),$$

(3.11)

where $0 \leq I_b \leq 255$.

(b) The shape of the cumulative distribution function has in general three "roughly linear parts" with different slopes, corresponding to umbra, penumbra and photosphere. Straight lines are fit to all three parts and the abscissas I_{u-pu} and I_u of their intersection points are taken to be the approximate umbra–penumbra and umbra–photosphere intensities.

(c) All pixels with intensities equal to those found in step (b) ±2 are marked in the displayed image. This forms contour bands which have widths varying with the local gradients. The intensity is varied interactively to ensure that the resulting contour follows the umbra–penumbra or penumbra–photosphere boundary in an optimal way. This usually means that the contour band is narrowest so that it coincides with the maximum intensity gradient.

(d) The umbra and penumbra areas are found by adding up the pixels between the appropriate boundaries.

3.2.3 Simulated Annealing

Simulated annealing is a technique which has proved useful in the numerical research field of optimization for solving a variety of global combinatorial extremization problems. At the heart of the approach is an analogy with the thermodynamics of the cooling or annealing of a metal to achieve the minimum energy state of the system. Cooling is slow to allow ample time for the atoms to redistribute, as they lose mobility, into the lowest energy configuration. In thermodynamic equilibrium at a given temperature the energy of a system is distributed over the available states according to the Boltzmann probability distribution, which allows a small probability of high energy states occurring at low temperatures and for the system to move out of a local energy minimum and find a better, more global one.

These ideas were first incorporated into numerical calculations for nonthermodynamic systems by Metropolis et al. (1953). This requires a description of the possible system configurations; a generator of random changes in the configuration; an objective function analogous to energy whose minimization is the goal of the procedure; and a control parameter T analogous to temperature and an annealing schedule specifying the number of random changes for each downward step in temperature and the size of the step.

An approach to segmenting sunspots in H_α images from the Sacremento Peak Observatory based on these ideas was proposed by Bratsolis and Sigelle (1998). Segmentation is separated into two stages in the first of which the image is sharpened and requantized to a reduced number of q intensity labels. In the second stage the requantized image is subjected to a relaxation labeling process, mean field fast annealing, which smoothes the boundaries between regions and removes small isolated regions. The steps involved in the implementation of the algorithm are as follows:

(a) A region of solar image containing a sunspot and of size 128×128 pixel is sharpened to enhance local contrast by convolving it with the following template

$$\begin{pmatrix} -1 & -2 & -1 \\ -2 & 13 & -2 \\ -1 & -2 & -1 \end{pmatrix} \qquad (3.12)$$

(b) The resulting image intensity I_1 is then requantized to intensity I_2 with q classes using the equation:

$$I_2(x,y) = \text{int}\left(\frac{I_1(x,y) - m}{M - m}(q - 1) \right) \qquad (3.13)$$

where M is the maximum and m the minimum intensity in the sharpened image. They present results for $q = 4$ and $q = 8$.

(c) The relaxation process is set up as follows. Each pixel site s in the image is now associated with a label l_s with probability distribution $P(l_s)$, mapped to the range 0–1 and initially given by I_2. Most images have the basic characteristic that the intensity values at neighboring locations are likely to be similar. To express the local properties of the image and penalize rough boundaries, a local energy function is introduced through which a pixel r in the neighborhood of s interacts with the pixel at s. Bratsolis and Sigelle (1998) chose to use the Potts potential interaction model U_{rs} defined by

$$U_{rs} = -K\frac{q-1}{q}(\hat{u}_s \cdot \hat{u}_r) - \frac{K}{q} + \frac{K}{2} \qquad (3.14)$$

where K is a constant and $\hat{u}_s$ and $\hat{u}_{rs}$ are random unit vectors at sites s and r under a $(q-1)$ dimensional space with values in R. Replacing the Potts model defined in (3.14) with its mean field approximation and using the Gibbs–Boltzmann probability distribution, the expected value of vector $\hat{u}_s$ at any pixel site s in the image S is calculated as:

$$E(\hat{u}_s) = \frac{\sum\limits_{k=0}^{q-1} \hat{u}_k \exp\left[\dfrac{\beta}{T} \sum\limits_{r \in N_s} \hat{u}_k \cdot E(\hat{u}_r) \right]}{\sum\limits_{k=0}^{q-1} \exp\left[\dfrac{\beta}{T} \sum\limits_{r \in N_s} \hat{u}_k \cdot E(\hat{u}_r) \right]}. \qquad (3.15)$$

The sum over r is taken over the four neighborhood of s. In the simplest case, replacing the potential coefficients by a scalar K, Bratsolis and Sigelle (1998) give the mean field equation expressed in a probabilistic iterative form as:

$$[P(l_s = i)]^{new} = \frac{\exp\left(\frac{1}{T}\left\{K\sum_{r\in N_s}[P(l_r = i)]^{old} + B\delta(l_s^0, i)\right\}\right)}{\sum_{k=0}^{q-1}\exp\left(\frac{1}{T}\left\{K\sum_{r\in N_s}[P(l_r = k)]^{old} + B\delta(l_s^0, k)\right\}\right)}, \quad (3.16)$$

$$\forall i \in I_2, \forall s \in S.$$

This is the equation on which the segmentation results are based. The superscript 0 indicates the initial label of a pixel site, *old* the previous label and *new* the next label during iteration of the equation. During the iteration, the parameter T starts at a high value T_0 and is then gradually reduced in value.

At iteration step k, the temperature $T(k) = \dfrac{T_0}{\log(k)}$ whereas in the fast cooling schedule the denominator is replaced by k. Bratsolis and Sigelle (1998) have tested both cooling schedules and have found they give very similar results.

The steps involved in the implementation of the mean field fast annealing algorithm are as follows:

(a) Define k_{limit}: number of sweeps and T_0: initial temperature.

(b) Initialize q buffers each of size N with $P(1_s = i) = 10^{-5} \; \forall i \in I_2, \forall s \in S$.

(c) for $k = 1 \dots k_{limit}$, $T = \dfrac{T_0}{k} T_0$, use the iterative mean field equation

(d) If $k = k_{limit}$ round off and display taking $l_s = \arg\max_{i \in I_2} P(l_s = i)$

When the system converges, the value displayed at each pixel site is the maximal probability from every state q. The parameter values used in there study were $B = 1.0$, $K = 1.0$, $T_c = 1.92$ for $q = 4$ and $T_c = 1.36$ for $q = 8$ where $T_0 = 2T_c$. With $q = 4$ the sunspot umbra only was segmented while with $q = 8$ both the sunspot umbra and penumbra were segmented. The number of sweeps $k_{limit} = 8$.

With N pixels and q gray levels, the MFFA algorithm needs $O(qN)$ updates at each fixed temperature. For comparison, the stochastic simulated annealing needs $O((q^N)^2)$ such updates that make this method computationally expensive if applied automatically to large archives. Although, the MFFA method appears to work well with noisy artificial data and with the sunspot images obtained with little distortions. Its behavior with a large number of sunspot image and other features needs investigation and it could perhaps be combined with a simpler method to perform an initial segmentation of sunspot regions.

3.3 Edge-Based Methods for Solar Feature Detection

V.V. Zharkova, S.I. Zharkov and S.S. Ipson

3.3.1   Lagrangian of Gaussian (LgOG) Method

General Description

Edge-based segmentation relies on discontinuities in the image data to locate the boundaries of the segments before assessing the enclosed region. The problem with this approach is that in practice the edge-profile is usually not known. Furthermore, the profile often varies heavily along the edge caused by, for example, shading and texture. Due to these difficulties usually a symmetrical simple step-edge is assumed and the edge-detection is performed based on a maximum intensity gradient. However, the resulting boundary is seldom complete and so edge linking is usually necessary to fill gaps. Region edges which are complete may also be defined by the zero-crossing of the Laplacian operator which provides a 2D isotropic measure of the second spatial derivative of an image (Jahne 1997). The Laplacian of an image has largest magnitudes at peaks of intensity and has zero-crossings at the points of inflection on an edge. Two common 3×3 convolution kernels used to calculate the digital Laplacian are as follows:

$$\begin{vmatrix} 0 & 1 & 0 \\ 1 & -4 & 1 \\ 0 & 1 & 0 \end{vmatrix} \tag{3.17}$$

$$\begin{vmatrix} 1 & 1 & 1 \\ 1 & -8 & 1 \\ 1 & 1 & 1 \end{vmatrix} \tag{3.18}$$

Because it is sensitive to noise, the Laplacian is often applied to an image that has first been smoothed with an approximation of a Gaussian smoothing filter and since convolution is a commutative operation, the two techniques are often combined as the Laplacian of Gaussian (LpOG) or Marr–Hildreth operator. Marr–Hildreth edge-detection (Marr and Hildreth 1980) is based on locating the zero-crossings of the LpOG operator applied to the image using various values for the standard deviation of the Gaussian. LgOG methods incorporate noise reduction and have potential for rugged performance (Huertas and Medioni 1986). However, the computation of the zero-crossings is complicated in general and although zero-crossing positions are correct for ideal edges, errors as large as the standard deviation of the Gaussian can occur in other cases. In practice, the zero-crossing detected edges often include many small closed loops because a threshold, if applied, is very small and the weak edges at the closed boundaries of many small minor regions are detected.

Inflection Point Method

A Laplacian based approach called the inflection point method (IPM) was proposed by Steinegger et al. (1997) for measuring the umbral and penumbral areas of sunspots. High-resolution sunspot images obtained with the 50 cm Swedish Vacuum Solar Telescope at the Observatorio del Roque de los Muchachos at La Palma were used. The images were captured using a Kodak Megaplus 8-bit CCD camera and filters centered at the wavelengths of 525.7 and 542.5 nm.

Starting with a rectangular region (possibly smoothed) which contains a sunspot umbra or whole sunspot, a map of contours is obtained by applying a 7×7 Laplacian operator and then searching for its zero-crossings. A mean intensity (formula 1), I_u, along the contour positions is computed and the isoline at this intensity level is taken as the umbra–penumbra boundary. The same method is used to obtain the penumbra–photosphere boundary. The steps involved in the implementation of the algorithm are as follows:

(a) To define the area of a sunspot umbra, a rectangular region containing only the umbra and a portion as small as possible of a surrounding penumbra is manually chosen for extraction from the whole frame. If necessary the selected regions is smoothed to reduce the effect of perturbing features prior to computation of the derivatives.

(b) The inflection points in the spatial distribution of intensity in the box are located by computing the second derivative Laplacian operator and by searching for its zero-crossing points, thus producing a map of contours.

(c) To compute the mean intensity along a given contour, their coded algorithm requires that the contour be closed and have no spurious internal contours. If necessary, a contour is closed and spurious internal contours removed manually by adding or deleting pixels in an interactive way directly on the displayed contour image, using as a guide both the smoothed image and the original image.

(d) The mean intensity in the image over the path defined by the zero-crossing contour is computed. The isoline at this intensity level is taken to be the definitive boundary of the umbra.

(e) The isoline at the mean intensity level may include small contours inside the umbra that nevertheless belong physically to the umbra. To include such small regions into the umbral area, an operation is necessary to interactively remove their contours from inside the umbral boundary. The value of the umbral area is calculated by summing up all pixels enclosed by the umbral boundary.

Following a similar scheme for the whole sunspot, both the umbra–penumbra boundary and the total spot area can be determined using the IPM.

The authors assert that methods for measuring sunspot areas based on fixed intensity thresholds cannot be justified because of the different characteristics of individual sunspots. In comparison with the difference and cumulative histogram based methods for measuring sunspot areas, the IPM is much less affected by image blurring due to the seeing conditions and more in accord with the physical structure of sunspots (Steinegger et al. 1997) but the methods are not fully autonomous.

3.3.2   Canny Edge Detector

Edge detectors attempt to find, in the presence of noise, the regions in intensity images where the intensity variation is a local maximum at the boundaries between adjacent features in the image such as the solar disk and the background sky in a solar image. Although a great many algorithms have been formulated for this purpose only one, devised by Canny (1986), which is very widely used, is described here. Briefly, Canny's approach to the problem was to propose a mathematical model of edges and noise, specify the performance criteria for the edge detector and on the basis of these two criteria formulate an optimal filter for the edge detector.

A practical implementation of his detector includes the following sequence of steps: application of a Gaussian smoothing filter to the intensity image; computation of the vertical and horizontal components of the gradient of the smoothed image; estimation of the gradient amplitude and direction at all pixel positions (gradient amplitude and direction maps); location of the local edge maxima; discarding of weak edges by hysteresis thresholding. Application of the algorithm requires the specification of the values of three parameters, the standard deviation of the Gaussian smoothing filter, the upper and the lower thresholds for the hysteresis thresholding.

To shorten execution time, the 2D Gaussian smoothing filter may be implemented as a horizontal 1D Gaussian convolution operation followed by a vertical 1D Gaussian convolution operation (or vice versa), since the 2D Gaussian convolution is separable. A suitable choice for the width of the Gaussian kernel of standard deviation σ is $5 \times \sigma$, since this includes 98.76% of the area under the Gaussian distribution and to adequately sample the Gaussian a minimum of five samples are required. The value of σ to be used depends on the noise level, the length of the connected edge contours of interest and the trade-off between edge localization and detection.

The local maxima of the gradient amplitude map may comprise wide ridges and these should be thinned to 1 pixel wide edges. A procedure to accomplish this approximates, at each pixel, the direction of the gradient as either 0°, 45°, 90°, or 135° and then forces the gradient amplitude to zero if it is smaller than at least one of its two neighbors in the identified direction. The resulting map of maximum edge points contains maxima caused by noise as well as the true maxima. Setting a low threshold above which maxima are accepted runs the risk of accepting maxima caused by noise as well as true weak edges, while setting a higher threshold runs the risk of fragmenting edge contours whose maxima fluctuate above and below the threshold. A solution to this is hysteresis thresholding which employs an upper and a lower threshold and scans through all the maxima in a fixed order. When a nonvisited before maximum above the upper threshold is encountered, the chain of maxima connected to that point is followed and marked as visited until a maxima below the lower threshold is encountered. This procedure reduces the likelihood of spurious edge contours because they must produce a response at least equal to the upper threshold to be detected while allowing continuous edge contours to be retained even when if they have weak sections with response no smaller than the lower threshold. Figure 3.4 shows the response of a Canny edge detector applied to a solar Ca K image from Meudon for three different values of Gaussian smoothing.

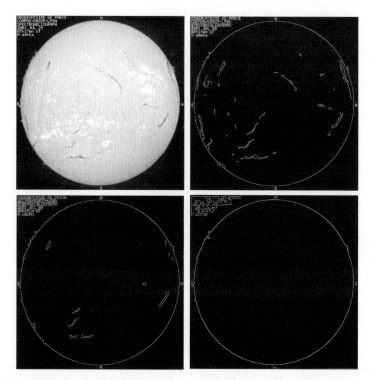

Fig. 3.4. Original CaII K3 solar image from Meudon at *top-left* to which Canny edge detectors have been applied with standard deviations of 1, 2 and 3 pixels (*top-right*, *bottom-left*, and *bottom-right*, respectively)

3.3.3 Automatic Sunspot Recognition in Full-Disk Solar Images with Edge-Detection Techniques

The first thresholding methods for the extraction of sunspot areas used an a priori estimated intensity threshold on white-light full-disk solar images (Chapman and Groisman 1984; Chapman et al. 1994a,b). Sunspots were defined as features with intensity 15% (CG84) or 8.5% (C94) below the quiet Sun background and simply counted all pixels below these values. Similar methods were applied to high-resolution nonfull disk images containing a sunspot or a group of sunspots using constant intensity boundaries for the umbra–penumbra and the penumbra–photosphere transitions at 59% and 85% of the photospheric intensity, respectively (Steinegger et al. 1990; Brandt et al. 1990).

The thresholding methods were significantly improved by using image histograms that help to determine the threshold levels in sunspots. Steinegger et al. (1996) used the DHM to determine the intensity boundary between a penumbra and the photosphere that was defined for each individual spot. Another method based on the cumulative sum of sunspot area contained in successive brightness bins of the histogram (Steinegger et al. 1997) that was applied to determine the umbral areas of sunspots observed with high-resolution nonfull disk observations (Steinegger et al. 1997). Another method using sunspot contrast and contiguity based on region growing technique was also developed (Preminger et al. 2001) that has produced more accurate results comparing to thresholding and histogram methods but was very time-consuming.

More accurate approach to sunspot area measurements was developed by utilizing edge-detection and boundary gradient intensity was suggested for high-resolution observations of individual sunspot groups, and/or nonfull disk segments (Győri 1998). The method is very accurate when applied to the data with sufficiently high-resolution. However, this method was not suitable for the automated sunspot detection on full-disk images of the low and moderate resolutions that are present in most archives. They needed some adjustments that have been implemented in the automated techniques described below.

SOHO/MDI White-Light Images (Space-Based)

The technique developed for the SOHO/MDI data relies on the good quality of the images evident from the top three graphs in Fig. 2.2 describing the variations of mean, curtosis and skewness during a long period of time (a year). This allows for a number of parameters, including the threshold values as percentages of the quiet Sun intensity, to be set constant for the whole data set. Since sunspots are characterized by strong magnetic field, the synchronized magnetogram data is then used for sunspot verification by checking the magnetic flux at the identified feature location.

Basic (binary) morphological operators such as dilation, closing and watershed (Matheron 1975; Serra 1988) are used in our detection code. Binary morphological dilation also known as Minkowski addition is defined as:

$$A \oplus B = \left\{ x : (\hat{B})_x \cap A \neq \varnothing \right\} = \bigcup_{x \in B} A_x , \qquad (3.19)$$

where A is the signal or image being operated on and B is called the "Structuring Element." This equation simply means that B is moved over A and the intersection of B reflected and translated with A is found. Dilation using disk structuring elements corresponds to isotropic swelling or expansion algorithms common to binary image processing.

Binary morphological erosion also known as Minkowski subtraction is defined as:

$$A \ominus B = \{x : (B)_x \subseteq A \neq \varnothing\} = \bigcap_{x \in B} A_x .$$ (3.20)

The equation simply means that erosion of A by B is the set of points x such that B translated by x is contained in A. When the structuring element contains the origin, Erosion can be seen as a shrinking of the original image.

Morphological closing us defined as dilation followed by erosion. Morphological closing is an idempotent operator. Closing an image with a disk structuring element eliminates small holes, fills gaps on the contours and fuses narrow breaks and long, thin gulfs.

The morphological watershed operator segments images into watershed regions and their boundaries. Considering the gray scale image as a surface, each local minimum can be thought of as the point to which water falling on the surrounding region drains. The boundaries of the watersheds lie on the tops of the ridges. This operator labels each watershed region with a unique index, and sets the boundaries to zero. We apply the watershed operator provided in the IDL library by Research Systems, Inc. to binary image where it floods enclosed boundaries and, thus, is used in a filling algorithm. For a detailed discussion of mathematical morphology see the references within the text and numerous books on Digital Imaging (see for example, Matheron 1975; Bow 2002).

The detection code is applied to a "flattened" full-disk SOHO/MDI continuum image, Δ, (Fig. 3.6a), with the estimated quiet Sun intensity, I_{QSun} (Fig. 3.5), image size, solar disk center pixel coordinates, disk radius, date of observation, and resolution (in arcsec per pixel). Because of the Sun's rotation around its axis, a SOHO/MDI magnetogram, M, taken at the time T_M, is synchronized to the continuum image time T_{WL} via a spatial displacement of the pixels to the position they had at the time T_{WL} in order to obtain the same point of view as those for the continuum.

The technique presented in the current paper uses edge-detection with threshold applied on the gradient image. This technique is significantly less sensitive to noise than the global threshold since it uses the background intensity in the vicinity of a sunspot. We consider sunspots as connected features characterized by strong edges, lower than surrounding quiet Sun intensity and strong magnetic field. Sunspot properties vary over the solar disk, so a two stage procedure is adopted. First, sunspot candidate regions are defined. Second, these are analyzed on the basis of their local properties to determine sunspot umbra and penumbra regions. This is followed by verification using magnetic information. A detailed description of the procedure is provided in the pseudocode below.

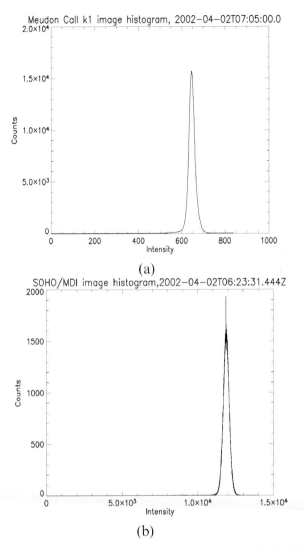

Fig. 3.5. Flat image histograms for the Meudon Observatory CaII K1 line (**a**) and SOHO/MDI white-light images (**b**)

Sunspot candidate regions are determined by combining two approaches: edge-detection and low intensity region detection (steps 1–3). First, we obtain a gradient gray-level image, Δ^P, from the original preprocessed image, Δ (Fig. 3.6a) by applying Gaussian smoothing with a sliding window (5×5) followed by Sobel gradient operator (step 1). Then (step 2) we locate strong edges via iterative thresholding of the gradient image starting from initial threshold, T_0, whose value is not critical but should be small. The threshold is applied followed by 5×5 median filter and the number of connected components, N_c, and the ratio of the number edge pixel to the total number of disk pixels, R, are determined. If the ratio is too large or the number of components is greater than 250, the threshold is incremented. The number 250 is based on the available recorded maximum number of sunspots which is around 170.[1] Since at this stage of the detection we are dealing with noise and the possibility of several features joined into a single candidate region, this limit is increased 250 to ensure that no sunspots are excluded. The presence of noise and fine structures in the original "flat" image will contribute many low gradient value pixels resulting in just a few very large connected regions, if the threshold is too close to zero. Imposing an upper limit of 0.7 on the ratio of numbers of edge pixels to disk pixels excludes this situation. A lower value increment in the iterative thresholding loop ensures better accuracy, at the cost of computational time. The increment value can also be set as a function of the intermediate values of N_c and R. The resulting binary image (Fig. 3.6b) contains complete and incomplete sunspot boundaries as well as noise and other features such as the solar limb.

Similarly, the original "flat" image is iteratively thresholded to define dark regions (step 3). The resulting binary image contains fragments of sunspot regions and noise. The two binary images are combined using the logical OR operator (Fig. 3.6c). The image will contain feature boundaries and blobs corresponding to the areas of high gradient and/or low intensity as well as the limb edge. After removing the limb boundary, a 7×7 morphological closure operator (Baxes 1994; Casasent 1992) is applied to close incomplete boundaries. Closed boundaries are then filled by applying a filling algorithm based on the IDL watershed function (Jackway 1996) to the binary image. A 7×7 dilation operator is then applied to define the regions of interest which possibly contain sunspots (step 4, Fig. 3.6d, regions masked in dark gray).

---

[1] The number varies depending on the observer.

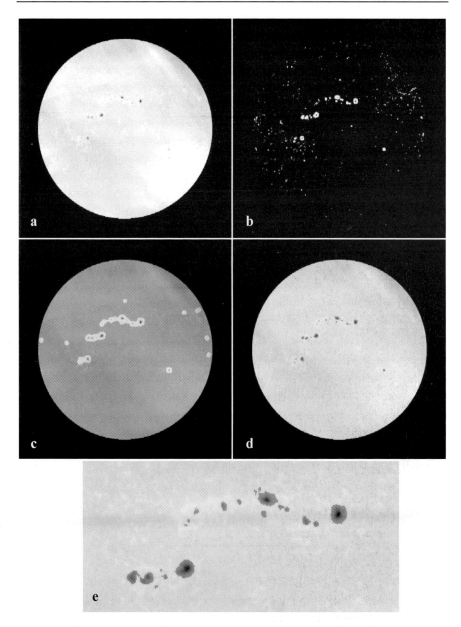

Fig. 3.6. The sunspot detection stages: **(a)** the standardized original image; **(b)** the detected edges; **(c)** the detected regions (dilated) superimposed on the original image; **(d)** the final detection results; **(e)** an extract from the final detection results

The pseudocode describing the sunspot detection algorithm in SOHO/MDI white-light images is given on the next page:

1. Apply Gaussian smoothing with sliding window 5×5 followed by the Sobel operator to a copy of Δ;
2. Using the initial threshold value, T_0, threshold the edge map and apply the median 5×5 filter to the result. Count the number of connected components, N_c, and the ratio of the number edge pixel to the total number of disk pixels, R. (Feature Candidates, Figs.3.6b and 3.7b). If N_c is greater than 250 or R is larger than 0.7, increase T_0 by set value (1 or larger depending on N_c and R) and repeat step 3 from the beginning.
3. Similarly, iteratively threshold original flat image to define less then 100 dark regions. Combine (using OR operator) two binary images into one binary Feature Candidate Map (Fig. 3.6c).
4. Remove the edge corresponding to the limb from Candidate Map and fill the possible gaps in the feature outlines using IDL's morphological closure and watershed operators (Figs. 3.6d and 3.7c).
5. Use blob coloring to define a region of interest, F_i, as a set of pixels representing a connected component on the resulting binary image, $\overline{B}_\Delta$
6. Create an empty Sunspot Candidate Map, B_Δ, a byte mask which will contain the detection results with pixels belonging to umbra marked as 2, penumbra as 1.
7. For every F_i extract a cropped image containing F_i and define $\mathbf{T_s}$ and $\mathbf{T_u}$:
i.   if $|F_i| =< 5$ pixels assign the thresholds:
     for penumbra $\mathbf{T_s = 0.91}\ I_{QSun}$; for umbra $\mathbf{T_u = 0.6}\ I_{QSun}$
i.   if $|F_i| > 5$ pixels assign the thresholds:
     for penumbra: $\mathbf{T_s = max\ \{\ 0.93}\ I_{QSun}\ ;\ (<F_i> - 0.5* \Delta\ F_i\ )\}$;
     for umbra: $\mathbf{T_u = max\ \{\ 0.55}\ I_{QSun}\ ;\ (<F_i> - \Delta\ F_i\ )\}$,
     where $<F_i>$ is a mean intensity and $\Delta\ F_i$ a standard deviation for F_i.
8. Threshold a cropped image at this value to define the candidate umbral and penumbral pixels and insert the results back into B_Δ (Figs. 3.6e and 3.7d). Use blob coloring to define a candidate sunspot, S_i as a set of pixels representing a connected component in B_Δ.
9. To verify the detection results, cross check B_Δ with the synchronized magnetogram, M, as follows:

for every sunspot candidate S_i of B_Δ extract

$$B_{max}(S_i) = max(M(p) \mid p \in S_i)$$

$$B_{min}(S_i) = min(M(p) \mid p \in S_i)$$

if $max(abs(B_{max}(S_i)), abs(B_{min}(S_i))) < 100$ then disregard S_i as noise.

10. For each S_i extract and store the following parameters: gravity center coordinates (Carrington and projective), area, diameter, umbra size, number of umbras detected, maximum–minimum–mean photometric intensity (as related to flattened image), maximum–minimum magnetic flux, total magnetic flux and total umbral flux.

In the second stage of detection, these regions are uniquely labeled using blob coloring algorithm (Rosenfeld and Pfaltz 1966) (step 5) and individually analyzed (steps 6 and 7, Fig. 3.6e). Penumbra and umbra boundaries are determined by thresholding at values T_u and T_s which are functions of the region's statistical properties and quiet Sun intensity defined in step 7. Practically, the formulae for determining T_u and T_s (step 7), including the quiet Sun intensity coefficients, 0.91, 0.93, 0.6, 0.55, are determined by applying the algorithm to a training set of about 200 SOHO/MDI WL images. Since smaller regions of interest (step 7i) carry less statistical intensity information lower T_s value reduces the probability of false identification. In order to apply this sunspot detection technique to datasets from other instruments, the values of the constants appearing in the formulas for T_u and T_s should be determined for each dataset. As mentioned in the Introduction, other authors (Chapman et al. 1994a,b) apply global thresholds at the different values 0.85 or 0.925 for T_s and 0.59 for T_u. The parameters are extracted and stored in the ASCII format in the final step 10.

Ground Based Images: CaII K1

The technique described above can also be also applied to CaII K1 images with the following modifications. First, these images contain substantial noise and distortions owing to instrumental and atmospheric conditions, so their quality is much lower than the SOHO/MDI white-light data (see Fig. 1.2, bottom three plots). Hence, the threshold in step 7 (i.e., item 7 in the pseudocode) has to be set lower, i.e., $T_s = max \{ 0.91 \, I_{QSun} ; (<F_i> - 0.25 * \Delta F_i)\}$

for penumbra and $T_u = \max\{ 0.55\ I_{QSun}\ ; (<F_i> - \Delta\ F)\}$, for umbra, where $\Delta\ F_i$ is the mean absolute deviation of the region of interest F_i.

The examples of sunspot detection with this technique applied to a CaII K1 image taken on 2 April 2002 are presented in Fig. 3.8. First, a full-disk solar image is preprocessed as described in Sect. 2.1.2 by correcting a shape of the disk to a circular one (via automated limb ellipse-fitting) and by removing the limb-darkening (Fig. 3.8a). Then Canny edge-detection (similar results were also achieved with morphological gradient operation (Serra 1988) defined as the result of subtracting an eroded version of the original image from a dilated version of the original image) is applied to the preprocessed image, followed by thresholding in order to detect strong edges (Fig. 3.8b).

This over segments the image; and then a 5×5 median filter and an 8×8 morphological closing filter (Matheron 1975; Serra 1988) are applied to remove noise, to smooth the edges and to fill in small holes in edges. After removing the limb edge, the watershed transform (Jackway 1996) is applied to the thresholded binary image in order to fill-in closed sunspot boundaries. Regions of interest are defined, similar to the WL case, via morphological closure and dilation (Matheron 1975; Serra 1988). Candidate sunspot features are then detected by local thresholding using threshold values in the previous paragraph. The candidate features' statistical properties such as size, principal component coefficients and eigenvalues, intensity mean and mean absolute deviation are then used to aid the removal of false identifications such as the artifacts and lines, often present in the Meudon Observatory images (Fig. 6d, e).

It can be seen that on the Ca K1 image shown in Fig. 3.8 the technique performs as well as on the white-light data (Fig. 3.7). However, in many other Ca images the technique can still produce a relatively large number of false identifications for smaller features under 10 pixel where there is not enough statistical information to differentiate between the noise and sunspots. This raises the problem of verification of the detected features that is discussed below.

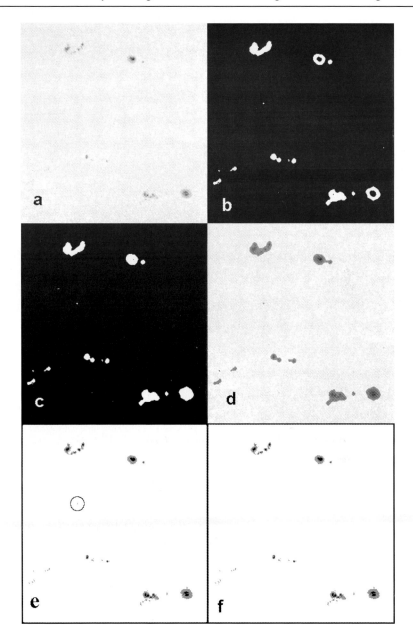

Fig. 3.7. A sample of the sunspot detection technique applied to the WL disk SOHO/MDI image from 19th April 2002 (presented for a cropped fragment for better resolution): (**a**) a part of the original image; (**b**) the detected edges; (**c**) candidate map; (**d**) the regions of interests after filtering as masks on original; (**e**) the detection results before magnetogram verification, false identification is circled; (**f**) the final detection results

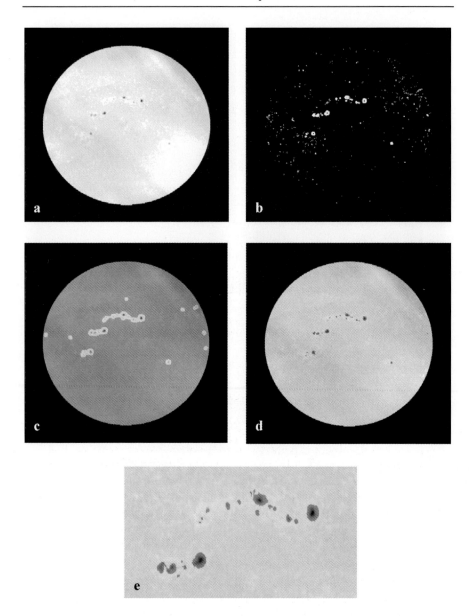

Fig. 3.8. Sunspot Detection on the CaII K1 line full-disk image obtained from Meudon Observatory on 01 April 2002: (**a**) the original image cleaned; (**b**) the detected edges; (**c**) the found regions (dilated); (**d**) the final detection results super-imposed on original image; (**e**) the extract of a single sunspot group from (**d**)

Verification and Accuracy

There are two possible means of verification. *The first option* assumes the existence of a tested well-established source of the target data that is used for a straightforward comparison with the automated detection results. In our case, such data would be the records (sunspot drawings) produced by a trained observer. However, the number (and geometry) of visible/detectable sunspots depend on the time (date) of observation (sunspot lifetime can be less than an hour), location of the observer, wavelength and resolution (in case of digital imaging). Therefore, this method works best when the input data for both detection methods is the same. Otherwise, a number of differences can appear naturally when comparing the two methods.

The second option is comparing two different datasets describing the same sunspots taken close together in time by different instruments, by extracting from each dataset a carefully chosen invariant parameter (or set of parameters), such as Sunspot Area, and looking at its correlation. For our technique verification both methods were applied and the outcome is presented below.

Verification with Drawings and Synoptic Maps

The verification of the automated sunspot detection results started by comparison with the sunspot database produced manually at the Meudon Observatory and published as synoptic maps in ASCII format. The comparison is shown in Table 3.1. The two cases presented in Table 3.1 correspond to two ways of accepting/rejecting detected features. In general, by considering feature size, shape (i.e., principal components), mean intensity, variance, quiet sun intensity and proximity to other features, one can decide whether the result is likely to be a true feature. In Table 3.1, case 1, we have included features with sizes over 5 pixels, mean intensities less than the quiet Sun's, mean absolute deviations exceeding 20 (which is about 5% of the quiet Sun intensity), principal component ratios less than 2.1. In case 2, we include practically all detected candidate features by setting the deviation and principal component ratio thresholds to 0.05.

The differences between the manual and automatic methods are expressed by calculating the false acceptance rate (FAR) (where we detect a feature and they do not) and the false rejection rate (FRR) (where they detect a feature and we do not). FAR and FRR were calculated for the available observations for the two different classifier settings described in the previous paragraph. The FAR is lowest for the classifier case 1 and does not exceed 8.8% of the total sunspot number detected on a day. By contrast, FRR is lowest for the classifier case 2 and does not exceed 15.2% of the total sunspot number.

The error rates in Table 3.1 are the consequences of several factors related to the image quality. First, different seeing conditions can adversely influence automated recognition results; for example, a cloud can obstruct a significant portion of the disk, thus greatly reducing the quality of that segment of the image, making it difficult to detect the finer details of that part of the solar photosphere. Second, some small (less than 5–8 pixels) dust lines and image artifacts can be virtually indistinguishable from smaller sunspots leading to false identifications.

Table 3.1. The accuracy of sunspot detection and classification for CaII K1 line observations in comparison with the manual set obtained from Meudon Observatory (see the text for description of defined cases 1 and 2)

Date	s/s Meudon	s/s case 1	FAR case 1	FRR case 1	s/s case 2	FAR case 2	FRR case 2
01 April 02	16	17	2	1	19	4	1
02 April 02	20	18	0	3	18	0	3
03 April 02	14	13	0	3	24	10	2
04 April 02	13	15	2	2	20	6	1
05 April 02	16	18	0	2	18	1	2
06 April 02	10	10	0	5	15	5	5
07 April 02	11	9	1	5	13	4	4
08 April 02	14	17	3	2	22	7	0
09 April 02	16	17	0	2	17	0	2
10 April 02	12	12	0	4	14	1	3
11 April 02	12	9	0	7	10	1	7
12 April 02	18	20	2	0	21	3	0
14 April 02	20	23	2	2	34	13	2
15 April 02	13	16	1	4	18	2	3
16 April 02	10	13	3	1	19	9	1
17 April 02	11	11	1	1	13	2	0
18 April 02	12	11	1	1	12	1	0
19 April 02	11	14	0	0	15	1	0
20 April 02	13	10	0	2	11	1	2
21 April 02	9	8	1	1	15	7	0
22 April 02	12	13	1	0	15	3	0
23 April 02	14	13	0	1	15	1	0
24 April 02	18	15	0	3	17	0	1
25 April 02	17	13	0	3	17	2	1
27 April 02	9	7	0	1	9	2	1
28 April 02	9	10	1	0	11	2	0
29 April 02	8	12	5	0	20	13	0

Also, in order to interpret the data presented in Table 3.1, the following points have to be clarified. Sunspot counting methods are different for different observatories. For example, a single large sunspot with one umbra is counted as a single feature at Meudon, but can be counted as three or more sunspots (depending on the sunspot group configuration) at the Locarno Observatory. Similarly, there are differences between the Meudon approach and our approach. For example, a large sunspot with several umbras is counted as one feature by us, but can be counted as several features by the Meudon observer. Furthermore, interpretation of sunspot data at Meudon is influenced by the knowledge of earlier data and can sometimes be revised in the light of the subsequent observations. Hence, for the instance, on 2 April 2002 there are 20 sunspots detected at Meudon Observatory. The automated detection applied to the same image yielded 18 sunspots corresponding to 17 of the Meudon sunspots with one of the Meudon sunspots detected as two. Thus, in this case FAR is zero, and FRR is 3.

Currently, our automated detection approach is based on extracting all the available information from a single observation and storing this information digitally. Further analysis and classification of the archive data is in progress that will allow us to produce sunspot numbers identical to the existing spot counting techniques.

For the verification of sunspot detection on the SOHO/MDI images, which have better quality (see Fig. 2.2, three upper plots), less noise and better time coverage (four per day) we used the daily sunspot drawings produced manually since 1965 at the Locarno Observatory in Switzerland. The Locarno manual drawings are produced in accordance with the technique developed by Waldmeier (1961) and the results are stored in the Solar Index Data Catalogue at the Royal Belgian Observatory in Brussels (SIDC). While the Locarno observations are subject to seeing conditions, this is counterbalanced by the higher resolution of live observations (about 1 arcsec). We have compared the results of our automated detection in white-light images with the available drawings in Locarno for June–July 2002, as well as for January–February 2004 along with a number of random dates in 2002 and 2003 (about 100 daily drawings with up to 100 sunspots per drawing). The comparison has shown a good agreement (~95–98%). The automated method detects all sunspots visually observable in the SOHO/MDI WL observations. The discrepancies are found at the level of sunspot pores (smaller structures), and can be explained by the time difference between the observations and by the lower resolution of the SOHO/MDI images.

Verification with the NOAA Dataset

Comparison of temporal variations of the daily sunspot areas extracted from the EGSO Solar Feature Catalogue in 2003 presented in Fig. 3.9 (bottom plot) with those available as ASCII files obtained from the drawings of about 365 daily images obtained in 2003 at the US Air Force/NOAA (taken from National Observatory for Astronomy and Astrophysics National

Giophysical Data Centre, US (Zharkov and Zharkova 2006), revealed a correlation coefficient of 96% (Fig. 3.9, upper plot). This is a very high accuracy of detection that ensures a good quality of extracted parameters within the resolution limits defined by a particular instrument.

Further verification of sunspot detection in WL images can be obtained by comparing sunspot area statistics with solar activity index such as sunspot numbers. Sunspot numbers are generated manually from sunspot drawings (SIDC) and are related to the number of sunspot groups. The first attempt to compare the sunspot area statistics detected by us with the sunspot number revealed a correlation of up to 86% (Zharkov and Zharkova 2004) or with the sunspot areas defined by NOAA up to 96% (Zharkov et al. 2005).

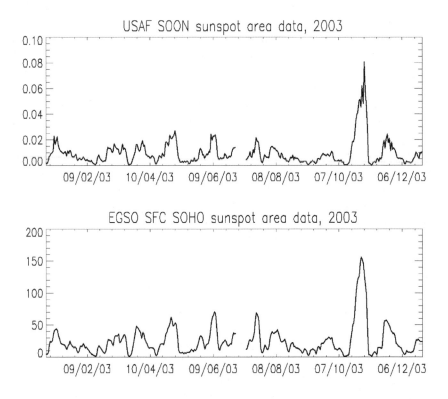

Fig. 3.9. A comparison of temporal variations of daily sunspot areas extracted in 2003 from the USAF/NOAA archive (US) and from the Solar Feature Catalogue with the presented technique

3.4 Region-Based Methods for Solar Feature Detection

V.V. Zharkova

3.4.1 Introduction

Region-based segmentation methods for pattern recognition in solar images rely on the homogeneity of spatially localized characteristics (e.g., intensity, multispectral values or texture values) within contiguous regions. One approach to region growing starts from one or more pixels, called seeds, and then incorporates neighboring pixels into regions around them according to specified homogeneity criteria (Fu and Mui 1981) which are often based on statistical tests. The process is continued until a predefined termination condition is reached.

An alternative approach is to subdivide the image into a set of arbitrary disjoint regions and then to merge or split the regions in an attempt to achieve contiguous regions satisfying the specified homogeneity criteria. Splitting may be achieved, for example, by dividing the image successively into smaller and smaller quadrants represented by a quad tree that is a tree with each node having exactly four descendents (Gonzalez and Woods 2001).

The preprocessed solar images with correct radial and nonradial illumination nonuniformities and dark lines resulting from dust, occasionally appearing in solar images can be used for region growing detection of active regions (Benkhalil et al. 2005), filaments (Fuller et al. 2005) or even coronal mass ejection movements (Hill et al. 2001). The efficiency of the method depends on the seed selection and on the intensity range definition for any types of features either brighter than background, i.e., active regions and plages, or darker, i.e., filaments.

3.4.2 Segmentation and Seed Selection

The important part of region growing method is a selection of a correct number of seeds because getting too many can lead to false detections while too few can lead to missed features.

In order to select the seeds as precisely as possible, we can use a threshold based on the mean intensity (see formula 1 in Sect. 3.1). The mean intensity can be either the global one, M_g, based on the whole solar disk pixel values as described in Sect. 3.2, or the windowed, or local, one, M_w, based on the pixel values only within a given window that can be adjacent to the feature to be detected. The global mean intensity M_g is subject to the image quality and contrast, since its value strongly depends on the image symmetry, or skewness, and the level of its flatness, or kurtosis. Therefore, it is much more accurate to use the windowed mean intensity that is better suited for comparison within the features within this window. Moreover, in order to further exclude random pixel errors, it is often more beneficial to use the second mean, M_{w2}, i.e., the mean without brightest and darkest regions within the window. Pixels with the values much less or more than the mean M_w are disregarded from the calculation of a second mean M_{w2}, which better reflects the "quiet Sun" intensity within each window.

Then the standard deviation σ_w is computed as per formula (2) in Sect. 3.1 from the same set of pixels, thus the window threshold is given by:

$$T_w = M_w \pm \alpha_1 \, \sigma_w \tag{3.21}$$

where the sign "+" is referred to the features brighter than background (i.e., active regions) and the sign "−" to darker than background features (i.e., filaments); α_1 is a constant obtained after testing a large set of images that also varies for different features. All pixels with the values lower (for darker features) or higher (for brighter features) than T_w are set to 1 and the rest of pixels are set to 0.

It can be noted that in finding seeds a lower threshold is preferential. The seeds are found as the centers of gravity of the regions with T_w equal to 1.

1. Segmentation

   (a) At first the solar disk is segmented into the regions of similar intensity using a region-growing algorithm that examines each of the four connected neighbors of a pixel and assigns them to the same region as the pixel if the difference in absolute intensity is less than a low, empirically determined threshold. This results in a large number of small segments.

   (b) Attributes, such as area, minimum, average, and maximum intensity, are computed and recorded for each segment component. The largest region, which is the area outside the disk, is labeled as "background" and the next largest region is labeled as "solar disk background."

   (c) A square adjacency matrix with one row and one column for each region is constructed. The image is scanned pixel by pixel and for each pixel that is in the ith segment and has a neighbor in the jth segment the (i, j) element of the matrix is incremented. When the matrix is complete it contains the lengths in pixels of the boundaries between each pair of regions.

   (d) The over-segmentation of the image is reduced by merging adjacent regions with similar properties.

2. Segment labeling

(a) A local brightness score S_l is computed for each segment based on its brightness relative to neighbors. Given n neighboring regions, where the ith neighbor shares a boundary of length P_i with the region being evaluated and has average intensity B_i, the weighted brightness of the set of local regions is computed as

$$\beta_l = \frac{\sum_{i=1}^{n} B_i P_i}{\sum_{i=1}^{n} P_i}. \qquad (3.22)$$

The differences Δ_l between the relative brightness of the current region β_c and the neighboring spots β_l, are computed as:

$$\Delta_l = \frac{[\beta_c - \beta_l]}{\beta_l}. \qquad (3.23)$$

A sigmoid function of the form:

$$\frac{1}{(1 + \exp(-2\Delta_l))} \qquad (3.24)$$

is used to produce a *local* intensity score, S_l between 0 and 1.

(b) A global brightness score S_g is generated by computing the mean intensity β_{bg} for the solar disk after removing outliers representing missing data and extremely dark or bright regions. The average intensity of each region is computed as:

$$\Delta_g = \frac{[\beta_{bg} - \beta_c]}{\beta_{bg}}. \qquad (3.25)$$

A sigmoid function of the form:

$$\frac{1}{(1 + \exp(-2\Delta_g))} \qquad (3.26)$$

is used to produce a *global* intensity score, S_g between 0 and 1.

(c) The local and global intensity score are combined using the function $S = S_l^\alpha S_g$, where α is a weighting factor defined empirically for a given instrument.

3.4.3   Automated Region Growing Procedure for Active Regions

The active regions (ARs) are detected automatically using the automated threshold and region growing (ATRG) technique (Benkhalil et al. 2005). The technique is applied to full-disk high-resolution solar images which have been preprocessed using the procedures indicated above. The technique include initial segmentation of the preprocessed image, noise reduction with median filtering, region labeling followed by region growing to define active region boundaries.

The Initial Segmentation

In order to define a suitable local threshold all images were first remapped into polar coordinates with origin at the solar disc center. Then an initial segmentation of the bright plages is carried out with a preliminary intensity threshold. The choice of these initial intensity threshold values is importation because if the value is too high this can lead to some real features being missed, whereas if the value is too low then binary images will contain more objects including noise and, hence, spurious features. Normally, this threshold is selected globally using the image histogram (Worden et al. 1996; Steinegger and Brandt 1998; Preminger et al. 2001). However, the optimum global threshold value varies with image brightness levels and the nonradial large-scale intensity variations which are particular problems in some Meudon images. In Fig. 3.10 the variations of mean intensity (μ) and standard deviation (σ) in Hα images are plotted for the whole of the year 2002, clearly indicating that these values change with the seeing conditions. These conditions also affect the local image quality over the solar disk leading to asymmetric intensity distributions in different quadrants (nonzero skewness).

This problem can be overcome sufficiently to complete an initial segmentation for seed generation purposes by using local thresholds (T_i) instead of a global one calculated from the image histogram. The local intensity threshold values are calculated for top-left, top-right, bottom-left and bottom-right quadrants of each image as follows:

$$T_i = \mu_i + (1 + \Delta_i) \times \sigma_i \tag{3.27}$$

where μ_i is the mean intensity value of quadrant i, σ_i is the standard deviation of the intensity of the same region and Δ_i is a constant that was set to

0.3 after investigating more than 30 Hα and Ca K II3 images. Then the pixels whose intensity values are greater than this intensity threshold have their values set to 1 and all other pixels have their values set to zero. Hence, we have defined regions where potential active regions may occur.

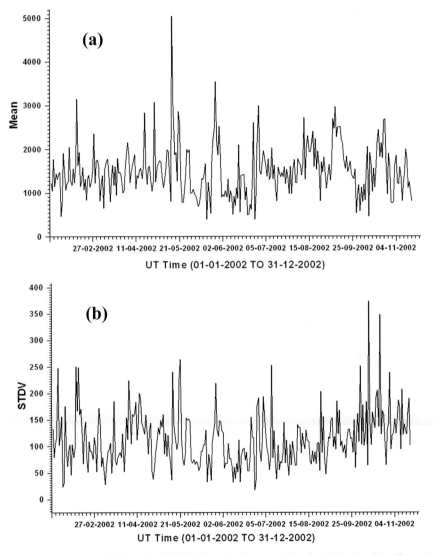

Fig. 3.10. Data statistics for full-disk solar images for the Meudon Hα-line, (**a**) mean (μ_i) and (**b**) standard deviation (STDV, σ_i)

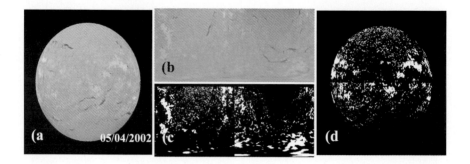

Fig. 3.11. The initial segmentation procedure stages: (**a**) an original preprocessed Hα image; (**b**) after transformation to Polar coordinates; (**c**) after initial thresholding; (**d**) after transformation back to Cartesian coordinates

The main stages of this initial process are illustrated in Fig. 3.11: a preprocessed image is shown in figure a, the result of image remapping into polar coordinates is shown in figure b, the initial segmentation-based on (3.27) is presented in figure c for the polar coordinates and in figure d for the Cartesian coordinates.

Noise Filtering and Region Labeling

The result of the initial segmentation generally includes noise and unwanted small features caused by over-segmentation (Fig. 3.12a). Over-segmentation is preferable to under-segmentation as the former can be remedied using median filtering and morphological operations (Matheron 1975; Serra 1988) whereas the latter could lose significant information. In order to remove the noise and small features, a 15×15 median filter is used (Fig. 3.12b) followed by a morphological opening operation of size 8×8 (Fig. 3.12c) that smoothes the features and fills in holes (the sizes having been chosen through experimentation with the standardized images). The detected regions are shown in Fig. 3.12d after their transformation back to Cartesian coordinates where every connected component is labeled, and its centroid is calculated for use as a seed in a region growing procedure. As can be seen, the noise and over-segmentation problems have been significantly reduced. Before using the seed, its location is checked to ensure that it is inside the associated region from which the region growing procedure can start. If found to be outside then its position is adjusted by searching the pixel values in the eight nearest neighbor directions until the new seed is found inside the region.

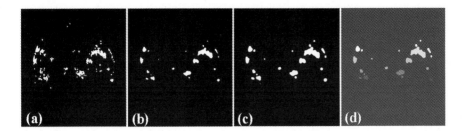

Fig. 3.12. Morphological, filtering, and labeling operations applied on the image resulting from the initial segmentation: (**a**) 8×8 morphological opening operation; (**b**) 15×15 median filter; (**c**) 7×7 dilatation to make the smaller features visible in the figure; (**d**) labeling the detected features with different gray values

The Region Growing Technique

The region-growing algorithm takes as input a standardized image and a corresponding set of seed points obtained by the procedures described in Sects. 2.1 and 2.2 on this book. The algorithm begins with seed pixels and scans the neighboring 8 pixels in a circular fashion, to determine a membership of the region around the central pixel that complies with the rules for belonging to the region grown described in the pseudocode (Fig. 3.13) and falls into a fixed threshold range defined by upper and lower pixel values. This allows us to give more accurate control in defining the outer boundaries of regions while also reducing the occurrence of holes in the regions. Similarly to the initial segmentation procedure (section "The Initial Segmentation"), the upper and lower threshold values within initially detected ARs are determined by exploiting again the statistical properties of the locally homogeneous background regions. In the case of Hα images the value of the upper threshold was found to noncritical and was set to the maximum intensity found in each solar disk. However, unlike the initial segmentation, the lower threshold value is defined as $\mu - 0.3\sigma$ (where μ is the mean and σ is the standard deviation of that region) because it provides more accurate threshold definition required at this stage.

As pixels are added, the process is repeated with the newly added pixels as the center pixels. Sometimes, more than one seed pixel is produced for a single AR, which results in two or more regions growing until they touch and merge into a single continuous AR. During the process, a list of the tested pixels is maintained, in order to avoid unnecessary duplication of the tests. A pseudocode procedure defining the region-growing algorithm

```
FOR j=1...n // Run over all n pixels in the image
  REPEAT
    FOR each pixel p at the border of Rᵢ
      FOR all neighbours (x,y) of p
        LOW_THⱼ = μⱼ – 0.3 ×σⱼ;// Let LOW_THⱼ be the lowest grey
                                            level of pixels in Rⱼ
        HIGH_THⱼ = max(R); // Let HIGH_THⱼ be the highest grey
                                            level of pixels in Rⱼ
      IF neighbour unassigned and
        (HIGH_THⱼ ≥ f(x,y) ≥ (LOW_THⱼ )
        Add neighbour to Rⱼ,
          update HIGH_THⱼ and LOW_THⱼ
      ENDFOR
    ENDFOR
  UNTIL no more pixels are assigned to the region
ENDFOR
```

Fig. 3.13. A pseudocode for the region-growing algorithm where R is an active region on the solar disk, R_1, R_2,...R_n are a set of regions selected for growing starting from an initial seed pixel and $f(x, y)$ is the gray level of pixel x, y

is given in Fig. 3.13. A set of selected seed locations and the active region detection results obtained using the region growing technique are shown in Fig. 3.14. For each AR automatically segmented, a number of physical parameters are extracted and populated into the AR feature table of the SFC database (Zharkova et al. 2005). These parameters include: center of gravity in pixel and heliographic coordinates; minimum, maximum, and mean intensities within the AR; area in square degrees; contrast ratios; bounding rectangle dimensions in pixels; chain-code representation of the AR boundary.

The example of an AR region bounding rectangle (cropped image) and a boundary chain-code saved in the SFC database, are shown in Fig. 3.15. The chain-code starts from any boundary pixel and lists the directions required to move from one boundary pixel to the next anticlockwise round the boundary until the starting pixel is reached. The directions are represented by the ASCII codes for the numbers from 0 to 7 as shown in Fig. 3.16.

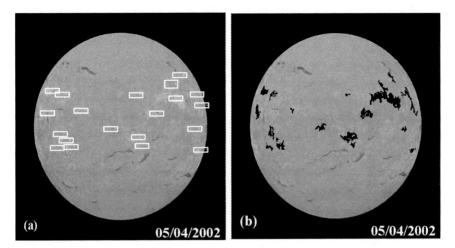

Fig. 3.14. (**a**) A standardized input image showing the selected seed locations (enclosed in white rectangles for visual identification), (**b**) the region growing results

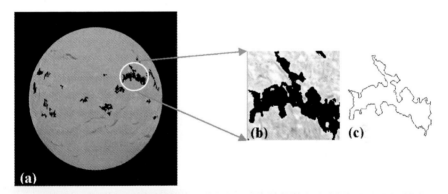

Fig. 3.15. (**a**) Detection results, (**b**) the cropped image, and (**c**) boundary represented by a chain-code

Verification of Detection with the Meudon and NOAA Synoptic Data

The ATRG procedure was tested on the synoptic image sequences obtained from full-disk solar images at the Meudon Observatory in January, February, March, April, and July 2002 and the overall comparison with Meudon and big bear solar observatory (BBSO) results are shown in Table 3.2. More detailed comparisons of our results with those from Meudon Solar Maps, BBSO Maps and NOAA AR Maps are provided in Table 3.3 and Figs. 3.17–3.23.

Table 3.2. A comparison of ARs detected per month by ATRG ($\Delta = 0.3$), Meudon and BBSO/NOAA

	ATRG	Meudon	NOAA	% ATRG/ Meudon	% NOAA/ ATRG
Jan. 02 (17 days)	256	364	182	70	71
Feb. 02 (18 days)	251	349	172	72	68
Mar. 02 (22 days)	351	408	212	86	86
Apr. 02 (28 days)	539	561	287	96	60
Jul. 02 (24 days)	311	479	202	64	65

A quantitative comparison of a subset of the results obtained using the ATRG technique with those obtained manually at the Meudon Observatory (Mouradian 1998) and at BBSO/NOAA (for the months of April and July) is shown in Table 3.3.

In comparison with ATRG and Meudon, NOAA detects about 50% fewer ARs on most days. For example, on 14 April, Meudon included 22 ARs in their results, the ATRG procedure detected 21 ARs and the NOAA Maps showed only 11 ARs. In order to quantify these differences the FAR (where we detect an AR and they do not) and the FRR (where they detect an AR and we do not) were calculated for each day and presented in the last four columns of Table 3.3. As can be seen, in most cases there is a larger number of ARs detected by ATRG than by NOAA with an average FAR of 8.3 per day in April (28 days) and only 4.5 per day in July (24 days). The FRR was very low at about 0.07 per day during April and about 0.125 per day during July. During this period the ATRG procedure failed to detect only 5 of those detected by NOAA. Comparing the ATRG result with that from Meudon produced a FAR of about 0.75 per day during April and about 0.58 per day during July. The FRR was larger at 1.9 per day during April and 4.16 per day during July.

One can note by comparing Meudon and NOAA results, that during this period the Meudon values for FAR and FRR are about 10 per day and 0.05 per day, respectively. Also it should be noted that the expected relationship between the results, ATRG number = Meudon number + FAR – FRR is not always satisfied in this table. For example, on 5th July in Table 3.2, ATRG detected 13 ARs and Meudon detected 19 ARs while the FAR and FRR values are 1 and 3, respectively, resulting in a discrepancy of four ARs. In this case, ATRG merges two regions detected by Meudon and Meudon separates one region detected by ATRG into four. Images from which results like these are obtained are presented and discussed below.

Table 3.3. A comparison of ARs detected with the ATRG technique with those from Meudon and BBSO/NOAA over April (4) and July (7) 2002

Date	Meudon		BBSO / NOAA		ATRG		FAR1 Meudon–ATRG		FRR1 Meudon–ATRG		FAR2 NOAA–ATRG		FRR2 NOAA–ATRG	
	4	7	4	7	4	7	4	7	4	7	4	7	4	7
01	19	**	8	**	15	**	0	**	4	**	7	**	0	**
02	20	21	8	6	20	18	0	0	0	3	12	12	0	0
03	20	20	9	5	14	8	0	0	3	9	5	4	0	1
04	22	20	9	9	21	16	0	0	1	4	12	7	0	0
05	21	19	8	9	20	13	0	1	1	3	12	4	0	0
06	22	**	10	**	16	**	0	**	4	**	6	**	0	**
07	23	**	13	**	24	**	2	**	1	**	11	**	0	**
08	22	20	13	8	21	12	3	1	2	7	8	5	0	0
09	22	**	14	**	17	**	2	**	1	**	3	**	0	**
10	25	20	12	9	29	10	2	1	1	4	17	1	0	0
11	22	19	13	9	18	14	0	2	2	3	5	5	0	0
12	21	18	13	6	19	14	0	0	2	4	6	8	0	0
14	22	17	11	6	21	12	1	2	2	4	11	6	0	0
15	18	**	12	**	16	**	0	**	2	**	4	**	0	**
16	19	18	12	9	16	14	0	0	3	4	4	6	0	0
17	16	19	10	9	14	13	0	0	2	2	4	4	0	0
18	19	17	7	6	21	13	3	2	1	3	13	7	0	0
19	20	18	8	4	16	7	0	0	4	6	7	3	0	0
20	17	19	9	4	22	9	2	0	2	6	13	5	0	0
21	17	19	8	4	17	10	0	0	2	5	8	6	0	0
22	16	16	8	5	18	9	1	0	2	4	8	2	0	0
23	20	20	10	9	18	14	2	0	3	6	8	6	0	1
24	22	**	11	**	19	**	0	**	3	**	6	**	0	**
25	21	21	13	12	27	21	3	1	0	1	14	9	0	0
26	**	21	**	13	**	14	**	0	6	**	1	**	0
27	20	23	10	12	25	12	0	0	0	7	15	0	0	0
28	18	24	12	13	14	15	0	0	4	3	2	2	1	0
29	18	22	7	13	18	16	0	0	3	3	9	3	1	0
30	19	24	9	12	23	14	0	0	0	5	11	2	0	0
31	**	24	**	10	**	13	**	4	**	4	**	4	**	1
Total	561	479	287	202	539	311	21	14	55	106	234	100	2	3

Verification and Classification Problems

In order to understand the results presented in Table 3.3 the total numbers of ARs per month are summarized for detection by ATRG Meudon and NOAA for the five month period. This comparison clearly indicates that, with the current threshold parameter setting $\Delta = 0.3$ (see (3.21)), the ATRG detection results are closer to Meudon than to NOAA. Increasing the initial threshold parameter value to $\Delta_i = 0.6$, then ATRG detects only the brighter regions and the results are in a closer agreement with those from NOAA On the other hand, if the initial threshold parameter is set lower to $\Delta_i = 0.2$, then ATRG detects more regions and the results are in closer agreement with those from Meudon. These results are illustrated in Fig. 3.17.

In Fig. 3.18a comparison is shown of the daily AR numbers extracted from the EGSO SFC with those available from the NOAA Space Environment Centre (Solar Region Report NOAA 2005, available as an ASCII file). Figure 3.18a shows a comparison of the AR numbers averaged over 50 days for the 5-year period (1998–2003). The cross-correlation analysis gives a correlation coefficient of 0.51879 and an average AR number per day of 11.7 for ATRG and 7.96 for NOAA. Figure 3.18b shows a similar comparison for the whole year 2002, to show the fluctuations of AR numbers in daily images.

In general, there are many factors that affect the quantitative comparison. In some cases, ATRG detects a single AR in CaII K3 (Fig. 3.19a,b) while Meudon splits it into four regions (on the basis of expert knowledge) (Fig. 3.19c) and NOAA splits it into two regions (on the basis of sunspot information) (Fig. 3.19d). A similar example in a Hα-line image is showed in Fig. 3.20 where ATRG detects a single AR (b) while NOAA splits it into two regions, on the basis of sunspot information (c). On the other hand in some cases, the ATRG procedure splits a region into two because of a low intensity channel between them (Fig. 3.21b) while the Meudon observer considered it to be single (CaII, Fig. 3.21c and Hα, Fig. 3.21e). A different example, presented in Fig. 3.21d–e, shows an AR in the Hα-line split by a filament into two regions. The Meudon observer recognizes the filament overlying a part of the AR but the ATRG detection currently does not take account the filament detection results and, thus, classifies them as two separate regions.

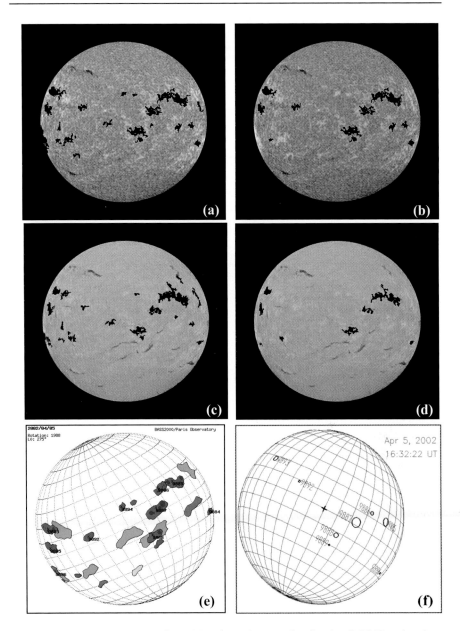

Fig. 3.17. The ATRG procedure ARs detection results for April 2002 using low value of T_i on the left, applied to Hα (**a**) and CaII K3 (**b**) and using a higher value of T_i on the right, applied to Hα (**d**) and CaII K3 (**e**). The Meudon and NOAA solar maps are shown for comparison in (**c**) and (**f**), respectively

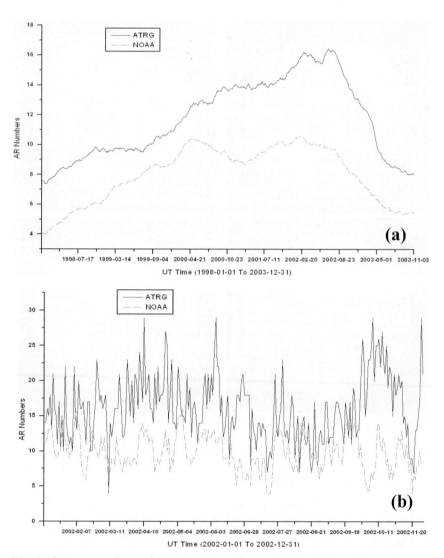

Fig. 3.18. A comparison of ATRG AR numbers with the NOAA AR numbers. (**a**) For 5 years 1998–2003 averaged over a 50 day window, (**b**) daily numbers for the year 2002

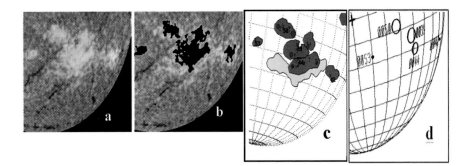

Fig. 3.19. An example of three ARs merging: (**a**) the Meudon CaII K3 input image (31 July 2002), (**b**) the same image after applying the region growing procedure, (**c**) the Meudon solar map image shown the same regions are divided into three ARs, and (**d**) the NOAA solar map shown the same region is divided into two ARs (NOAA0039, NOAA0044)

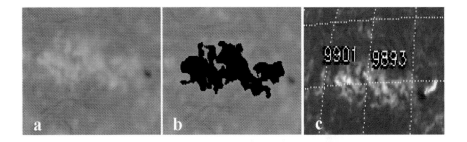

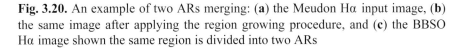

Fig. 3.20. An example of two ARs merging: (**a**) the Meudon Hα input image, (**b**) the same image after applying the region growing procedure, and (**c**) the BBSO Hα image shown the same region is divided into two ARs

The ATRG as currently implemented does not detect very small regions or regions of very low brightness and this is illustrated in Fig. 3.22a,c where the Meudon observer has detected a small AR (b) and a weak decaying AR (d), respectively. It also happens, although rarely, that the ATRG procedure detects a region not detected by Meudon or NOAA. This is illustrated in Fig. 3.23 where the region indicated by the arrow was detected only by ATRG despite the fact that it is similar in intensity to the neighboring AR which was detected by Meudon and NOAA. There are also some cases where NOAA detects regions which are not detected by Meudon and an example is given in Fig. 3.23d where NOAA detected the region labeled NOAA 41. The Meudon Map for the corresponding region of the sun is shown in Fig. 3.23e.

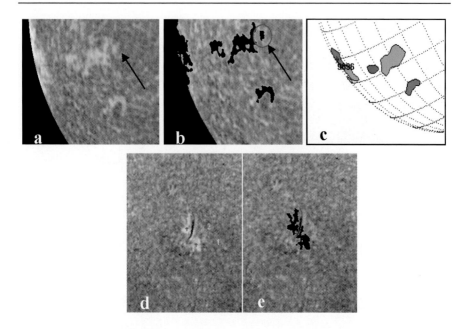

Fig. 3.21. An example of two ARs splitting: (**a**) the Meudon CaII K3 input image, (**b**) the same image after applying the region growing procedure, the arrow indicates the split region, (**c**) the Meudon solar map shown the same regions are merged into one A, (**d**) the input Hα image, and (**e**) our detection results

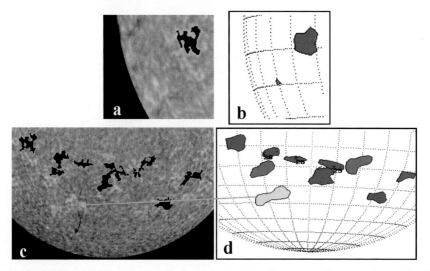

Fig. 3.22. The ATRG misdetection cases: (**a, b**) of very small regions, (**c, d**) of weak regions

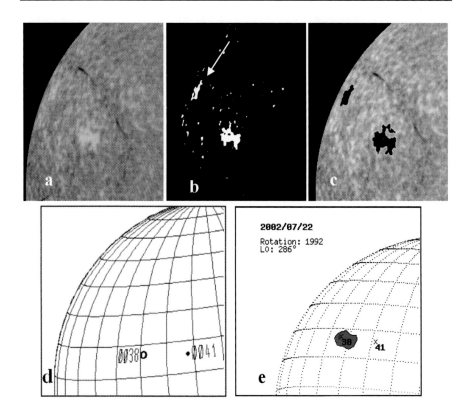

Fig. 3.23. Region detected by ATRG but not detected by NOAA and Meudon, (**a**) input CaII K3 image (22 July 2002), (**b**) the thresholding result, (**c**) the detection results, (**d**) NAAA map, and (**e**) Meudon solar map

The main reason for these differences is believed to be due to the different definitions of ARs. At Meudon all bright regions (plages) are detected and are defined as the regions in the chromosphere which are brighter than the normal "quiet" Sun background. At NOAA an AR is defined as a bright area on the Sun with a large concentration of magnetic field, containing sunspots. However, not all plages contain a strong magnetic field as they might be decaying ARs with a weakening magnetic field (van Driel-Gesztelyi 2002).

Figure 3.24 clearly illustrates this case by showing the results of ARs detection at NOAA, Meudon and by using the ATRG technique with Hα, CaII K3 (Meudon) and Fe XII 195Å (SOHO/EIT) solar images on the same day (30 July 2002). In general, the agreement with both Meudon and NOAA is thought to be understood, considering that in NOAA the decisions are taken using more information than the ATRG does.

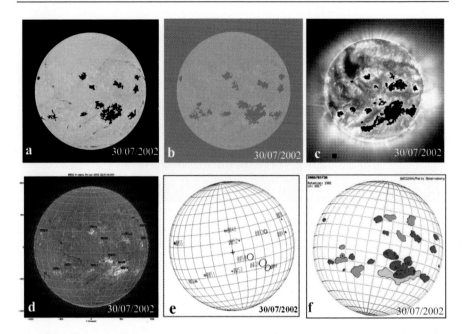

Fig. 3.24. A comparison of AR detection results: (**a**) ATRG procedure applied to Hα; (**b**) ATRG procedure applied to CaII K3; (**c**) ATRG procedure applied to Fe XII 195 Å; (**d**) BBSO solar image showing 12 ARs; (**e**) BBSO solar map showing 12 ARs; (**f**) Meudon solar map showing 24 ARs

3.4.4   Region Growing Procedure for Filaments

Procedure Description

The first step after image standardization and cleaning is to find the seeds as described in Sect. 3.4.1. The next major step is to grow the seed regions into larger ones by adding the pixels, which comply with the intensity criteria within the applied windows. These windows are centered on the seed and their sizes depend on the seed dimensions in both Ox and Oy directions. The lowest value of the intensity range is set to 0 and the highest T_{max} is obtained the same way as for seeds:

$$T_{max} = M_w - \alpha_2 \times \sigma_w \quad with \quad \alpha_2 < \alpha_1 \qquad (3.28)$$

The value of each pixel adjacent to a seed is checked and, if it is in the range [0, T_{max}], it is appended to the seed.

Figure 3.25 shows an example of the final segmentation of the region growing for filaments in one full-disk image in Hα-line obtained by Meudon Observatory. Note that it is advisable to define a minimum region size in order to remove the regions that did not grow into large enough ones.

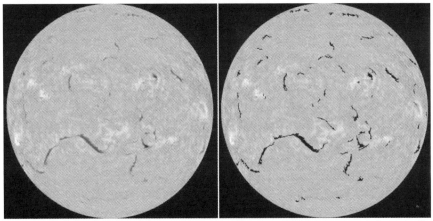

Original cleaned image   Detected filaments superimposed

Fig. 3.25. The results of filament detection (*right*) with region growing technique from the original preprocessed Hα-image (Meudon Observatory, *left*)

Verification of Filament Detection

The automated detection results were compared with those produced manually at the Meudon Observatory by defining a shape descriptor corresponding as close as possible to that from the manual detection process (Fuller et al. 2005). The latter involves choosing a few points along the path of the filament and computing the full path by linking the points together. The best way to represent the automatically detected filaments is thus to compute their pruned skeletons. This iteratively removes pixels from the boundary of a filament region, using a morphological operator, until the region is only one pixel thick. The resulting skeleton is then pruned to remove short branches and leave only two end points (see Fig. 3.26).

The automatic procedure for the detection of filaments has been applied to the observations between May 1995 and May 2005 and compared with manually detected filaments (see examples from manual detection can be seen on the Bass2000 web site (http://bass2000.obspm.fr) and their synoptic maps available in ASCII format.

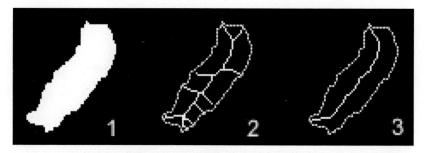

Fig. 3.26. Original shape – 1, full skeleton – 2 and, pruned skeleton – 3

A comparison of the automatically and manually detected filaments (with the size greater than 5°) over this period showed: 89% of the "automatic" filaments matched the "manual" ones; 4% of the "automatic" filaments do not correspond to "manual" ones (they are present but are faint and were missed by the observer); 11% of "manual" filaments have not been automatically detected. The difference between the total lengths of all "automatic" and "manual filaments" is 7% resulting from the filaments, which were not automatically detected because they were too faint and felt below the limit of the filament definition (threshold). Hence, one can conclude that the differences between the two methods come from the subjective interpretation of the original image but not from the accuracy of automated technique.

3.4.5   Region Growing Methods for Different Classes of Features

Preminger et al. (2001) used the region growing approach for semimanual detection of active regions in the three-trigger method by splitting pixels into three contiguous classes whose contrast is greater than the given trigger criteria for each feature type (sunspots, active regions, and quiet Sun background). These are then used as seeds, from which the regions are grown by adding adjacent pixels that satisfy a less stringent threshold criterion.

The values for triggers and thresholds are estimated from the statistical properties of the quiet Sun. They are set to give what are considered to be acceptably low levels of false identifications of sunspots in H_α images and faculae in CaII K. Successful application of the three-trigger method over the whole solar surface does require very clean images. The cleaning techniques they used to prepare the "contrast" images are discussed in Walton et al. (1998).

The steps involved in the implementation of the algorithm are as follows:

(a) A contrast image is produced from a calibrated image by dividing each pixel in the disk by the appropriate value from the mean limb-darkening curve and then subtracting one. This procedure produces an image that contains some vertical or horizontal artifacts mainly due to a small nonuniformity in the CFDT filters and noise. These artifacts are removed from the image by subtracting from each column of the image the median value in that column and then repeating the procedure for each row (Walton et al. 1998).

(b) The probability of a false positive identification of a feature by the three-trigger method is $p_+ = p - p(1-p)^8 - 8p^2(1-p)^7)$, where p is the probability of a randomly chosen pixel satisfying the trigger criteria. The expected number of false positives for an image containing N pixels is $Np_+ / 9$. Using the H_α contrast image histogram, they set the trigger threshold for dark features to be the contrast (-1.9%) at which $p = 0.003$ which should result in only about one false positive per 100 images. For the much more numerous bright features they set the trigger threshold to the contrast (-1.4%) for which $p = 0.01$, giving an expectation of one false positive per three images. For the CaII K images only bright features were considered. Because there are typically hundreds of bright features on these images they preferred to err on the side of missing no potentially real features. They set the trigger threshold to the contrast (2%) at which $p = 0.02$, giving an expected two false positives per image.

(c) An isolated feature is identified by searching the solar disk for a pixel whose contrast is greater than the given trigger contrast criterion. Then adjacent pixels are examined to check if they are a part of the same feature. If three contiguous pixels meeting the trigger criterion are found, it is concluded that a real feature has been found. The search was limited to pixels whose distance from the solar center was less than 0.95 of the solar radius. This limit was imposed because of the reduction in signal-to-noise ratio near the limb.

(d) For each detected three-pixel region, adjacent pixels are added to the region providing they satisfy a threshold criterion. In the case of H_α images, for both dark and bright features, the region growing threshold was set to the contrast (1.4%) at which the $p = 0.01$. In the case of the CaII K images, the region growing threshold was set to the contrast (2%) at which the $p = 0.02$.

The authors report that the three-trigger method is fast and accurate except very close to the limb. By comparison a straightforward threshold method employing the same low contrast thresholds would produce many spurious features on each image. However, the method does require very clean images with no artifacts and Gaussian noise whose distribution is constant across the solar image.

3.5   Other Methods for Solar Feature Detection

V.V. Zharkova and S.S. Ipson

3.5.1   Bayesian Inference Method for Active Region Detection

In the context of image segmentation, Bayesian inference is a tool for determining the likelihood that a particular object x is present in a scene, given that sensor data y (i.e., image data) is observed (Duda and Hart 1973). In Bayesian terms, this likelihood is called the *posterior model*, $P(x \mid y)$, and is computed using Bayes' Rule:

$$P(x \mid y) = P(y \mid x)P(x)/P(y) \propto P(y \mid x)P(x), \tag{3.29}$$

where $P(x)$ is a *prior model*, which expresses the probability that object x is present (regardless of the sensor data), and $P(y \mid x)$ is a *sensor model*, which expresses the likelihood that the sensor will generate data y given that x is present. The required priors may be computed by assuming image pixels belong to an appropriate small number of classes, and analyzing a set of training images to estimate the likelihood priors and by assuming an appropriate general form (Geman and Geman 1984) for the prior probability of a given classification. In practice, the first prior enforces fidelity to the data while the second enforces smoothness. Segmentation is achieved by adjusting the classification to maximize the posterior probability.

An application of Bayesian methods for automatically identifying various surface structures on the Sun is described in a series of papers by Turmon et al. (Pap et al. 1997; Turmon et al. 1998). The primary source of data for their study was a set of CaII K (at wavelength 393.4nm) full-disk spectroheliograms obtained from the Sacramento Peak National Solar Observatory. For their purposes they divided the surface structures into three types:

(a) Plages, which are high-intensity, clustered regions that spatially coincide with active regions. Sunspots also coincide with these regions, although sunspots are features of the photosphere.
(b) Network, which is a lower-intensity cellular structure formed by the hot boundaries of convection cells.

(c) Background, which is the lowest-intensity region formed by the cooler cell interiors

In order to partition the image into plage, network and background components, they adopted the Bayesian framework of inference of underlying pixel classes based on the observed intensity. Denoting pixel sites by $s = [s_1 s_2]$, in an image domain N, observed intensities by $\mathbf{y}$, and matrices of class labels by $\mathbf{x} = \{x_s\}_{s \in N}$, a small-integer is associated with each pixel of the observed image defining its class.

The maximum a posteriori rule maximizes the posterior probability of the labels given data:

$$\hat{\mathbf{x}} = \arg \max_{\mathbf{x}} \log P(\mathbf{y} \mid \mathbf{x}) + \log P(\mathbf{x}). \tag{3.30}$$

The prior probability of a given labeling (prior models) is specified by the Markov field smoothness priors:

$$P(\mathbf{x}) = Z^{-1} \exp[-\beta \sum_{s \sim s} 1(x_s \neq x_s)], \tag{3.31}$$

where Z is a constant normalizing the distribution, $s \sim s'$ represents neighborhood relationship (each interior pixel has eight neighbors). For $\beta = 0$ the distribution is uniform on all labeling, while as β is increased, smoother labeling is favored.

The likelihood function, assuming that intensities are independent conditional on the labels being known, is given by $P(\mathbf{y} \mid \mathbf{x}) = \prod_{s \in N} P(y_s \mid x_s).$

The densities $P(y \mid x = k)$ are estimated from labeled data (supplied by scientists). The lognormal distribution is suggested as a good model for the per-class intensities in case of segmentation of CaII K images into plage, network and quiet sun areas (Kolmogorov–Smirnov and Cramer–von Mises distributional tests confirm this, according to Pap et al. (1997). Thus, in this case the objective function becomes,

$$-\sum_{s \in N} \left(\frac{(\log y_s - \mu_{x_s})^2}{2\sigma_{x_s}^2} + \log \sigma_{x_s} \right) - \beta \sum_{s \sim s} 1(x_s \neq x_s) \tag{3.32}$$

If $\beta = 0$ and the class variances are identical, the threshold rule used in practice is recovered. With $\beta > 0$, the optimization becomes coupled across the sites, and is entirely intractable for the three-class problem. The

Gibbs sampler is used to tackle this problem, cycling through each site, computing $P(x_s \mid y_s, x_{N(s)})$ for each class, and choosing the next label from this distribution. For finite label spaces the resulting random sequence converges in distribution to the posterior. To extremize the posterior, the distribution is sharpened by decreasing a scale parameter slowly to zero, and the resulting labeling is the maximum a posteriori estimate. Turmon et al. (1998) presented segmentation results with $\beta = 0.7$ obtained after 800 sweeps through the image.

The authors show a small region of an original image and its segmentation into network and plage regions via simple thresholding and via their Bayesian approach (Pap et al. 1997; Turmon et al. 1998). In their example the segments produced by the thresholding approach have rough boundaries while the boundaries produced by the Bayesian approach are smooth. Small isolated segments produced by the thresholding approach are rarely produced by the Bayesian approach. However, the Bayesian method is very time-consuming and should be compared in terms of speed and accuracy with other quicker methods that take into account spatial structure as well as the intensity histogram. In reference Turmon et al. (2002) the Bayesian method is applied to combined high-resolution (1,024×1,024 pixel) magnetograms and quasi-white-light photograms (taken by MDI) and demonstrates that the combined data can potentially be used to accurately segment the solar images into four specific regions: sunspot umbra, sunspot penumbra, faculae, and background.

3.5.2   Detection of Flares on H$_\alpha$ Full-Disk Images (Veronig Method)

Flares and coronal mass ejections (CMEs) are transient events accompanied by a rapid release of energy and the transport of energy material as downwards to the solar surface so away from the surface of the Sun, contributing to interplanetary particles, emission, and shocks. We review recent techniques for the detection of these features in solar images using combinations of edge-based, region-growing, and Hough transform techniques.

A combination of region-based and edge-based image segmentation methods (cf. Sects. 3.3 and 3.4), applied to H$_\alpha$ full-disk images, is described by Veronig et al. (2001). In this technique region growing is started at seed pixels having intensities greater than a threshold, chosen as twice the quiet sun level and is stopped when either the intensity falls below the threshold or a region edge is encountered. The Canny edge-detection method (Canny 1986) is used to identify edges and involves the following steps:

1. Smooth the image with a Gaussian filter
2. Compute the gradient-magnitude and orientation using finite-difference approximations for the partial derivatives
3. Apply nonmaxima suppression, with the aid of the orientation image, to thin the gradient-magnitude edge image
4. Track along edges starting from any point exceeding a higher threshold as long as the edge point exceeds the lower threshold
5. Apply edge linking to fill small gaps

The method as described attempts to follow boundaries between poorly defined objects as well as hard edges and ignores weak edges which are not connected to stronger edges.

The full steps involved in the implementation of the algorithm of Veronig et al. are as follows:

(a) An H_α full-disk image is first preprocessed to remove nonuniform illumination effects. A median filter with a large mask is first applied to a solar image to smooth the image without blurring the edge. A Sobel operator, computing the magnitude of the gradient from the vertical and horizontal obtained by convolving the image with following Kernels

$$\begin{vmatrix} -1 & -2 & -1 \\ 0 & 0 & 0 \\ 1 & 2 & 1 \end{vmatrix} \qquad \begin{vmatrix} -1 & 0 & 1 \\ -2 & 0 & 2 \\ -1 & 0 & 1 \end{vmatrix}, \qquad (3.33)$$

is then used to detect the limb edge. After thresholding the gradient-magnitude to detect edge points, a least squares fit to a circle is performed giving the sun radius and center. The large-scale variation of intensity over the solar disk is obtained by dividing the solar disk into concentric rings and fitting polynomials to them. These are used to correct the solar image for nonuniform illumination. Further processing is only done on images that pass a quality criterion. This is based on the standard deviation of the intensity histogram of the corrected image, which should be normally distributed about the mean.

(b) Pixels in the corrected image that have intensities greater than twice the quiet sun level are marked as possible flare regions.

(c) The area of a flare region is obtained by region growing. Adjacent pixels are added to the region provided intensities are large enough and providing they do not coincide with a region boundary. The boundaries are obtained by first applying a Canny edge detector and then using a separate edge tracing algorithm to mark the individual closed boundaries.

The authors developed this procedure to identify solar flares in H_α full-disk images automatically and in a quasi-real time. They applied it to archived the data and found that it produced appropriate results. The authors suggest that the combination of region and edge-based segmentation should work irrespective of the size and intensity of the flare, which is a crucial requirement of an automatic flare identification system that cannot be achieved by a simple threshold based segmentation scheme.

3.5.3   Detection of Coronal Mass Ejections

An autonomous procedure for detecting CMEs in image sequences from LASCO C2 and C3 data was reported by Berghmans et al. (2002). This procedure uses a Hough transform approach, which, in general, is used to detect the presence of edge pixels in an image which fall on a specified family of curves and uses an accumulator array whose dimension equals the number of parameters in the curves to be detected. For example, in the case of a straight line expressed in the normal form $\rho = x\cos\theta + y\sin\theta$, the parameters are ρ and θ and the two indices of the accumulator array correspond to quantized values of ρ and θ over the range of values of these parameters which are of interest. The accumulator array is used as follows. Each edge pixel in the image with coordinates x, y is considered in turn and those elements in the accumulator array which satisfy the straight line $\rho = x\cos\theta + y\sin\theta$ with quantized values of ρ and θ are incremented. After all the edge pixels in the image have been processed in this way, the accumulator array is examined to locate those elements with largest values. The largest elements indicate the parameters of the straight lines within the image which have the most supporting evidence. Using this information, the image can be searched to find the points lists of the line segments with corresponding parameter values. The implementation of the Hough transform is similar for curves with more parameters but the execution time increases very rapidly with the number of model parameters. Fast variants of the Hough transform exist, which take account of the gradient of edge about each edge point (O'Gorman and Clowes 1976; Burns and Hanson 1986). Bergman et al. utilize the Hough transform to find bright ridges in height-time maps generated from an image sequence. Because the original data has much greater resolution than required for the detection of CMEs and the LASCO C2 and C3 data have different spatial and temporal resolutions a number of preprocessing steps are also applied to reformat the images.

Hough Transform Method

The steps involved in the procedure are as follows:

(a) Each image is read in, bright point like sources remove, and intensity normalized to a standard exposure time.
(b) A mapping from Cartesian (x, y) to polar (r, θ) coordinates is applied and simultaneously the resolution is reduced from 1,024×1,024 to 200×360.
(c) The C2 and C3 (r, θ) images are combined into one composite image by rescaling and matching the different spatial and temporal resolutions.
(d) A background image is estimated as a running average over one day (see Sect. 2.8) and subtracted from the original sequence resulting in a prepared (r, θ, t) data stream in which most of the non-CME signal is removed or strongly attenuated.
(e) A CME produces a ridge inclined at an angle θ_R corresponding to its speed v in a (t, r) slice from the prepared (r, θ, t) data which is detected using the Hough transform. A (v, θ, t) data cube is set up and integrated over v and CMEs detected by locating clusters in the resulting (θ, t) map.

The reported success rate of the automatic procedure compared with the human observer is about 75%.

Multiple Abstraction Level Mining Method

Images from a sequence of SOHO images are preprocessed to preextract and index spatio-temporal objects (bright and dark regions) using the region growing method described in Sect. 3.4.2. Objects are tracked from image to image in the sequence by assuming that objects that have the maximum spatial overlap in temporally adjacent images are the same object, since the rotational speed of the sun causes the objects to move very little in the time interval. After a time series of objects have been identified, the time series for area, minimum intensity, average intensity and maximum intensity are assembled. Each time series is then segmented using the multiple abstraction level mining (MALM) technique (Li et al. 1998) using a set of seven labels including "sharply rising," "rising," "flat," etc. To query the database for CME events, component classes of brightening objects and darkening objects are defined with specified time constraints and a CME class is defined to consist of both brightening and darkening objects next to each other. The authors reported that over a period of two months of SOHO data from 1997 to 1998 they detected 11 out of 13 CME events identified in the images by astronomers. The system returned between 20 and 30 images, out of about 2,500 images for an expert user to inspect and identify the 11 CMEs.

3.5.4   Magnetic Inversion Line Detection

Magnetic inversion lines feature in many papers related to solar magnetic field activity ranging from long term changes through the 11 (22) year solar cycle to rapid changes with the time span of flare events. We review four methods, which have been proposed for detecting magnetic inversion lines with and without the application of smoothing. Smoothing is thought to provide estimates of inversion line positions, which correspond to higher altitudes in the solar atmosphere but eliminates spatial detail, associated with flaring phenomena for example. The fourth method was originally developed by the current authors and uses a distance transform approach simplifying the structure of magnetic inversion lines in the regions of weak magnetic fields but not in the regions of strong ones.

Contour Method

Many papers define a magnetic neutral or inversion line as the region corresponding to the zero contour of the line-of-sight magnetic field or the region between corresponding positive and negative contours having the lowest thresholds based on the instrument errors. This method is applied to unsmoothed magnetic field measurements possibly with a high cadence of up to 1 min relevant to fast dynamic events like flares, microflares or bright points as illustrated in Fig. 3.27.

An example applying this approach is the work of Falconer et al. (1997) with Marshal Space Flight Center Vector Magnetograph data (Hagyard et al. 1984) aligned with Yohkoh SXT images. They examined the magnetic structure at 94 sites in the bright coronal interiors of five active regions that are not flaring while exhibiting persistent pronounced coronal heating. The MSFC magnetograms used have uncertainties of 50 G in line-of-sight magnetic fields and uncertainties of 75 G for transverse fields. They plotted line-of-sight contours for positive and negative fields with magnitudes of 25, 500, and 1,500 G. The neutral line was defined to be between the -25 G and +25 G contours, which in the regions of interest had a width of two pixels or less.

Falconer et al. determined magnetic shear at the point on a neutral line as the difference between the azimuthal direction of the observed transverse field at that point and the direction of the potential field calculated to satisfy the boundary condition of the observed line-of-sight field. The results revealed that the most persistent enhanced heating of coronal loops requires (a) the presence of strong polarity inversion near at least one of the loop footpoints, (b) is greatly aided by the presence of strong shear along the neutral line, and (c) is controlled by some variable process that acts in the magnetic environment, which was suggested to be a low-height reconnection accompanying magnetic flux cancellation.

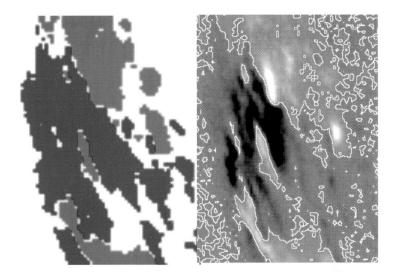

Fig. 3.27. Line-of-sight magnetic field in the vicinity of the July 23rd, 2002 flare seen near the limb in a SOHO MDI magnetogram. On left ±25 G contour region in white, with higher field in red and lower field in blue. On the right field is displayed with zero gauss as mid gray and with positive and negative fields increasing and decreasing in brightness, respectively. The zero contour regions are shown as white lines

Spatial Smoothing Method

This approach is related to the suggestion that magnetic loops of various magnitudes produce different spatial signatures in the LOS magnetic field signals measured in the photosphere and by smoothing the magnetic data one can approximate the magnetic field from higher loops (Priest 1984). This approach was applied by Durrant (2002) who investigated the magnetic neutral lines in the north and south polar regions using magnetograms from the National Solar Observatories, Kitt Peak (NSOKP), Mt. Wilson Observatory (MWO) and SOHO MDI archives for Carrington rotation 1955 (the period 10 October to 5 November 1999). The magnetic data were compared with each other in order to establish their reliability, and then were smoothed before locating the zero field contours. These locations were checked by overlaying with the positions of solar polar filaments observed in H$_\alpha$ images at the BBSO and the Meudon Solar Observatory (Durrant 2002). With increasing smoothing, the resulting structures approximate the magnetic field at increasing heights in the solar atmosphere. The behavior of these structures at high latitudes is important in the investigation of the evolution of the solar magnetic field, but had previously been little studied because of the observational difficulties and the substantial computing time required executing smoothing.

The Procedure

The source solar magnetograms have the following specifications:

> NSOKP, 1,788×1,788 pixel, (resolution ~1 arcsec) using the 868.8 nm Fe I line.
> MDI, 1,024×1,024 pixel, (~2 arcsec) using the 676.8 nm Ni I line.
> MWO, 512×512 pixel, (~3.75 arcsec) using the 525.0 nm Fe I line.

The magnetograms are mapped to equal interval latitude–longitude grids (Plate Care projection, −90 to +90 (latitude) and 0–360 longitude) of size 2,560×5,120 pixel for NKSOKP and MDI data and 512×1,024 pixel for MWO data as illustrated in Fig. 3.28. The NSOKP and MDI maps were smoothed to the effective resolution at the center of the MWO data and then reduced in resolution to a 512×1,024 pixel grid. A second stage of smoothing over circles of diameter 60, 120, and 180 arcsec is then applied to the 512×1,024 pixel grid data from all three sources before remapping back to a 512×512 pixel image showing the solar disc as north or south polar plots.

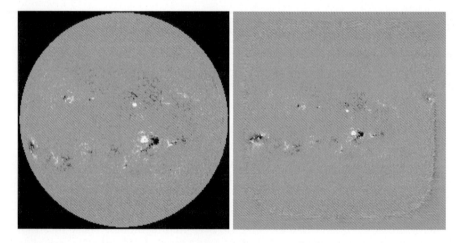

Fig. 3.28. The SOHO MDI magnetogram of 23rd July 2002 on the left shown transformed to latitude longitude projection on the right (Zharkova et al. 2005)

Temporal Averaging Method

Ulrich et al. (2002) reported a method of averaging spatial features over time to remove fast temporal variations of magnetic field while retaining stable long-lived structures like large equatorial loops. They extracted the east–west and meridional components of the slowly evolving large-scale magnetic field from Mount Wilson Solar Observatory data. The observation noise was reduced by temporal averaging and the resulting neutral line detection over a narrow strip about the sun's center meridian at the time of observation were plotted onto a synoptic map where the horizontal axis is the Carrington longitude and the vertical axis is the heliographic latitude. To avoid introducing smearing caused by the differing rates of solar rotation at different latitudes, a Carrington coordinate is considered to apply to a point at the time of its central meridian crossing and all other observations are differentially rotated in time backward or forward to be overlaid at the specified Carrington coordinate. To keep the distorting effect at high latitude minimal, this averaging procedure is limited to including the observations over less than one-quarter of the solar period and the zero point is reset for each new rotation.

The Procedure

The analysis is based on the approach developed by Shrauner and Scherrer (1994). A measurement B_i of the line-of-sight component of the sun at longitude L_i is resolved into local meridional and east–west transverse components B_v and B_t, respectively, as follows.

$$B_i = B_v \cos(L_i) + B_t \sin(L_i) \tag{26}$$

The two equations given by the following weighted sums:

$$\sum_i B_i \sin(L_i) = B_v \sum_i \sin(L_i)\cos(L_i) + B_t \sum_i \sin^2(L_i)$$
$$\sum_i B_i \cos(L_i) = B_v \sum_i \cos^2(L_i) + B_t \sum_i \sin(L_i)\cos(L_i) \tag{27}$$

for all observations i of the same heliographic coordinate (after taking differential rotation at different latitudes into account) are solved for B_v and B_t. Further, the tilt angle φ of the magnetic field in the east–west direction is defined by $\varphi = \tan^{-1}(B_t / B_v)$. Ulrich et al. produced two types of synoptic maps. The first show the synoptic variation of B_v, with the neutral lines where B_v is zero superimposed. The second type show the synoptic variation of φ, with both the neutral lines and the regions where φ is zero superimposed. The authors refer to regions where φ is zero and therefore the averaged field is meridional, as meridional neutral lines to distinguish them from the ones where B_v is zero.

Distance Transform Method

Ipson et al. (2004) applied this method to line-of-sight magnetic data from the SOHO MDI level 1.5 magnetograms to smooth the data outside strong magnetic field features while retaining the original data within the features. The method is suitable for the detection of medium-term variations of magnetic field in filaments outside the polar crown.

The SOHO/MDI 1,024×1,024 pixel magnetograms have a spatial resolution of about 2 arcsec and an uncertainty in magnetic field of about ±20 G. In a typical MDI magnetogram the boundaries between adjacent positive and negative magnetic 20 G contours are narrow only at comparatively few locations between local concentrations of opposite fields associated with sunspots. Elsewhere the positive and negative unipolar regions are characterized by the presence of mainly positive (or negative) fluctuations from the ±20 G, near zero, range of magnetic field values. Rather than apply a spatial averaging using a large kernel, which smoothes all the details, this approach uses a distance transform method that retains the fine structure in the high field regions while reducing it elsewhere.

The Procedure

The magnetogram data is then segmented into three types of magnetic region namely, negative, neutral, and positive by applying a two-level magnetic field threshold ±T, where for example, T would be set at the uncertainty level 20 G. A fast Euclidean transform (Cuisenaire and Macq 1999), is applied to label the pixels in the neutral region with their distances from the nearest positive or negative boundary pixel. On completion of the distance transform, each pixel in the neutral region is labeled positive or negative according to whether the pixel is nearest to an original negative or positive boundary. The boundaries between the positive and negative labeled regions (forming a Voronoi tessalation) are the estimated neutral lines and can be marked explicitly by applying an edge detector.

By applying the distance transform approach with values of T larger than 20 G, Ipson et al. produced neutral line maps containing decreasing amounts of spatial details in the low-field regions. The method gives the similar results over the T = 20–40 G tested, to those obtained by applying increasing amounts of smoothing using Gaussian filters as illustrated in Fig. 3.29.

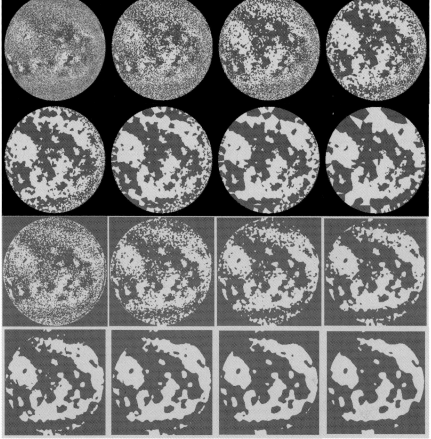

Fig. 3.29. Magnetic neutral lines indicated by the boundary between dark (negative) and light (positive) regions in MDI magnetogram for 23 July 2002. Upper set using distance transform method and thresholds varying from 20 to 90 G from *top-left* to *bottom-right*. Lower set using Gaussian smoothing with standard deviations varying from 4 to 32 arcsec from *top-left* to *bottom-right* (from Ipson et al. 2004)

3.6 Skin Lesion Recognition with Region Growing Methods

Ilias Maglogiannis

3.6.1 Introduction

The field of medical image processing is continually evolving. During the past ten years, there has been a significant increase in the level of interest in medical image morphology, full-color image processing, image data compression, image recognition, and knowledge based medical image analysis systems. One major area of accelerated growth is the field of dermatology where a lot of research has been done involving the development of diagnostic tools, which are designed to serve as diagnostic adjuncts for medical professionals in skin lesion assessment.

Dermatologists and general physicians base the conventional diagnosis of skin lesions on visual assessment of pathological skin and the evaluation of skin macroscopic features. Therefore the correct assessment is highly dependent on the physician's experience and on his or her visual acuity. Moreover, the human vision lacks accuracy, reproducibility and quantification in gathering visual information, elements which are substantial during follow-up studies in diseases monitoring. During the last years however, computer vision-based diagnostic systems for the assessment of digital skin lesion images have demonstrated significant progress. Such systems have been used in several hospitals and dermatology clinics, aiming mostly at the early detection of malignant melanoma tumor, which is among the most frequent types of skin cancer, versus other types of nonmalignant cutaneous diseases. The significant interest in melanoma is due to the fact that its incidence has increased faster than that of almost all other cancers and the annual incidence rates have increased on the order of 3–7% in fair-skinned populations in recent decades (Marks 2000). The advanced cutaneous melanoma is still incurable, but when diagnosed at early stages it can be cured without complications. However, the differentiation of early melanoma from other nonmalignant pigmented skin lesions is not trivial even for experienced dermatologists. In several cases primary care physicians underestimate melanoma in its early stage (Pariser and Pariser 1987).

The main design issues, needing to be addressed during the development of a machine vision system for skin lesion assessment, concern the image acquisition set up, the image processing and the classification methodology. More specifically, the following questions have to be addressed:

1. How reproducible and reliable images can be acquired using existing image acquisition devices and how we process them to increase their quality?
2. Which image features assist diagnosis, i.e., what are we looking for?
3. How are these features calculated by the image matrix, in space or frequency domains using image processing techniques?
4. How many of the defined features should be used for optimal results? (feature selection).
5. Which classification tools may be used and how is the "importance" of each feature to classification determined?

6. How may the performance of a classifier be assessed?
7. What are the reported accuracies of the results from existing systems?

Section 3.6.2 provides answers to the above questions 1–6, while Sect. 3.6.3 covers question 7. Section 3.6.4 describes a case study application, measuring the effectiveness of the techniques presented in this chapter, for discriminating malignant melanoma tumors versus dysplastic naevi lesions and Sect. 3.6.5 concludes our presentation.

3.6.2 Building a Computer Vision System for the Characterization of Pigmented Skin Lesions

The complete architecture of a computer vision system for the characterization of pigmented skin lesions is depicted in Fig. 3.30.

The main modules include: Image acquisition, Image Processing (Preprocessing, Segmentation, and Registration), Feature Extraction and Selection, use of Classifiers for skin lesion classification. The final result is the recognition of the skin lesion type. The following subsections present all the design issues concerning the implementation of each distinct module.

Image Acquisition

The first step in a skin machine vision-based expert system involves the acquisition of digital images of tissue, which answers question 1. Although this seems quite simple, the construction of a device with the ability to capture reliable and reproducible images of skin is rather challenging due to equipment and environmental constraints including image resolution, image noise, illumination, skin reflectivity and pose uncertainty. The reproducibility is considered essential for image classification and for the comparison of sequential images during follow-up studies.

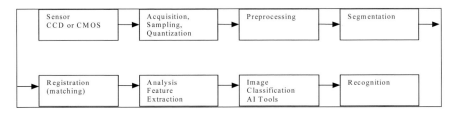

Fig. 3.30. Schematic representation of Computer Vision System for the Characterization of Pigmented Skin Lesions

The main techniques used for this purpose are epiluminence microscopy (ELM or dermoscopy) and image acquisition using still or video cameras. By placing a thin layer of oil on a lesion and then pressing a special hand-held microscope against the oil field on the patient's skin, ELM provides for a more detailed inspection of the surface of pigmented skin lesions and renders the epidermis translucent, making many features become visible. The biggest drawback of this acquisition method is that the pressure exerted on the skin tissue is not measurable or reproducible, and stimulates local hematoma, which changes the color and the morphology of the skin.

The use of commercially available photographic cameras is quite common in skin lesion inspection systems, particularly for telemedicine purposes (Loane et al. 1997). However, the poor resolution of very small skin lesion images, i.e., lesions with diameter of less than 0.5 cm and variable illumination conditions are not easily handled and therefore high-resolution devices with low-distortion lenses have to be used. However, the requirement for constant image colors, (necessary for image reproducibility) remain unsatisfied, as it requires real time, automated color calibration of the camera, i.e., adjustments and corrections to operate within the dynamic range of the camera and to measure always the same color regardless of the lighting conditions. The problem can be addressed by using video cameras that are parameterizable online and can be controlled through software (Maglogiannis 2003; Gutenev et al. 2001) at the price of higher complexity and costs.

In order to solve the problem of light reflections, skin image acquisition systems use appropriate lighting geometry and lighting sources. In the past incandescent lamps have had a significant advantage over fluorescent lamps in the area of color rendering. However newer fluorescent lamps have excellent color rendering ability. To optimize color rendering ability, lamp temperature should be within the 2,900–3,300 K range, with a Color Rendering Index (degree of ability to produce light output optimal for true color rendering) of 85 or higher. The light of the lamp should be transmitted and delivered onto the surface at an angle of 45° and the reflected light should be collected at 0° to the surface normal. This illumination and capturing geometry is internationally established for color measurements, because it eliminates the shadows and the reflections. However it is very difficult to create uniform illumination at an angle over 45° over the entire field of view and over curved surfaces as it is depicted in Fig. 3.31. Therefore in most acquisition systems polarizing filters are also used for eliminating the remaining reflections.

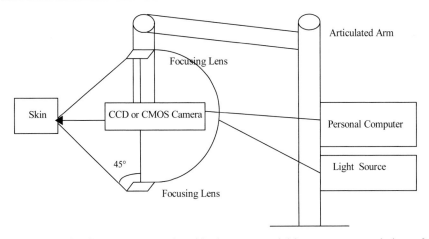

Fig. 3.31. Lighting geometry of a skin image acquisition system consisting of a CCD or CMOS sensors, focusing lenses (mounted on an articulated arm) that face the target at an angle of 45°, a light source that sends lights to the lenses through fiber optics and a personal computer for image processing

Recently new techniques have been presented, that use multispectral images. The chosen wavelengths interact preferentially with constituents of the skin and are able to reveal the structure of the skin lesion. An example is the work presented in Moncrieff et al. (2002).

Image Processing

Image acquisition is usually followed by a preprocessing stage used for image enhancement and noise removal. Simple low-pass filters such as Median or Average filters may be used for this purpose (Maglogiannis et al. 2005). Removal of noise due to dark hairs surrounding the skin lesion is achieved by identifying the hair location with morphological operators and replacing the hair pixels by the nearby nonhair pixels (Lee et al. 1997). After the preprocessing stage the set of pixels corresponding to the skin lesion must be separated from the healthy skin. This task is called segmentation and it can be implemented either manually by an expert dermatologist, or automatically by the computer with the help of unsupervised segmentation algorithms (Zhang et al. 2000; Chung and Sapiro 2000; Xu et al. 1999).

Unsupervised segmentation methods of skin lesion images are based on one of two basic properties: discontinuity and similarity. The first group of methods focuses on the detection of discontinuities between the skin lesion and the surrounding healthy skin. This is accomplished by the use of operators and filters capable of finding edges in an image. Thus edge-detection

algorithms are followed by linking and other boundary detection procedures designed to assemble edge pixels into meaningful boundaries. The
second group deals with the fact that pixels, which belong in skin lesion,
have different color attributes from pixels corresponding to healthy skin.
This criterion is used for the image segmentation into parts with similar
color attributes. Typical methods used in this approach are thresholding,
clustering, and region growing. Thresholding is implemented by choosing
an upper and a lower value and then isolate the pixels which have values in
this range. Region Growing is a procedure that groups pixels or subregions
into larger regions. It starts with a set of "seed" pixels and from these
grows regions by appending neighboring pixels that have similar color
properties. Finally clustering initially divides the image into rectangular
regions small enough to be considered as having only a single color. This
is followed by conservative merging, where adjacent regions whose colors
are similar are coalesced. Iterative optimal merging processes then complete the segmentation. A review and comparison of several segmentation
methods for digital skin images exists in Maglogiannis (2003). Figure 3.32
depicts examples of image segmentation algorithms, applied to dermatological images.

A significant problem in capturing skin images in vivo is the fact that the
target is moving, even if only slightly. Furthermore during follow up examinations images are captured in different time spaces and it is almost impossible to have the same capturing scene, identical patient position, and main-
tain same distance from the camera. Therefore another application from the
field of digital image processing is introduced in computer vision systems
for skin lesion inspections: image registration. This is a procedure for finding correspondences between two different images in order to correct trans-
positions caused by changes in camera position. The input of an image registration algorithm consists of the two images and the output is the values of
four parameters: Magnification, Rotation, Horizontal Shift and Vertical
Shift. The implementation involves the selection of a similarity criterion,
which measures the similarity of the two images. The most commonly used
criteria are the correlation coefficient, the correlation function, and the sum
of the absolute values of the pixel differences. The second step is an optimization algorithm, which maximizes the similarity criterion.

A lot of algorithms have been proposed for the solution of the medical/skin
image registration problem in general. According to a classification made by
Maintz and Viergever (1998) the most common approaches are the (a)
landmark-based, (b) segmentation-based, (c) voxel property-based, and (d)
transformation based registration methods. Landmark-based algorithms
require the selection of a number of corresponding landmarks; landmarks are
anatomical, i.e., salient and accurately locatable points of the morphology

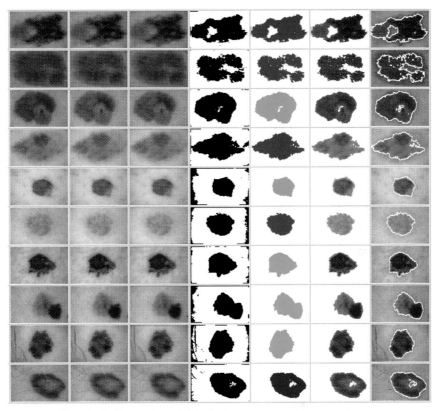

Fig. 3.32. Examples of skin lesion segmentation procedure

of the visible skin anatomy, usually identified interactively by the user and secondly by the utilization of an interpolating transformation model, i.e., thin-plate splines (Rohr et al. 2001). Segmentation-based methods rely on prior image data reduction achieved by segmentation and voxel property-based algorithms rely on the correlation between original images or extracted feature images or the maximization of mutual information (relative entropy) of properties, such us the image histogram (Wells et al. 2001).

A method that uses the cross-correlation of the log-polar Fourier spectrum for achieving registration of skin lesion digital images is presented in Maglogiannis et al. (2005). The magnitude of the Fourier spectrum is independent of horizontal and vertical shifting, while the use of the log-polar transform eliminates the dependency on magnification and rotation. If the two images are f(x,y) and f'(x,y) then they are related through a four-parameter geometric transformation, described by the following equation:

$$f'(x, y) = f(a(x\cos b + y\sin b) - \Delta x, a(-x\sin b + y\cos b) - \Delta y) \qquad (3.34)$$

where Δx and Δy are the shifts, a is the magnification factor and b is the rotation angle. The Fourier spectra of the two images are related through the following equation:

$$\left|F'(u,v)\right| = \frac{1}{a^2}\left|F\left(\frac{(u\cos b + v\sin b)}{a}, \frac{(-u\sin b + v\cos b)}{a}\right)\right| \qquad (3.35)$$

which displays the independence from horizontal and vertical shifts. Then the log-polar transformation is performed by two equations:

$$\left|F'(r,\theta)\right| = \frac{1}{a^2}\left|F\left(\frac{r}{a}, \theta + b\right)\right| \qquad (3.36)$$

where:

$$r = \sqrt{u^2 + v^2}, \quad \theta = \tan^{-1}\frac{v}{u} \qquad (3.37)$$

and

$$\left|F'(\rho,\theta)\right| = \frac{1}{a^2}\left|F(\rho - \ln(a), \theta + b)\right| \qquad (3.38)$$

where

$$\rho = \ln(r). \qquad (3.39)$$

This mapping of the Fourier magnitudes into polar coordinates (r, θ) achieves the decoupling of the rotation and scale factors; rotation maps to a cyclic shift on the θ-axis and scaling maps to a scaling of the r-axis. A logarithmic transformation of the r-axis further transforms scaling into a shift. The next step is the use of the cross-correlation function to find the scale factor and the rotation angle that maximizes the corresponding criterion:

$$XC(R,T) = \sum_{\rho=\rho_{min}}^{\rho_{max}} \sum_{\theta=0}^{2\pi} F(\rho + R, \theta + T)F'(\rho,\theta) \qquad (3.40)$$

where the parameters are the log-difference of the scale factors R and the difference of the rotation angle T. The problem is then reduced to finding the horizontal and vertical shifting, which is simpler and we resolved it using classical comparison methods as the aforementioned sum of the Absolute Values of the Differences. After the completion of image segmentation and registration, features are extracted and analyzed for building classification tools, as discussed in the next subsection.

Definition of Features for Detection of Malignant Melanoma

This section examines the features, i.e., the visual cues that are used for skin lesion assessment, providing answers to question 2. Similarly to the traditional diagnosis procedure, the computer-based systems look for features and combine them to characterize the lesion. The features employed have to be measurable and of high sensitivity, i.e., high correlation of the feature with the corresponding lesions and high probability of true positive response. Furthermore, the features should have high specificity, i.e., high probability of true negative response. Although in the typical classification paradigm both factors are considered important (a trade-off expressed by maximizing the area under the receiver–operating-characteristic curve), in the case of malignant melanoma the suppression of false negatives (i.e., increase of true positives) is obviously more important.

In the conventional procedure, the following diagnostic methods mainly used by dermatologists are (a) *ABCD rule* of dermoscopy (b) *Pattern Analysis*; (c) *Menzies* method; and (d) *7-Point Checklist* (Argenziano et al. 2003). The features used for these methods are presented below:

The *ABCD rule* examines the *asymmetry* (A), *border* (B), *color* (C) (Fig. 3.33), and *differential structures* (D) (Fig. 3.34) of the lesion and defines the basis for a diagnosis by a dermatologist. More specifically:

- *Asymmetry*: the lesion is bisected by two axes that are positioned to produce the lowest asymmetry possible, in terms of borders, colors, and dermoscopic structures.
- *Border*: the lesion is divided into eight pie-piece segments. Then it is examined if there is a sharp, abrupt cut-off of pigment pattern at the periphery of the lesion or a gradual, indistinct cut-off.
- *Color*: the number of colors present is determined. They may include: Light Brown, Dark Brown, Black, Red (red vascular areas are scored), White (if whiter than the surrounding skin), Slate-blue.
- *Differential structures*: the number of structural components present is determined, i.e., Pigment Network, Dots (scored if three or more are present), Globules (scored if two or more are present), Structureless Areas (counted if larger than 10% of lesion), Streaks (scored if three or more are present).

The *Pattern Analysis* method seeks to identify specific patterns, which may be global (Reticular, Globular, Cobblestone, Homogeneous, Starburst, Parallel, Multicomponent, Nonspecific) or local (Pigment network, Dots/globules, Streaks, Blue-whitish veil, Regression structures, Hypopigmentation, Blotches, Vascular structures).

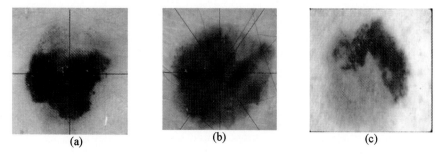

Fig. 3.33. Asymmetry Border Color Features; (**a**) Asymmetry Test, (**b**) Border Test, (**c**) Color variegation (source: http://www.dermoncology.com/)

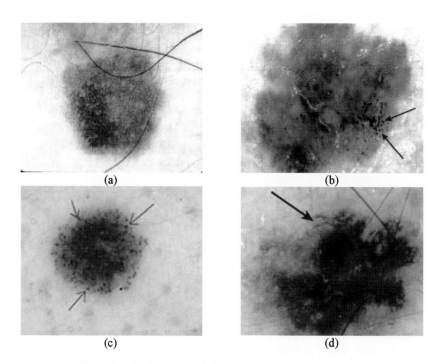

Fig. 3.34. Differential Structures; (**a**) Pigmented network, (**b**) Dots, (**c**) Brown globules, (**d**) Branched streaks (source: http://www.dermoncology.com/)

The *Menzies* method looks for negative features (Symmetry of pattern, Presence of a single color) and positive (Blue–white veil, Multiple brown dots, Pseudopods, Radial streaming, Scar-like depigmentation, Peripheral black dots/globules, Multiple (5–6) colors, Multiple blue/gray dots, Broadened network).

The 7-point checklist seeks for Atypical pigment network, Blue-whitish veil, Atypical vascular pattern, Irregular streaks, Irregular dots/globules, Irregular blotches, Regression structures.

The researchers that seek to identify automatically malignant melanoma exploit the available computational capabilities by searching for many of the above, as well as, for additional features. The main features used for skin lesion image analysis are summarized below, as well as their calculation method, answering question 3.

Asymmetry Features

The asymmetry is examined with respect to a point, one or more axes. Color, texture and shape must be taken into account. It is represented by features such as the asymmetry index. The asymmetry index is computed by first finding the principal axes of inertia of the tumor shape in the image (e.g., Nachbar et al. 1994; Rubegni et al. 2002). It is obtained by overlapping the two halves of the tumor along the principal axes of inertia and dividing the nonoverlapping area differences of the two halves by the total area of the tumor.

Apart from the area-based symmetry features, other features can be extracted by dividing symmetric subregions of the lesion image and by extracting for each subregion, local features similar to the ones described in the following subsections. If for each region i the measurement is represented by the quantity Q_i then the symmetry may be evaluated through examination of the ratios:

$$R_i = Q_i / \Sigma Q_i \qquad (3.41)$$

Another approach is to examine pairs of symmetric lesion pixels, with respect to the principal axes, and to calculate the sum of their Euclidean distances in the color space (e.g., Ganster et al. 2001).

Border Features

The most popular border-based features are Greatest Diameter, Area, Border Irregularity, Thinness Ratio, and Border Asymmetry. More specifically Irregularity is defined as:

$$Irregularity = \frac{Perimeter}{Area} \qquad (3.42)$$

or

$$Irregularity = \frac{Perimeter}{GreatestDiameter} \qquad (3.43)$$

where the second expression was found more efficient as it is independent of the size of the skin lesion. Thinness Ratio measures the circularity of the skin lesion and is defined as:

$$ThinnessRatio = 4\pi \frac{Area}{(Perimeter)2} \qquad (3.44)$$

This measure has a maximum value of 1, which corresponds to a circle. It can be used for the detection of skin diseases, which have the property to spread radial on the human skin. Apart from regarding the border as a contour, emphasis is also placed on the features that quantify the transition from the lesion to the skin. Such features are the minimum, maximum, average, and variance responses of the gradient operator, applied on the intensity image along the lesion border.

Color Features

Typical color images consist of the three-color channels RGB (red, green, and blue). The color features are based on measurements on these color channels or other color channels such as CMY (Cyan, Magenta, Yellow), HSV (Hue, Saturation, Value), YUV (Y-luminance, UV chrominance components) or various combinations of them, linear or not. Color variegation may be calculated by measuring minimum, maximum, average, and standard deviations of the selected channel values and by measuring chromatic differences inside the lesion (Maglogiannis et al. 2001).

Differential Structures

The differential structures as described in the ABCD method, as well as most of the patterns that are used by the pattern analysis, the Menzies method and the 7-points checklist are very rarely used for automated skin lesion classification, obviously due to their complexity.

Skin Lesion Kinetics

Several efforts concern the kinetics of skin lesions (e.g., Hansen et al. 1997; Herbin et al. 1993). In Hansen et al. (1997) the ratio of variances RV has been defined as

$$RV = \frac{SD_{B^2}}{SD_{B^2} + SD_{I^2} + SD_{A^2}} \tag{3.45}$$

SD_B^2 (standard deviation between days) is between day variance of the color variable computed using the mean values at each day of all wound sites and subjects.

SD_I^2 (standard deviation intra day) is the intra day variance of the color variable estimated from the computations at each day of all wound sites and subjects.

SD_A^2 (standard deviation analytical) is the variance of the color variable computed using normal skin sites of all subjects and times.

Features Selection and Skin Lesion Classifiers

The success of image recognition depends on the correct selection of the features used for the classification, which answers question 4. This is a typical optimization problem, which may be resolved with heuristic strategies, greedy or genetic algorithms or other computational intelligence methods (Handels et al. 1999). The use of feature selection algorithms is motivated by the need for highly precise results, by computational reasons and by a peaking phenomenon often observed when classifiers are trained with a limited set of training samples. If the number of features is increased the classification rate of the classifiers decreases after a peak (Jain 1986; Jain and Waller 1978). Feature selection may be also dealt by more sophisticated algorithms such as sequential floating forward selection (SFFS) or generalized sequential forward selection (GSFS) presented in Kudo and Sklansky (2000).

This section answers questions 5 and 6 as well by examining the most popular methods for skin lesion classification. The task involves mainly two phases after feature selection, learning, and testing, which are analyzed in the following:

During the learning phase typical feature values are extracted from a sequence of digital images representing classified skin lesions. The most classical recognition paradigm is statistical (Duda and Hart 1973). Covariance matrices are computed for the discriminative measures, usually under the multivariate Gaussian assumption. Parametric discriminant functions are then determined, allowing classification of unknown lesions (discriminant analysis).

The main aim of discriminant analysis (Mardia et al. 1979) is to allocate an individual to one of two or more known groups, based on the values of certain measurements **x**. The discriminant procedure identifies that combination (in the commonest case, as applied here, the linear combination) of these predictor variables that best characterizes the differences between the groups. The procedure estimates the coefficients, and the resulting discriminant function can be used to classify cases. The analysis can also be used to determine which elements of the vector of measurements **x** are most useful for discriminating between groups. This is usually done by implementing stepwise algorithms, as in multiple regression analysis, either by successively eliminating those predictor variables that do not contribute significantly to the discrimination between groups, or by successively identifying the predictor variables that do contribute significantly.

One important discriminant rule is based on the likelihood function. Consider k populations or groups $\Pi_1,\ldots,\Pi_k$, $k \geq 2$ and suppose that if an individual comes from population Π_j, it has probability density function $f_j(\mathbf{x})$. The rule is to allocate **x** to the population Π_j giving the largest likelihood to **x**

$$L_j(\mathbf{x})=\max L_i(\mathbf{x}) \tag{3.46}$$

In practice, the sample maximum likelihood allocation rule is used, in which sample estimates are inserted for parameter values in the pdf's $f_j(\mathbf{x})$. In a common situation, let these densities be multivariate normal with different means μ_i but the same covariance matrix Σ. Unbiased estimates of $\mu_1,\ldots,\mu_g$ are the sample means $\overline{\mathbf{x}}_1,\ldots,\overline{\mathbf{x}}_k$, while

$$S_u= \Sigma \, n_i \, S_i / (n-k) \tag{3.47}$$

is an unbiased estimator of Σ, where S_i is the sample covariance matrix of the ith group. In particular when k=2 the sample maximum likelihood discriminant rule allocates **x** to Π_1 if and only if

$$\mathbf{a}'\{\mathbf{x}-\frac{1}{2}(\overline{\mathbf{x}}_1 +\overline{\mathbf{x}}_2)\} > 0, \text{ where } \mathbf{a}' = S_u^{-1}(\overline{\mathbf{x}}_1 - \overline{\mathbf{x}}_2). \tag{3.48}$$

Another important approach is Fisher's Linear Discriminant Function (Mardia et al. 1979). In this method, the linear function **a'x** is found that maximizes the separation between groups in the sense of maximizing the ratio of the between-groups sum of squares to the within-groups sum of squares,

$$a'Ba/ a'Wa \tag{3.49}$$

The solution to this problem is the eigenvector of $\mathbf{W}^{-1}\mathbf{B}$ that corresponds to the largest eigenvalue. In the important special case of two populations, Fisher's LDF becomes:

$$(\overline{x}_1 - \overline{x}_2)'\,W^{-1}\left\{x - \tfrac{1}{2}(\overline{x}_1 + \overline{x}_2)\right\} \tag{3.50}$$

The discrimant rule is to allocate a case with values $\mathbf{x}$ to Π_1 if the value of the LDF is greater than zero and to Π_2 otherwise. This allocation rule is exactly the same as the sample ML rule for two groups from the multivariate normal distribution with the same covariance matrix. However, the two approaches are quite different in respect of their assumptions. Whereas the sample ML rule makes an explicit assumption of normality, Fisher's LDF contains no distributional assumption, although its sums of squares criterion is not necessarily a sensible one for all forms of data. The major problem of the statistical approach is the need for large training samples.

The methodology of neural networks involves mapping a large number of inputs into a small number of outputs and it is therefore frequently applied to classification problems in which the predictors $\mathbf{x}$ form the inputs and a set of variables denoting group membership represent the outputs. It is thus a major alternative to discriminant analysis and a comparison between the results of these two entirely different approaches is interesting. Neural networks are very flexible as they can handle problems for which little is known about the form of the relationships. Furthermore, they can fit linear, polynomial and interactive terms without requiring the kind of modeling that would be necessary in the statistical approach. Neural Networks are covered in a different chapter of this book.

Finally, Support Vector Machines (SVMs) is a popular algorithm for data classification into two classes (Burges 2001) (Christianini and Taylor 2000). SVMs allow the expansion of the information provided by a training data set as a linear combination of a subset of the data in the training set (support vectors). These vectors locate a hypersurface that separates the input data with a very good degree of generalization. The SVM algorithm is based on training, testing, and performance evaluation, which are common steps in every learning procedure. Training involves optimization of a convex cost function where there are no local minima to complicate the learning process. Testing is based on the model evaluation using the support vectors to classify a test data set. Performance evaluation is based on error rate determination as test set data size tends to infinity.

Answering question 6 the performance of each classifier is tested using an ideally large set of manually classified images. A subset of them, e.g., 80% of the images is used as training set and the rest 20% of the samples are used for testing using the trained classifier. The training and test images are exchanged for all possible combinations to avoid bias in the solution.

3.6.3   Reported Experimental Results from Existing Systems

The development of machine vision systems for the characterization of skin images preoccupies many biomedical laboratories. There are several research projects dealing with the feature extraction issue and the implementation of diagnostic tools for clinical dermatological help. In University of Illinois USA S. Umbaugh, R. Moss, and W. Stoecker identified successfully six skin different states by using features vectors (Umbaugh et al. 1993; Umbaugh et al. 1997). The six states were tumor, crust, hair, scale, shiny, and ulcer. The average success rates were 85% with the chromaticity coordinates to overpass in performance the remaining color spaces. Of course it is expected that a combination of color spaces will result in a better identification. The same team used neural networks to identify variegated coloring in skin tumors with 89% correct results (Umbaugh et al. 1991). Nischic and Forster in University of Erlangen Germany used the CIELAB color space for the analysis of skin erythema (Nischic and Forster 1997). This research was conducted to test the effect of specified drugs to the human skin. In University of South Florida a method based on feature extraction is proposed for the objective assessment of burn scars (Tsap et al. 1998). The results illustrated the ability to objectively detect differences in skin elasticity between normal and abnormal tissue. In Europe, in Italy the RGB and the HIS color planes are used for an skin lesion computer-assisted diagnosis application specialized to the detection of melanoma (Tomatis et al. 1998). The general trend of the data was in agreement with the clinical observations according to which melanoma is usually darker, more variegated and less round than a benign nevus. This research has shown that the use of both color planes had better results than each one separately. The same feature extraction techniques are used even for skin images obtained by microscope with good results (Sanders et al. 1999). The results from the computer-based system were compared with the results from traditional methods and the mean absolute difference was about 5%. The most common installation type in the above systems seems to be the video camera, obviously due to the control features that it provides. The still camera is of use in some installations, while infrared or ultraviolet illumination (in situ or in vivo) using appropriate cameras is a popular choice, e.g., Bono et al. (1996) and Chwirot et al. (1998) correspondingly. Microscopy (or epiluminence microscopy) installations are applied in the works of Sanders et al. (1999) and Ganster et al. (2001).

More specifically, regarding images produced from the fluorescence of human skin in the nonvisible spectra, the feature, which is used for identification is the intensity of the fluorescence. The first attempt on using fluorescence methods for in situ detection of melanoma was made in 1988 by Lohman and Paul (Lohman and Paul, 1988). The authors excited in vivo autofluorescence of skin tissues with ultraviolet light and recorded the spectra of light emitted by healthy tissues, naevi, and melanomas. They found that melanomas generated specific patterns of variation in the fluorescence intensity. Specifically they noticed local maxima in the transition zone between the melanomas and the healthy skin, which was not found for naevi. Bono and others implemented an image analysis system base on imaging of pigmented skin lesions in infrared. The results were very promising with 77% of lesions correctly diagnosed against 81% of correct clinical diagnoses. Chwirot and associates used the ratio I_{max}/I_{min} where I_{max} and I_{min} are the maximum and minimum value of fluorescence intensity in regions located up to 40 mm from the lesions (Chwirot et al. 1998). This ratio had average values 14.3 for melanoma, 5.7 for naevi and 6.1 for other skin lesions. Similar techniques are described in Mansfield et al. (1998) and Jones (1998) for the infrared imaging.

The most common features that are used for automated lesion characterization are the ones that are associated with color in various color spaces (RGB, HIS, CIELab). Some of them combine features in more than one color spaces for better results, e.g., HIS and RGB. Asymmetry and border features as defined in the subsection "Image Processing" are quite common, e.g., Ganster et al. (2001) while features based on differential structures are very rare.

The most common classification methods are the statistical ones. More advanced techniques such as neural networks are presented in works like Umbaugh et al. (1997), while the k-nearest neighborhood classification scheme is applied in Ganster et al. (2001). The success rates for the methods presented in the literature indicate that the work toward automated classification of lesions and melanoma in particular may provide good results. These rates along with the other system features are summarized in Table 3.4. We should note here that the results are not comparable but rather indicative, mainly due to the fact that different images from different cases are used. Moreover, the classification success rates are not applicable to the methods calculating healing indexes.

Table 3.4. Computer-based systems for the characterization of digital skin images

Reference	Detection goal	Installation type	Visual features	Classification method	Success rates
Umbaugh et al. (1997)	Tumor, crust, hair, scale, shiny ulcer of skin lesions	Video RGB Camera	Color (chromaticity) coordinates (more)	Neural networks	85–89% in average
Nischic and Forster (1997)	Skin erythema	Video RGB Camera	Color - CIE L*a*b* color space	Statistical	Monitoring indexes for Follow ups
Tsap et al. (1998)	Burn scars	Video RGB Camera	Image Intensity, Skin Elasticity	Finite element analysis,	Monitoring indexes for Follow ups
Tomatis et al. (1998)	Melanoma Recognition	Video RGB Camera	Color in RGB and HIS (more)	Statistical	5% deviation from manual diagnosis
Sanders et al. (1999)	Melanoma Recognition	Tissue microscopy	Epidermal and dermal features	Statistical	Difference in epidermal features was 5.33% , for dermal features it was 2.76%
Herbin et al. (1993)	Wound Healing	Still CCD Camera	Ratio of variances, in HIS and RGB	Healing indexes measuring, the wound area	Monitoring indexes for Follow ups
Bono et al. (1996)	Melanoma Recognition	In situ, ultraviolet illumination	Auto fluorescence of skin tissues	Statistical	77% (81% manual diagnoses)
Chwirot et al. (1998)	Melanoma Recognition	Ultraviolet illumination	Imax/Imin, (fluorescence intensity)	Statistical	Sensitivity of 82.5%, specificity of 78.6% positive predictive value of 58.9%
Ganster et al. (2001)	Melanoma Recognition	Epilumines cence microscopy (ELM)	RGB/HIS/Border	Statistical (k-nearest-neighbor)	sensitivity of 87% and a specificity of 92%

3.6.4 Case Study Application

In our study we tried to quantify the success rates of the classification methods described in this chapter for the distinction of malignant melanoma from dysplastic nevus (Maglogiannis et al. 2005). An integrated image analysis system with the modules depicted in Fig. 3.35 was implemented.

Regarding the acquisition of images we applied the lighting engineering described in section "Image Acquisition" to ensure as constant illumination as possible. With regard to the software corrections, three main types were implemented: Color Calibration incorporating the Gretag-McBeth

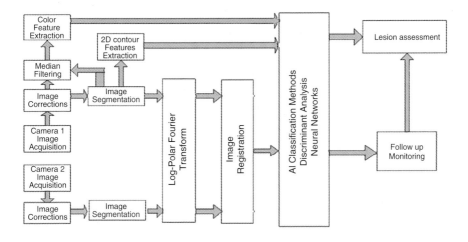

Fig. 3.35. Schematic representation of the complete digital image analysis architecture implemented in our system

Color Calibration chart, Shading correction and Median filtering. Shading correction is performed by division of an image with all pixels having R=G=B=255 by the image of a perfect diffuser (BaSO$_4$) and then multiplying a captured image with the look-up table generated by the division. Median Filtering is used for noise reduction due to hair, scales, and light reflections. This procedure ends in blurring the image, which visually seems a poorer result but on the other hand assists the segmentation and analysis procedure. Image was then segmented and registrated using the techniques discussed in section "Image Processing", while the features listed in section "Definition of Features for Detection of Malignant Melanoma" were extracted.

Three groups of data were considered. The first group (denoted VGP – vertical growth phase) consists of cases of malignant melanoma, with measurements taken on the entire extent of the lesion. The second group (RGP – radial growth phase) also refers to the malignant melanomas, but measurements are restricted to the dark area of the melanoma. The third group (DSP – dysplastic) comprises cases of dysplastic nevus (see Fig. 3.36). Separate analyses were carried out, one between VGP and DSP, and the other between RGP and DSP. Both comparisons are made by linear discriminant analysis, by fitting a neural network model and by utilizing the SVM algorithm. The training data set was taken from 34 cases at the Dept of Plastic Surgery and Dermatology in Athens General Hospital collected within a period of six months. The total number of lesions captured was: 14 melanomas and 20 dysplastic naevi (although the set is small

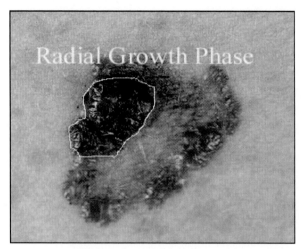

Fig. 3.36. The RGP phase of melanoma is the circled area

it can provide some insight about the classifier performance and how the number of features influences the results). The mean thickness of melanomas lesions was measured after biopsy at approximately 1.5 mm penetration through the skin. The sensitivity and specificity rates using discriminant analysis, neural networks and SVMs are presented in Table 3.5. and Table 3.6. for VGP–DSP and RGP–DSP classification. We have used the cross-validation or "leaving-one-out" estimator of the rate of correct classifications, obtained by seeing how each observation is classified according to a function recalculated after omitting that observation from the analysis.

Table 3.5. Sensitivity and Specificity Indexes of the VGP–DSP classification

Method	Total correct classification	Sensitivity (%)	Specificity (%)
Discriminant Analysis using four Features as the most significant for discrimination	33/34 or 97%	93	100
Discriminant Analysis using two Features as the most significant for discrimination	32/34 or 94%	86	100
Neural Networks using four principal components as input	33/34 or 97%	93	100
Neural Networks using two principal components as input	29/34 or 85%	79	90
SVM (Gaussian RBF Kernel, sigma=4, 7 support vectors)	94% 32/34	86	100

Table 3.6. Sensitivity and specificity indexes of the RGP–DSP classification

Method	Total correct classification	Sensitivity (%)	Specificity (%)
Discriminant Analysis using four Features as the most significant for discrimination	33/34 or 97%	93	100
Discriminant Analysis using two Features as the most significant for discrimination	30/34 or 88%	86	100
Neural Networks using four principal components as input	34/34 or 100%	100	100
Neural Networks using two principal components as input	32/34 or 94%	86	90
SVM (First order polynomial kernels, 5 support vectors)	97%	93	100

3.6.5 Conclusions

The most significant systems for the automated skin lesion assessment have been surveyed. These systems employ a variety of methods for the image acquisition, the feature definition and extraction as well as the lesion classification from features. The most promising image acquisition techniques appear to be those that reveal the skin structure through selected spectral images.

Regarding the features, it is clear that the emphasis has been on assessment of lesion size, shape, color, and texture. These statistical parameters were chosen primarily for computational convenience; they can be acquired with well-established analytic techniques at a manageable computational cost. However, they do not correspond to known biological phenomena and do not model human interpretation of dermoscopic imagery. On the contrary, the structural patterns that are considered essential for manual lesion categorization seem to have been neglected by the computational intelligence community, due to their complexity, although their exploitation could provide crucial information.

However, from our survey and the case study application it is clear that the implementation of a machine vision system for dermatology is feasible and useful. Such a system may serve as diagnostic adjunct for medical professionals and for the training of dermatologists. The use of such computer-based systems is intended to avoid human subjectivity and to perform specific tasks according to a number of criteria. However the presence of an expert dermatologist is considered necessary for the overall visual assessment of the skin lesion and the final diagnosis.

References

Argenziano, G. et al. (2003). Dermoscopy of pigmented skin lesions: Results of a consensus meeting via the Internet. *J. Am. Acad. Dermatol., 48*(5), 680–693

Baxes, G. (1994). *Digital image processing: principles and applications.* New York: Wiley

Benkhalil, A.K., Zharkova, V., Ipson, S., & Zharkov, S. (2003). *Proceedings of the AISB'03 symposium on biologically-inspired machine vision, theory and application.* Aberystwyth: University of Wales, p. 66

Bentley, R.D. (2002). *Proceedings of the second solar cycle and space weather Euro-conference,* 24–29 September 2001. Huguette, S-L. (Ed.), Vico Equense, Italy: ESA Publication SP-477

Berghmans, D., Foing, B.H., & Fleck, B. (2002). Automated detection of CMEs in LASCO data. In *Proceedings of the SOHO-11 workshop, from solar min to max: Half a solar cycle with SOHO.* Switzerland: Davos

Bono, A., Tomatis, S., Bartoli, C. (1996). The invisible colors of melanoma. A telespectrophotometric diagnostic approach on pigmented skin lesions. *Eur. J. Cancer, 32A,* 727–729

Bow, S.-T. (2002). *Pattern recognition and image processing.* New York, USA: Dekker

Brandt, P.N., Schmidt, W., & Steinegger, M. (1990). On the umbra–penumbra area ratio of sunspots. *Solar Phys., 129,* 191

Bratsolis, E., & Sigelle, M. (1998). Solar image segmentation by use of mean field fast annealing. *Astron. Astrophys. Suppl. Ser., 131,* 371–375

Burges, C. (2001). A tutorial on support vector machines for pattern recognition. URL: http://www.kernel-machines.org

Burns, J.R., Hanson, A.R., & Riseman, E.M. (1986). Extracting straight lines. *IEEE Trans. Pattern Anal. Mach. Intell., 8,* 425–455

Canny, J. (1986). A computational approach to edge detection. *IEEE Trans. Pattern Anal., 6,* 679–698

Casasent, D. (1992). New advances in correlation filters. *Proc. SPIE Conf. Intell. Robots Comput. Vision, 1825*(XI), 2–10

Chapman, G.A., & Groisman, G. (1984). A digital analysis of sunspot areas. *Solar Phys., 91,* 45

Chapman, G.A., Cookson, A.M., & Dobias, J.J. (1994a). Observations of changes in the bolometric contrast of sunspots. *Astrophys. J., 432,* 403–408

Chapman, G.A., Cookson, A.M., & Hoyt, D.V. (1994b). Solar irradiance from Nimbus-7 compared with ground-based photometry. *Solar Phys., 149,* 249

Christianini, N., & Shawe-Taylor, J. (2000). *An introduction to support vector machines.* Cambridge: Cambridge University Press

Chung, D.H., & Sapiro, G. (2000). Segmenting skin lesions with partial-differential-equations-based image processing algorithms. *IEEE Trans. Med. Imaging, 19*(7), 763–767

Chwirot, W., Chwirot, S., Redziski, J., & Michniewicz, Z. (1998). Detection of melanomas by digital imaging of spectrally resolved ultraviolet light-induced autofluorescence of human skin. *Eur. J. Cancer, 34,* 1730–1734

Cuisenaire, O., & Macq, B. (1999). Fast and exact signed Euclidean distance transformation with linear complexity. In *Proceedings of the IEEE international conference on acoustics, speech and signal processing,* Vol. 6, pp. 3293–3296. Phoenix, AZ: ICCASSP99

van Driel-Gesztelyi, L. (2002). *Proceedings of the SOLMAG, the magnetic coupling of the solar atmosphere euroconference & IAU colloquium 188,* Greece, June 2002, p. 113

Duda, R.O., & Hart, P.E. (1973). *Pattern classification and skin analysis.* New York: Wiley

Durrant, C.J. (2002). Polar magnetic fields – filaments and the zero-flux contour. *Solar Phys. 211,* 83–102

Falconer, D.A., Moore, R.L., Porter, J.G., & Gary, G.A. (1997). Neutral-line magnetic shear and enhanced coronal heating in solar active regions. *Astrophys. J., 482,* 519–534

Fuller, N., Aboudarham, J., & Bentley, R.D. (2005). Filament recognition and image cleaning on Meudon Halpha Spectroheliograms. *Solar Phys., 227*(1), 61–73

Ganster, H., Pinz, P., Rohrer, R., Wildling, E., Binder, M., & Kittler, H. (2001). Automated melanoma recognition. *IEEE Trans. Med. Imaging, 20*(3), 233–239

Geman, S., & Geman, D. (1984). Stochastic relaxation, Gibbs distribution and the Bayesian restoration of images. *IEEE Trans. Pattern Anal. Mach. Intell., 6,* 721–741

Gonzalez, R.C., & Woods, R.E. (2001). *Digital Image Processing.* New Jersey: Prentice-Hall

Gonzalez, R.C., & Woods, R.E. (2002). *Digital image processing,* 2nd edn., Upper Saddle River, NJ: Prentice-Hall

Gutenev, A., Skladnev, V.N., & Varvel, D. (2001). Acquisition-time image quality control in digital dermatoscopy of skin lesions. *Comput. Med. Imaging Graph., 25,* 495–499

Győri, L. (1998). Automation of area measurement of sunspots. *Solar Phys., 180,* 109–130

Hagyard, M.J., Smith, J.E., Teuber, D., & West, E.A. (1984). A quantitative study relating to observed shear in photospheric magnetic fields to repeated flaring. *Solar Phys., 91,* 115–126

Handels, H., Roß, T., Kreusch, J., Wolff, H.H., & Pöppl, S.J. (1999). Feature selection for optimized skin tumor recognition using genetic algorithms. *Artif. Intell. Med., 16,* 283–297

Hansen, G., Sparrow, E., Kokate, J., Leland, K., Iaizzo, P. (1997). Wound status evaluation using color image processing. *IEEE Trans. Med. Imaging, 16*(1), 78–86

Herbin, M., Bon, F., Venot, A., Jeanlouis, F., Dubertret, M., Dubertret, L., Strauch, G. (1993). Assessment of healing kinetics through true color image processing. *IEEE Trans. Med. Imaging, 12*(1), 39–43

Hill, M., Castelli, V., Chung-Sheng, L., Yuan-Chi, C., Bergman, L., Smith, J.R., & Thompson, B. (2001). *International conference on image processing*, Thessaloniki, Greece, 7–10 October 2001, Vol. 1, p. 834

Huertas, A., & Medioni, G. (1986). Detection of intensity changes with sub-pixel accuracy using Laplacian-Gaussian masks. *IEEE Trans. Pattern Recognit. Mach. Intell.*, 8(5), 651–664

Ipson, S.S, Zharkova, V.V, Benkhalil, A.K, Zharkov, S.I, Aboudarham, J. (2002). SPIE Astronomical telescopes and instrumentation, astronomical data analysis II (AS14) conference, Hawaii, USA, August 2002. Innovative telescopes and instrumentation for solar astrophysics. In Stephen, L.K., & Sergey, V.A. (Eds.). *Proceedings of the SPIE, 4853*, 675

Jackway, P. (1996). Gradient watersheds in morphological scale-space. *IEEE Trans. Image Process., 5*, 913–921

Jahne, B. (1997). *Digital image processing*. Berlin Heidelberg New York: Springer

Jain, A.K. (1986). Advances in statistical pattern recognition. In Devijer, P.A., & Kittler, J. (Eds.). *Pattern recognition, theory and applications*. Berlin Heidelberg New York: Springer

Jain, A.K., & Waller, W.G. (1978). On the optimal number of features in the classification of multivariate gaussian data. *Pattern Recogt., 10*, 365–374

Jones, B. (1998). Reappraisal of the use of infrared thermal image analysis in medicine. *IEEE Trans. Med. Imaging, 17*(6), 1019–1027

Kittler, J., & Illingworth, J. (1985). On threshold selection using clustering criteria. *IEEE Trans. Syst. Man Cyber., SMC-15*, 652–655

Kudo, M., & Sklansky, J., (2000). Comparison of algorithms that select features for pattern classifiers. *Pattern Recognit., 33*, 25–41

Kullback, S. (1959). Information theory and statistics. New York: Wiley

Lee, T., Ng, V., Gallagher, R., Coldman, A., & McLean, D. (1997). DullRazor: A software approach to hair removal from images. *Comput. Biol. Med., 27*, 533–543

Lee, S.U., Chung, S.Y., & Park, R.H. (1990). A comparative performance study of several global thresholding techniques for segmentation. *Comput. Vis. Graph. Image Process., 52*(2), 171–190

Li, C.S., Yu, P.S., & Castelli, V. (1998). MALM: a framework for mining sequence databases at multiple abstraction levels. In: *Proceedings of the 7th International Conference Information Knowledge Management, CIKM'98*, MD, USA, pp. 267–272

Loane, M., Gore, H., Corbet, R., & Steele, K. (1997). Effect of camera performance on diagnostic accuracy. *J. Telemed. Telecare, 3*, 83–88

Lohman, W., & Paul, E. (1988). In situ detection of melanomas by fluorescence measurements. *Naturewissenschaften, 75*, 201–202

Maglogiannis, I., Pavlopoulos, S., & Koutsouris, D. (2005). An integrated computer supported acquisition, handling and characterization system for pigmented skin lesions in dermatological images. *IEEE Trans. Inform. Technol. Biomed., 9*(1), 86–98

Maglogiannis, I. (2003). Automated segmentation and registration of dermatological images. *J. Math. Model. Algorithms, 2,* 277–294

Maglogiannis, I., Caroni, C., Pavlopoulos, S., & Karioti, V. (2001). Utilizing artificial intelligence for the characterization of dermatological images, 4th international conference neural networks and expert systems in medicine and healthcare. Greece: NNESMED, pp. 362–368

Maintz, J.B., & Viergever, M.A. (1998). A survey of medical image registration. *J. Med. Image Anal., 2*(1), 1–36

Mansfield, J., Sowa, M., Payette, J., Abdulrauf, B., Stranc, M., & Mantsch, H. (1998). Tissue viability by multispectral near infrared imaging: A fuzzy C-means clustering analysis. *IEEE Trans. Med. Imaging, 17*(6), 1011–1018

Mardia, K.V., Kent, J.T., & Bibby, J.M. (1979). *Multivariate analysis.* London: Academic

Marks, R. (2000). Epidemiology of melanoma. *Clin. Exp. Dermatol., 25,* 459–463

Marr, D., & Hildreth, E. (1980). Theory of edge detection. *Proc. R. Soc. Lond., B207,* 187–217

Matheron, G. (1975). *Random sets and integral geometry.* New York: Wiley

Metropolis, N., Rosenbluth, A., Rosenbluth, M., Teller, A., & Teller, E. (1953). *J. Chem. Phys., 21,* 1087–1092

Moncrieff, M., Cotton, S., Claridge, E., & Hall, P. (2002). Spectrophotometric intracutaneous analysis – a new technique for imaging pigmented skin lesions. *Br. J. Dermatol., 146*(3), 448–457

Mouradian, Z. (1998). *Synoptic Solar Phys. ASP Conf. Ser., 140,* 181

Nachbar, F., Stolz, W., Merkle, T., Cognetta, A.B., Vogt, T., Landthaler, M., Bilek, P., Braun-Falco, O., & Plewig, G. (1994). The ABCD rule of dermatoscopy: High prospective value in the diagnosis of doubtful melanocytic skin lesions. *J. Am. Acad. Dermatol., 30*(4), 551–559

Nischic, M., & Forster, C. (1997). Analysis of skin erythema using true color images. *IEEE Trans. Med. Imaging, 16*(6)

NOAA National Geophysical Data Center. URL: http://www.ngdc.noaa.gov./stp/SOLAR /ftpsunspotregions.html

NOAA (2005). http://www.solar.ifa.hawaii.edu/ARMaps/armaps.html

O'Gorman, F., & Clowes, M.B. (1976). Finding picture edges through collinearity of feature points. *IEEE Trans. Comput., C25,* 449–454

Otsu, N. (1979). A threshold selection method from gray level histograms. *IEEE Trans. Syst. Man Cyber., SMC-9,* 62–66

Pap, J.M., Turmon, M., Mukhtar, S., Bogart, R., Ulrich, R., Fröhlich, C., Wehrli, Ch. (1997). Automated recognition and characterization of solar active regions based on the SOHO/MDI images. In: Wilson, A. (Ed.). *31st ESLAB Symposium,* Held 22–25 September 1997, Vol. 477, ESTEC, Noordwijk, The Netherlands, European Space Agency, ESA SP-415

Pariser, R.J., & Pariser, D.M., (1987). Primary care physicians errors in handling cutaneous disorders. *J. Am. Acad. Dermatol., 17,* 239–245

Pettauer, T., Brandt, P.N. (1997). On novel methods to determine areas of sunspots from photoheliograms. *Solar Phys., 175,* 197

Preminger, D.G., Walton, S.R., & Chapman, G.A. (2001). Solar feature identification using contrast and contiguity. *Solar Phys., 202,* 53

Priest, E.R. (1984). Solar magneto-hydrodynamics. *Geophys. Astrophys.* Monographs. Dordrecht: Reidel

Priest, E.R. (1984). *Geophys. Astrophys. Monographs.* Dordrecht: Reidel

Rohr, K. et al. (2001). Landmark based elastic registration using approximating thin plate splines. *IEEE Trans. Med. Imaging, 20*(6), 526–534

Rosenfeld, A., & Pfaltz, J.L. (1966). Sequential operations in digital image processing. *J. Assoc. Comp. Mach., 13,* 471–494

Rubegni, P., Cevenini, G., Burroni, M., Perotti, R., Dell'eva, G., Sbano, P., Miracco, C., Luzi, P., Tosi, P., Barbini, P., & Andreassi, L. (2002). Automated diagnosis of pigmented skin lesions. *Int. J. Cancer, 101,* 576–580

Sanders, J., Goldstein, B., Leotta, D., & Richards, K. (1999). Image processing techniques for quantitative analysis of skin structures. *Comput. Methods Programs Biomed., 59,* 167–180

Serra, J. (1988). *Image analysis and mathematical morphology,* Vol. 1. London: Academic

Shrauner, J.A., & Scherrer, P.H. (1994). East–west inclination of large-scale photospheric magnetic fields. *Solar Phys., 153,* 131–141

Solar Index Data Catalogue, Belgian Royal Observatory. http://sidc.oma.be/products

Steinegger, M., & Brandt, P.N. (1998). *Solar Phys., 177,* 287

Steinegger, M., Bonet, J.A., & Vazquez, M. (1997). Simulation of seeing influences on the photometric determination of sunspot areas. *Solar Phys., 171,* 303

Steinegger, M., Brandt, P.N., Pap, J., & Schmidt, W. (1990). Sunspot photometry and the total solar irradiance deficit measured in 1980 by ACRIM. *Astrophys. Space. Sci., 170,* 127–133

Steinegger, M., Vazquez, M., Bonet, J.A., & Brandt, P.N. (1996). On the energy balance of Solar Active Region. *Astrophys. J., 461,* 478. 24/10/0624/10/06

Tomatis, S., Bartol, C., Tragni, G., Farina, B., & Marchesini, R. (1998). Image analysis in the RGB and HS colour planes for a computer assisted diagnosis of cutaneous pigmented lesions. *Tumori, 84,* 29–32

Tsap, L., Goldgof, D., Sarkar, S., & Powers, P. (1998). Vision-based technique for objective assessment of burn scars. *IEEE Trans. Med. Imaging, 17,* 620–633

Turmon, M., Pap, J.M., & Mukhtar, S. (1998). *Proceedings of the SoHO 6/GONG 98 workshop, structure and dynamics of the interior of the sun and sun-like stars,* Boston

Ulrich, R.K., Evens, S., Boyden, J.E., & Webster, L. (2002). Mount Wilson synoptic magnetic fields: Improved instrumentation, calibration and analysis applied to the 2000 July 14 flare and to the evolution of the dipole field. *Astrophys. J. Suppl. Ser., 139*(1), 259–279

Umbaugh, S., Wei, Y., & Zuke, M. (1997). Feature extraction in image analysis. *IEEE Eng. Med. Biol., 16,* 62–73

Umbaugh, S.E., Moss, R.H., & Stoecker, W.V. (1993). Automatic color segmentation of images with application to skin tumor feature identification. *IEEE Eng. Med. Biol. Mag., 12*(3), 75–82

Umbaugh, S.E., Moss, R.H., & Stoecker, W.V. (1991). Applying artificial intelligence to the identification of variegated coloring in skin tumors. *IEEE Eng. Med. Biol. Mag.,* 57–62

Veronig, A., Steinegger, M., Otruba, W., Hanslmeier, A., Messerotti, M., & Temmer, M. (2001). *HOBUD7, 24*(1), 195–2001

Waldmeier, M. (1961). *The sunspot activity in the years 1610–1960, sunspot tables.* Zurich: Schulthess

Walton, S.R., Chapman, G.A., Cookson, A.M., Dobias, J.J., & Preminger, D.G. (1998). Processing photometric full-disk solar images. *Solar Phys., 179*(1), 31–42

Wells III, W.M., Viola, P., Atsumi, H., Nakajima, S., & Kikinis, R. (1997). Multimodal volume registration by maximization of mutual information. *Med. Image Anal., 1,* 35–51

Worden, J.R., White, O.R., & Woods, T.N. (1996). *Solar Phys., 177,* 255

Xu, L., Jackowski, M., Goshtasby, A., Roseman, D., Bines, S., Yu, C., Dhawan, A., & Huntley, A. (1999). Segmentation of skin cancer images. *Image Vis. Comput., 17,* 65–74

Zhang, Z., Stoecker, W.V., & Moss, R.H. (2000). Border detection on digitized skin tumor images. *IEEE Trans. Med. Imaging, 19*(11), 1128–1143

Zharkov, S.I., & Zharkova, V.V. (2004). Statistical analysis of sunspot area and magnetic flux variations in 2001–2003, IAU Symposium 223. Multi-Wavelength Investigations of Solar Activity, St. Petersburg, Russia, 14–19 June 2004

Zharkov, S.I., & Zharkova, V.V. (2006). Statistical analysis of the sunspot area and magnetic flux variations in 1996–2005 extracted from the Solar Feature Catalogue. *Adv. Space Res.* 38(5), 568–575

Zharkov, S.I, Zharkova, V.V., & Ipson, S.S. (2005a). Statistical properties of sunspots and their magnetic field detected from full disk SOHO/MDI images in 1996–2003. *Solar Phys., 228*(1), 401–423

Zharkov, S., Zharkova, V., Ipson, S., & Benkhalil, A. (2005b). The technique for automated recognition of sunspots on full disk solar images. *EURASIP J. Appl. Signal Process. (EURASIP JASP), 15,* 2573–2584

Zharkova, V.V., Aboudarham, J., Zharkov, S., Ipson, S.S., Benkhalil, A.K., & Fuller, N. (2005). *Solar Phys., 228,* 61

Zharkova, V.V., Ipson, S.S., Zharkov, S.I., Benkhalil, A.K., Aboudarham, J., & Bentley, R.D. (2003). *Solar Phys., 214,* 89

4 Advanced Feature Recognition and Classification Using Artificial Intelligence Paradigms

V. Schetinin and V.V. Zharkova

4.1 Neural-Network for Recognizing Patterns in Solar Images

This section describes a neural-network technique developed for an automated recognition of solar filaments visible in H-α hydrogen line full disk spectroheliograms. An atmospheric heterogeneity and instrumental noise affect these observations. Besides, the brightness of backgr ound and filament elements is dependent on their location on the full disk image. Under such conditions, the automated recognition of filament elements is still a difficult task. The technique is based on an artificial neural network (ANN) capable of learning to recognize the filament elements in the solar fragments.

4.1.1 Introduction

Solar filaments are seen as dark elongated features seen in absorption which are the projections on a solar disk of prominences seen on the solar limb as very bright and large-scale features (Zharkova et al. 2003a, b; Zharkova and Schetinin 2005). Their location and shape does not change very much for a long time and, hence, their lifetime is likely to be much longer than one solar rotation. However, there are visible changes seen in the filament elongation, position with respect to an active region and magnetic field configuration. For this reason the automated detection of solar filaments is a key task in understanding the physics of prominence formation, support, and disruption.

Recently, two techniques were applied for the feature detection such as the rough detection with a mosaic threshold technique and region growing techniques (Qahwaji and Green 2001; Bader et al. 1996; Gao et al. 2001).

V. Schetinin et al.: *Advanced Feature Recognition and Classification Using Artificial Intelligence Paradigms,* Studies in Computational Intelligence (SCI) **46**, 151–338 (2007)
www.springerlink.com   © Springer-Verlag Berlin Heidelberg 2007

However, these techniques are strongly dependent on the background cleaning procedures (Zharkova 2003a, b) and for this reason they are not suitable for an automated feature classification.

The standard neural network techniques (Bishop 1995; Duda and Hart 2001) applied to the filament recognition problem should be trained on a representative data set. The data have collected for training should represent the image fragments depicting filaments in different locations on the solar disk and background with different intensity, and for this reason the number of the training examples should be large. However, the image fragments are still to be labelled manually, which leads to an actual number of the labelled fragments being rather limited, increasing the learning process accuracy requirement.

The proposed technique described in Zharkova and Schetinin (2003; 2005) is based on an ANN consisting of two hidden neurons and one output neuron. These neurons learn separately to distinguish the contributions of a variable background and filaments. Their activation function can be linear or parabolic.

4.1.2 Problem Description

The filaments observable on a solar disk can be fragmented into separate rectangular images with smaller numbers of pixels. Since these image fragments come from different parts of a solar disk, their brightness varies from fragment to fragment or even within a fragment depending on the observing conditions and instrumental errors.

Let us introduce an image fragment data as an $(n \times m)$ matrix $\mathbf{X} = \{x_{ij}\}$, $i = 1, \ldots, n, j = 1, \ldots, m$, consisting of pixels with brightness ranged between 1 and 255. This image fragment depicts a filament, which is a dark elongated feature observable on a solar disk with higher background brightness. Within the fragment, a pixel x_{ij} can belong either to a filament region, class Ω_0, or to a non-filament region, class Ω_1. Hence, the recognition problem is to distinguish pixels of the given matrix $\mathbf{X}$ between the classes Ω_0 and Ω_1.

Now let us define a background function $u_{ij} = \varphi(\mathbf{X}; i, j)$, which reflects the brightness u_{ij} of pixel x_{ij} taken from a given image $\mathbf{X}$ in a position $<i, j>$. As the brightness of pixel x_{ij} may depend on its position in the image $\mathbf{X}$, so this function has to satisfy either of the following two properties:

$$\varphi(\mathbf{X}; i, j) = \varphi_0 \ \forall \ i = 1, \ldots, n, \ j = 1, \ldots, m, \tag{4.1}$$

$$| \Delta \varphi_{ij} | > 0, \qquad\qquad (4.2)$$

where $\varphi_0 \geq 0$ is a constant and $\Delta \varphi_{ij}$ is a finite difference of the function φ.

The first property is referred to a case when and the variability of background elements is non-distinguishable, i.e. the contribution of a background to any pixel value x_{ij} in the image $\mathbf{X}$ is a constant. The second property is referred to an alternative case when in a vicinity of the pixel x_{ij}, the background elements make different contribution to the brightness of the pixels.

A structure of the background function φ can be predefined and its parameters fitted to a given image matrix $\mathbf{X}$. Calculating the output of this function, one can evaluate the brightness u_{ij} for all the pixels of $\mathbf{X}$ and reduce the contribution of a variable background to the filament recognition assuming this contribution is additive. Usually, it is assumed that the influence of neighbouring pixels on a central pixel x_{ij} is restricted to k elements.

Therefore, for recognition of solar filaments the influence of variable background on the brightness of pixels involved in window $\mathbf{P}$ has to be eliminated. This requires evaluating an output u_{ij} of the background function φ for each pixel $x_{ij}. \in \mathbf{X}$ and then subtract the value of u_{ij} from the brightness of pixels in the window.

4.1.3 The Neural-Network Technique for Filament Recognition

In order to estimate a background function, one can introduce a rectangular window of pixels as $(t \times t)$ matrix $\mathbf{P}(i, j)$, with the central pixel x_{ij} and $(t - 1)$ nearest pixels. Then by sliding this window $\mathbf{P}$ through the image $\mathbf{X}$, every central pixel x_{ij} in the window can be assigned to one of the two classes: Ω_0 and Ω_1. This can be done by transforming the original image $\mathbf{X}$ to a matrix $\mathbf{Z} = (\mathbf{z}^{(1)}, ..., \mathbf{z}^{(q)})$, where $q = (n - t + 1)(m - t + 1)$ is a number of columns. Each column consists of $r = t^2$ elements taken from matrix $\mathbf{P}$ so that the central element of $\mathbf{P}$ is located in the $(r + 1)/2$ position of the column in the matrix $\mathbf{Z}$.

Figure 4.1 schematically depicts a case when $\mathbf{X}$ is a (4×5)-matrix, $t = 3$ and $\mathbf{P}$ is a (3×3)-rectangle matrix. For this case $r = 9$, $q = 6$ and $\mathbf{Z}$ is a (9×6)-matrix of which columns $\mathbf{z}^{(1)}, ..., \mathbf{z}^{(6)}$ are the elements of matrix $\mathbf{P}$ sliding from a position $<2, 2>$ to $<3, 4>$.

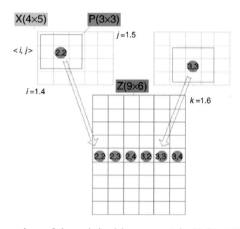

Fig. 4.1. A transformation of the original image matrix $\mathbf{X}$ (4×5) into the matrix $\mathbf{Z}$ (9×6) using a sliding window $\mathbf{P}$, a (3×3)-matrix. The circles are the central pixels of the window sliding from the position <2, 2> to <3, 4>

For the matrix $\mathbf{Z}$, the output of a background function can be rewritten as $u_k = \varphi(\mathbf{X}; k)$, $k = 1, \ldots, q$. Note that, for a given image $\mathbf{X}$, this function is described as $u_k = \varphi(k)$, i.e. the contribution of the background elements to pixel z_k is dependent only on a position of this pixel in the image. Therefore, the background function properties (4.5) and (4.6) can be rewritten as follows:

$$\varphi(\mathbf{X}; k) = \varphi_0 \ \forall \ k = 1, \ldots, q, \tag{4.3}$$

$$|\Delta\varphi_k| > 0, \tag{4.4}$$

In general, there are several functions, which can match the properties (4.3) and (4.4). First one is a linear function $u_k = \alpha_0 + \alpha_1 \cdot k$, where α_0 and α_1 are constants. Indeed, the property (4.11) is satisfied, if the contribution of the background elements is described as $u_k = \alpha_0$, and it is not dependent on the position k. The property (4.4) is satisfied by the contribution being linearly dependent on k. In this case the finite difference $\Delta\varphi_k = \alpha_1$.

The second function can be a parabolic one, which is more applicable to large filaments with the brightness of pixels varying in a wider range than that for the small ones. In this case the contribution of background elements is described as $u_k = U(\alpha; k) = \alpha_0 + \alpha_1 k + \alpha_2 k^2$. In both cases the coefficients of the linear and parabolic approximations are fitted to the data so that the squared error e is minimal:

$$e = \sum_{k=1}^{q} [U(\mathbf{a};k) - \beta_0 - \sum_{i=1}^{r} \beta_i z_i^{(k)}]^2, \tag{4.5}$$

where $\beta_i > 0$ are some coefficients given from the pixel brightness in a sliding window and β_0 is the bias.

A weighted sum of the pixels $z_i^{(k)}$ over all r elements of the kth column vector $\mathbf{z}^{(k)}$ acts as a mean filter of the image that suppresses the influence of a background noise in every single pixel and produces the average contribution of k neighbouring elements in the kth column of vector $\mathbf{z}^{(k)}$. The coefficients of the approximation can be fitted, for example, by a least square method (Duda 2001; Farlow 1984; Madala and Ivakhnenko 1994; Müller and Lemke 2003).

The process described earlier can be implemented within a feed-forwards ANN consisting of the two hidden neurons and one output neuron as depicted in Fig. 4.2. The first hidden neuron is fed by r elements of the kth column vector $\mathbf{z}^{(k)}$. The second hidden neuron evaluates the value u_k of the background elements. Considering these two neurons, the output neuron, f_3, makes a decision, $y_k = \{0, 1\}$, on the central pixel in the kth column vector $\mathbf{z}^{(k)}$.

Assuming that the first hidden neuron is fed by r elements of column vector $\mathbf{z}^{(k)}$, its output s_k is calculated as follows:

$$s_k = f_1(w_0^{(1)}, \mathbf{w}^{(1)}; \mathbf{z}^{(k)}), \ k = 1,\ldots, q, \tag{4.6}$$

where $w_0^{(1)}$, $\mathbf{w}^{(1)}$, and f_1 are the bias term, weight vector and an activation function of the neuron, respectively.

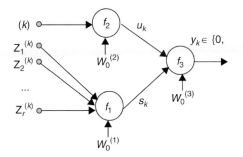

Fig. 4.2. The ANN used for detection consisting of the two hidden neurons and one output neuron

The activity of the second hidden neuron reflects the contribution of the background elements to the central pixel of the kth vector $\mathbf{z}^{(k)}$ and can be written as follows:

$$u_k = f_2(w_0^{(2)}, \mathbf{w}^{(2)}; k), k = 1,\ldots, q, \qquad (4.7)$$

where f_2 is the activation function, a linear or parabolic function described earlier.

The bias term $w_0^{(2)}$ and weight vector $\mathbf{w}^{(2)}$ of the hidden neuron are fitted to make the approximation error (4.5) minimal.

Taking into account the values of s_k and u_k, the output neuron makes a final decision, $y_k \in \{0, 1\}$, for a central pixel of vector $\mathbf{z}^{(k)}$ as follows:

$$y_k = f_3(w_0^{(3)}, \mathbf{w}^{(3)}; s_k, u_k), k = 1,\ldots, q. \qquad (4.8)$$

Depending on activities of the hidden neurons, the output neuron assigns a central pixel of the vector $\mathbf{z}^{(k)}$ either to the class Ω_0 or Ω_1. Therefore, the ANN is trained to recognize all the pixels of the given image $\mathbf{X}$ and distinguish between filament and background elements.

4.1.4 Training Algorithm

For training the ANN depicted in Fig. 4.2, one can use the standard back-propagation technique aimed to minimize an error function for a given number of the training epochs. During training this technique repeatedly updates the weights of the hidden neurons and then calculates their outputs for all q columns of matrix $\mathbf{Z}$. Clearly, this technique is computationally expensive. However, there are some local solutions, which can be found if the hidden and output neurons are trained separately. Providing an acceptable accuracy, such solutions may be found more easily than the global ones. Based on these local solutions, our training algorithm can be described as follows.

First we need to fit the weight vector of the second hidden, or a "background" neuron, which evaluates the contribution of the background elements. The contribution of these elements to a brightness of the kth pixel is dependent on weather conditions and instrumental errors. As an example, the top left plot in Fig. 4.3 depicts the image matrix $\mathbf{X}$ presenting a filament on the unknown background, and the bottom left one depicts the output of u_k summed with the weights $\beta_i = 1$ and $\beta_0 = 0$. This filament is relative small and the number of pixels in the image is 6,936.

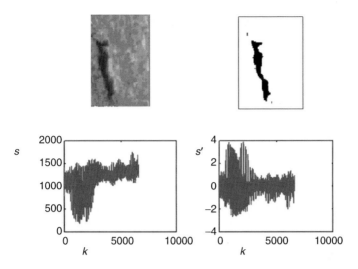

Fig. 4.3. The example of a small filament depicted (the left top plot) with a noisy background (the left bottom plot), the filament recognized with linear approximation of a background contribution (the right top plot), and the contribution of filament elements (the right bottom plot)

Observing the image in Fig. 4.3, we can see that the brightness of a background varies from the lowest level at the left bottom corner to the highest level at the right top corner. These variations increase the output value s_k calculated over the q columns of matrix $\mathbf{Z}$ as depicted in the left bottom plot. The increase of the s_k shows that the contribution of background elements is changed over k and, in the first approximation, it can be approximated by a linear function.

For larger filaments as depicted at the left top plot in Fig. 4.4, one can see that the background varies more widely. In this case the better approximation of the curve u_k depicted at the left bottom plot can be achieved with a parabolic approximation.

Based on these observations, we can define a parabolic activation function of the "background" neuron as follows:

$$u_k = f_2\left(w_0^{(2)}, \mathbf{w}^{(2)}; k\right) = w_0^{(2)} + w_1^{(2)}k + w_2^{(2)}k^2. \tag{4.9}$$

The weight coefficients $w_0^{(2)}$, $w_1^{(2)}$, and $w_2^{(2)}$ of this neuron is fitted to the data $\mathbf{Z}$ so that the squared error e between the outputs u_k and s_k became minimal, that changes (4.5) to the following:

$$e = \Sigma_k(u_k - s_k)^2 = \Sigma_k\left(w_0^{(2)} + w_1^{(2)}k + w_2^{(2)}k^2 - s_k\right)^2 \to min, \tag{4.10}$$

where s_k is the output of the first hidden neuron for the kth vector $\mathbf{z}^{(k)}$.

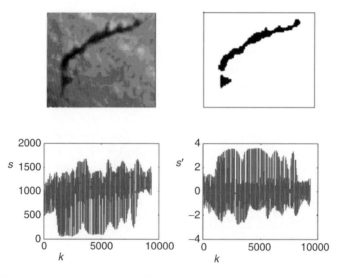

Fig. 4.4. The example of a large filament depicted (the left top plot) with a background not taken into account (the left bottom plot), the filament recognized with the parabolic approximation of a background (the right top plot) and the normalized output s' (the right bottom plot)

The desirable weight coefficients are fitted by the least deviation method. So the "background" neuron evaluates the contribution of background elements for each column vector $\mathbf{z}^{(k)}$.

The right bottom plots in Figs. 4.3 and 4.4 depicts the normalized values s' which are no longer affected by the background component. Comparing the left and right bottom plots in these figures, one can see that after the background neuron correction the background function in the right plot does not increase from the bottom to the top of a fragment, as it was before (left plot). Hence, we conclude that the "background" neuron in both cases has been successfully trained, and the contribution of background pixels was excluded. It can be seen that, as a result, the recognized filament elements, marked in black in the right top plots of Figs. 4.3 and 4.4, match the original filaments rather well; a classification accuracy of the detection is discussed later.

For the first neuron a local solution is achieved by setting its weights $w_i^{(1)}$ equal to 1 and bias $w_0^{(1)}$ to 0. Within such values the ANN performed well in our experiments.

Having defined the weights for both hidden neurons, it is possible now to train the output neuron which assigns the central pixel of the kth vector $\mathbf{z}^{(k)}$ to one of two classes. The output y_k of this neuron is described as follows:

$$y_k = 0, \text{ if } w_0^{(3)} + w_1^{(3)}s_k + w_2^{(3)}u_k < 0, \tag{4.11}$$

$$y_k = 1, \text{ otherwise.}$$

The weights of the output neuron can be fitted so that the recognition error rate e was minimal:

$$e = \Sigma_i |y_i - t_i| \rightarrow min, \, i = 1,\ldots, h, \tag{4.12}$$

where $|\cdot|$ is an absolute value, $t_i \in \{0, 1\}$ are the target elements, and h is the number of the training examples.

The desirable values of $w_0^{(3)}$ and $\mathbf{w}^{(3)}$ can be achieved by using the standard learning methods, including the perceptron rule (Bishop 1995; Duda 2001). When the noise in image data is not Gaussian, the learning algorithm described in Schetinin (2003) can be applied.

4.1.5 Experimental Results and Discussion

An ANN consisting of two hidden neurons and one output neuron, depicted in Fig. 4.2, was used for learning to recognize the filaments visible in H-α hydrogen line full disk spectroheliograms. These images were obtained at the Meudon Observatory (France) during March–April 2002. The 55 fragments of filaments were taken from several regions on the full disk images with different brightness caused by instrumental errors and varying observing conditions.

Because of a large amount of pixels, an expert was able to visually label only one fragment for training the ANN. The remaining 54 fragments were labelled by using a semi-automated technique described in Zharkova (2003b), which has revealed acceptable classification accuracy. The ANN trained on 6,936 labelled pixels correctly recognized 87.2% of 365,180 pixels in the testing fragments. The misclassifications were mainly caused by reducing the sizes of the recognized filaments as depicted in Figs. 4.3 and 4.4 at the top right plots.

The activation function of the hidden neuron can be assigned either liner for small fragments or parabolic for large fragments. Let us now compare the performance of the ANNs exploiting these functions on the image fragment depicted at the left plot of Fig. 4.5. This is a relatively large fragment including multiple filaments.

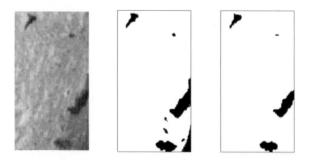

Fig. 4.5. The original fragment with multiple filaments (the left plot), detection results for the linear background function (the middle plot), and for the parabolic function (the right plot)

The ANN performing the linear approximation fails to recognize the filaments properly as it detects more candidates than those are really present. This happens because the background brightness, which is widely varied within the fragment, affects the recognition of the filaments located at the right bottom edge in Fig. 4.5.

Meanwhile, the ANN performing the parabolic approximation recognizes the multiple filaments visually better than that is achieved with a linear approximation, see the right plot in Fig. 4.5. The quantitative estimates of the performance on all the 54 test fragments shows that the classification accuracy of the ANN with a parabolic approximation is better on 17.5% than that achieved with the linear approximation.

Thus, we conclude that the proposed neural-network technique can be effectively used for an automated recognition of filaments in solar images.

4.2 Machine Learning Methods for Pattern Recognition in Solar Images

V. Schetinin, V.V. Zharkova and A. Brazhnikov

In this section we describe the basic ideas of machine learning methods allowing experts to learn models from solar data under the lack of a priori information. For examples, experts can be uncertain about the importance of features representing patterns. To deal with the uncertainty, we consider some feature selection approaches and discuss possible ways of improving the performance of the machine learning methods.

4.2.1 Introduction

Solar activity is characterized by complex patterns of which most interesting features are instrumentally observed and registered on ground-based and space stations. These patterns may have a complex dynamics represented by features changing in the time and location. For this reason, experts can be uncertain about the importance of the features for the pattern analysis.

The observable solar data are stored and further processed in order to gain information on the solar activity models and forecasts. The volume of these data grows quickly and now to handle them it is required to develop new effective approaches (Zharkova 2003a, b, 2005). Such approaches can be developed on the base of machine learning paradigms allowing experts to recognize and analyze the observable patterns (Galant 1993; Bishop 1995; Duda 2001).

When experts are uncertain about a priori information, the machine learning methods allow desired models to be induced from data within an acceptable accuracy (Bishop 1995; Duda 2001). These methods can be implemented within an artificial neural network paradigm which is described in this chapter. However, such methods are difficult to apply to real large-scale problems (Schetinin and Schult 2005). In this chapter we discuss how some of these difficulties can be overcome on the base of selection of relevant features.

4.2.2 Neural-Network-Based Techniques for Classification

In this section we briefly describe the standard feed-forwards neural-network techniques developed for classification and pattern recognition. Then we discuss some difficulties of using these techniques for real problems.

Neural-Network-Based Classifiers

In general, classification refers to learning *classification models* or *classifiers* from data presented by *labelled examples*. The aim of learning is to induce a classifier which is able to assign an unknown example to one of the given classes with acceptable *accuracy*. A typical classification problem is presented by a *data set* of labelled examples which are characterized by *variables* or *features.* Experts developing a classifier may assume features which make the distinct contribution to the classification problem. Such features are called *relevant*. However, among these features may be assumed *irrelevant* and/or *redundant*: the first can seriously hurt the classification accuracy whereas the second are useless for the classi-fication and can obstruct understanding how decisions are arrived at. Both the irrelevant and redundant features have to be discarded (Galant 1993; Bishop 1995).

Solving a classification problem, the user has to induce a classification model from the *training data* set and test its performance on the *testing data* set of the labelled examples. These data must be disjoint in order to objectively evaluate how well the classification model can classify unseen examples (Galant 1993; Bishop 1995; Duda 2001).

In practice, the user applying a neural network technique often cannot properly assume relevant features and avoid irrelevant and redundant. Moreover, some features become relevant being taken into account in the combination with others. In such cases the machine learning techniques exploit a special learning strategy capable of *selecting* relevant features during the induction of classification model (Duda 2001; Sethi and Yoo 1997; Quinlan 1993; Salzberg et al. 1998). Such a strategy allows users to learn classification models more accurately than strategies selecting features before learning.

Another practical problem is that most of classification methods suggested for inducing comprehensible models imply a *trade-off* between classification accuracy and representation complexity (Duda 2001). Much less work has been undertaken to develop the methods capable of discovering the comprehensible models without decreasing their classification accuracy (Avilo et al. 2001).

Standard Neural-Network Technique

A standard neural-network technique exploits a *feed-forwards* fully connected network consisting of the *input nodes*, *hidden* and *output* neurons which are connected each other by the adjustable *synaptic weights* (Galant 1993; Bishop 1995). This technique implies that a *structure* of neural network has to be predefined properly. This means that users must preset an appropriate number of the input nodes and hidden neurons and apply a suitable *activation function*. For example, the user may apply a *sigmoid* activation function described as:

$$y = f(\mathbf{x}, \mathbf{w}) = 1/(1 + \exp(-w_0 - \Sigma_i^m w_i x_i)), \tag{4.13}$$

where $\mathbf{x} = (x_1, ..., x_m)^\mathrm{T}$ is a $m \times 1$ input vector, $\mathbf{w} = (w_1, ..., w_m)^\mathrm{T}$ is a $m \times 1$ synaptic weight vector, w_0 is a bias term and m is the number of input variables.

Then the user has to select a suitable learning algorithm and then properly set its parameters such as the *learning rate* and the number of the *training epochs.* Note that when the neural networks include at least two hidden neurons, the learning algorithms with *error back-propagation* usually provide the best performance in term of the classification accuracy (Galant 1993; Bishop 1995).

Within the standard technique first the learning algorithm initializes the synaptic weights **w**. The values of w are updated whilst the *training error* decreases for a given number of the training epochs. The resultant classification error is dependant on the given learning parameters as well as on the initial values $\mathbf{w}^0$ of neuron weights. For these reasons neural networks are trained several times with random values of initial weights and different learning parameters. This allows the user to avoid local minima and find out a neural network with a near minimal classification error.

After training the user expects that the neural network can classify new inputs well and its classification accuracy is acceptable. However the learning algorithm may fit the neuron weights to specifics of training data, which are absent in new data. In this case neural networks become to be *over-fitted* and cannot generalize well. Within the standard technique the generalization ability of the trained network is evaluated on a *validation* subset of the labelled examples, which have not been used for training the network.

Figure 4.6 depicts a case when after $k*$ training epochs the validation error starts to increase whilst the training error continues to decrease. This means that after $k*$ training epochs the neural network becomes over-fitted. To prevent over-fitting, we can update the neuron weights whilst the validation error decreases.

When classification problems are characterized in the m-dimensional space of input variables, the performance of neural networks may be radically improved by applying the principal component analysis (PCA) to

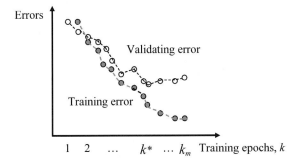

Fig. 4.6. Learning curves for the training and validating sets

training data (Galant 1993; Bishop 1995). The PCA may significantly reduce the number of the input variables and consequently the number of synaptic weights, which are updated during learning. A basic idea behind the PCA is to turn the initial variables so that the classification problem might be resolved in a reduced input space.

Figure 4.7 depicts an example of a classification problem resolved in a two-dimensional space of the input variables x_1 and x_2 by using a separating function $f_1(x_1, x_2)$. However we can turn the x_1 and x_2 so that this problem might be solved in one-dimensional input space of a principal component $z_1 = a_1 x_1 + a_2 x_2$, where a_1 and a_2 are the coefficients of a linear transformation. In this case a new separating function is $f_2(z_1)$ that is equal to 0 if $z_1 < \vartheta_1$ and 1 if $z_1 \geq \vartheta_1$, where ϑ_1 is a threshold learnt from the training data represented by the new variable z_1.

As we see, the new components z_1 and z_2 make different contribution to the variance of the training data: the first component contributes much greater than the second. This example demonstrates how the PCA can rationally reduce the input space. However users using PCA must properly define a variance level and the number of components making the contribution to the classification.

Using the earlier technique, we may find out a suitable neural-network structure and then fit its weights to the training data whilst the validation error decreases. Each neural network with a given number of input nodes and hidden neurons should be trained several times, say 100 times.

Thus, we can see that the standard technique is computationally expensive. For this reason, users use fast learning algorithms, for instance, a back-propagation algorithm by Levenberg–Marquardt (Bishop 1995).

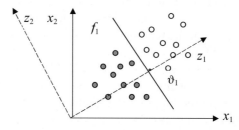

Fig. 4.7. An example of two-dimensional classification problem

4.2.3. Neural-Network Decision Trees

This section describes neural network decision tree techniques which exploit the multivariate linear tests and the algorithms searching for relevant features. Such linear tests are easily observable for experts. We also describe an algorithm which is capable of selecting relevant features.

Multivariate Decision Trees

Decision tree (DT) methods have successfully been used for inducing multi-class models from real-world data represented when experts are uncertain about the importance of features (Duda 2001; Sethi and Yoo 1997; Brodley and Utgoff 1995; Quinlan 1993; Salzberg et al. 1998). Experts find that a DT is easy-to-observe by tracing the route from its entry point to the outcome one. This route may consist of the subsequence of questions which are useful for the classification and understandable for medical experts.

The conventional DTs consist of the *nodes* of two types. One is a *splitting* node containing a *test*, and other is a *leaf* node assigned to an appropriate class. A *branch* of the DT represents each possible outcome of the test. An example is presented to the *root* of the DT and follows the branches until the leaf node is reached. The name of the class at the leaf is the resulting classification.

The node can test one or more of the input variables. A DT is a *multivariate* or *oblique* one, if its nodes test more than one of the features. Multivariate DTs are in general much shorter than those which test a single variable. These DTs can test threshold logical units (TLU) or *perceptrons* which perform a weighted sum of the input variables. Users can interpret such tests just as a weighted sum of the questions where weights usually mean the significance of the feature for the test outcome.

In order to learn from the numerical features extracted from the data, multivariate DTs are suggested that are capable of distinguishing linear separable patterns by means of linear tests (Duda 2001; Salzberg et al. 1998). However using the algorithms described in (Sthi 1997; Brodley and Utgoff 1995) the multivariate DTs can be also used for classifying non-linearly separable patterns.

In general, the DT algorithms require computational time that grows proportionally to the number of training examples, input features and classes. Nevertheless, the computational time, which is required to induce multi-class models from large-scale data sets, becomes overwhelming, especially, if the number of training examples is tens of thousands.

Let us now consider a linear machine (LM) which is a set of *r linear discriminant* functions calculated to assign a training example to one of the

$r \geq 2$ classes (Duda 2001). Each node of the LM tests a linear combination of m input variables $x_1, x_2, \ldots, x_m$ and $x_0 \equiv 1$.

For a m-input vector $x = (x_0, x_1, \ldots, x_m)$ and a discriminant function $g(x)$, the linear test at the jth node has the following form:

$$g_j(x) = \Sigma_i w_i^j x_i = w^{jT}x > 0, \tag{4.14}$$

where $w_0^j, \ldots, w_m^j$ are the real valued coefficients also known as a weight vector w^j of the jth TLU, $i = 0, \ldots, m, j = 1, \ldots, r$.

The LM assigns an example x to the j class if and only if the output of the jth node is larger than the outputs of the other nodes:

$$g_j(x) > g_k(x), \ k \neq j, k = 1, \ldots, r. \tag{4.15}$$

This strategy of making a decision is known as winner take all (WTA).

During learning the LM, the weight vectors w^j and w^k of the discriminant functions g_j and g_k are updated on each example x that the LM misclassifies. A learning rule increases the weights w^j, where j is the class to which the example x actually belongs, and decreases the weights w^k, where k is the class to which the LM erroneously assigns the example x. This is done using the following error correction rule:

$$w^j := w^j + cx, \ w^k := w^k - cx, \tag{4.16}$$

where $c > 0$ is a given amount of correction.

If the training examples are linearly separable, earlier procedure can yield a desirable LM giving maximal classification accuracy in a finite number of steps (Duda 2001). If the examples are non-linearly separable, this training procedure may not provide predictable classification accuracy. For this case the other training procedures have been suggested some of them we will discuss later.

A Pocket Algorithm

To learn DTs from data which are non-linearly separable, it has been suggested a pocket algorithm (Galant 1993). This algorithm searches weights of multivariate tests that minimize the classification error. The pocket algorithm uses error correction rule (4.16) to update the weights w^j and w^k of the corresponding discriminant functions g_j and g_k. The algorithm saves in the pocket the best weight vectors W^P that are seen during learning.

In addition, Gallant has suggested the "ratchet" modification of the pocket algorithm. The idea behind this algorithm is to replace the weight W^P by current W only if the current LM has correctly classified more training examples than was achieved by W^P. The modified algorithm finds the optimal weights if sufficient training time is allowed.

To implement this idea, the algorithm cycles train the LM for the given number of epochs, n_e. For each epoch, the algorithm counts the current number of input series of correctly classified examples, L, and evaluates accuracy A of the LM on the training set.

As searching time that the algorithm requires grows proportional to the numbers of the training examples as well as of the input variables and classes, the number of epochs must be large enough to achieve an acceptable classification accuracy. For example, in our case the number of the epochs is set to the number of the training examples. The best classification accuracy of the LM is achieved if c is equal to 1.

When the training examples are not linearly separable, the classification accuracy of LMs may be unpredictable large. There are two cases when the behaviour of the LM is destabilized during learning. In the first case, a misclassified example is far from a *hyperplane* dividing the classes. In such a case the dividing hyperplane has to be substantially readjusted. Such relatively large adjustments destabilize the training procedure. In the second case, the misclassified example lies very close to the dividing hyperplane, and the weights do not converge.

Feature Selection Strategies

In order to induce accurate and understandable DT models, we must eliminate the features that do not contribute to the classification accuracy of DT nodes. To eliminate irrelevant features, we use the sequential feature selection (SFS) algorithms (Galant 1993; Duda 2001) based on a *greedy* heuristic, called also the *hill-climbing* strategy. The selection is performed whilst the DT nodes learn from data. This avoids over-fitting more effectively than the standard methods of feature pre-processing.

The SFS algorithm exploits a *bottom up search* method and starts to learn using one feature. Then it iteratively adds the new feature providing the largest improvement in the classification accuracy of the linear test. The algorithm continues to add the features until a specified stopping criterion is met. During this process the best linear test T_b with the minimum number of the features is stored.

The stopping rule is satisfied when all the features have been involved in the test. In this case $m + (m - 1) + \ldots + (m - k)$ linear tests have been made, where k is the number of the steps. Clearly if the number of the features, m, as well as the number of the examples, n, is large, the computational time needed to terminate may be unacceptable.

However, the classification accuracy of the resulting linear test depends on the order in which the features have been included in the test. For the SFS algorithm, the order in which the features are added is determined by their contribution to the classification accuracy. As we know, the accuracy depends on the initial weights as well as on the sequence of the training examples selected randomly. For this reason the linear test can be non-optimal, i.e. the test can include more or fewer features needed for the best classification accuracy. The chance of selecting the non-optimal linear test is high, because the algorithm compares the tests that differ by one feature only. This hypothesis has been experimentally proved as described in Schetinin and Schult (2005).

4.2.4 Conclusion

An automated classification of solar data is a difficult problem because of the variable background caused by the instrumental errors and inhomo-geneities in the terrestrial atmosphere. Nevertheless, the classification techniques based on machine learning methods presented in the chapter have revealed promising results in our experiments (Zharkova 2003b). We have found that the standard neural network techniques can learn classification models from real data well, however domain experts cannot easily understand such classification models.

Obviously in practice a priori information is not always available for domain experts. When this is the case, the machine learning methods allow desired models to be induced from data within an acceptable accuracy by using, for example, an artificial neural-network paradigm described in this chapter. However, such techniques are difficult to apply to real large-scale problems. In this chapter we discussed how to overcome these difficulties by selecting those features which are most relevant to the classification. Such a selection has allowed the classification models to achieve a better performance. Thus, we conclude that machine learning methods enhanced by the feature selection can be applied to large-scale astrophysical data under the lack of a priori information.

4.3 The Methodology of Bayesian Decision Tree Averaging for Solar Data Classification

V. Schetinin, V.V. Zharkova and S. Zharkov

Bayesian methodology uses a priori knowledge in order to achieve the maximal accuracy in estimating the class posterior probabilities and distributions which are important for evaluating the uncertainty in decisions. In this section we discuss how models learnt from data can be probabilistically interpreted and made understandable for human experts. Although the methodology of Bayesian averaging over classification models is computationally expensive, Markov Chain Monte Carlo technique of stochastic sampling makes this methodology feasible. Based on such a technique, we describe a new methodology of Bayesian averaging over decision trees (DT) which can be easily understood by human experts and providing acceptable classification accuracy. The proposed Bayesian DT technique is used for the classification of real solar flares.

4.3.1 Introduction

The methodology of Bayesian model averaging is used for applications such as predictions when the evaluation of uncertainty in decision is of crucial importance. In general, uncertainty is a trade-off between the amount of data available for training, the classifier diversity and the classi-fication accuracy (Duda 2001; Denison et al. 2002; Chipman 1998). The interpretability of classification models can also give useful information to experts responsible for making reliable classifications. For this reason DTs seem attractive classification models for experts (Breiman et al. 1984; Chipman et al. 1998a; Denison 2002; Dietterich 2000; Duda 2001; Kuncheva 2004).

The main idea of using DT classification models is to recursively partition data points in an axis-parallel manner. Such models provide natural feature selection and uncover the features which make the important contribution the classification. The resultant DT classification models can be easily interpretable by users.

By definition, DTs consist of splitting and terminal nodes, which are also known as tree leaves. DTs are said to be binary if the splitting nodes ask a specific question and then divide the data points into two disjoint subsets, say the left or the right branch. Figure 4.8 depicts an example of the DT consisting of two splitting and three terminal nodes.

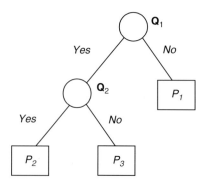

Fig. 4.8. An example of decision tree consisting of two splitting and terminal nodes are depicted by the *circles* and *rectangles*. The split nodes ask the questions Q_1 and Q_2 and an outcome is assigned to one of the terminal nodes with the probabilities P_1, P_2, and P_3

Note that the number of the data points in each split should not be less than that predefined by a user. The terminal node assigns all data points falling in that node to a class of majority of the training data points residing in this terminal node. Within a Bayesian framework, the class posterior distribution is calculated for each terminal node (Duda 2001; Denison 2002; Chipman 1998; Breiman et al. 1984).

The required diversity of the DTs can be achieved on the base of Bayesian Markov Chain Monte Carlo (MCMC) methodology of sampling from the posterior distribution (Denison 2002; Chipman 1998). This technique has revealed promising results when applied to some real-world problems. For sampling from large DTs, the MCMC technique has been extended with reversible jumps (RJ) (Green 1995). The RJ MCMC technique making such moves as *birth* and *death* allows the DTs to be induced under the priors given on the shape or size of the DTs. Exploring the posterior distribution, the RJ MCMC should keep the balance between the birth and death moves under which the desired estimate of the posterior can be unbiased (Chipman 1998; Denison 2002; Green 1995).

Within the RJ MCMC technique the proposed moves for which the number of data points falling in one of splitting nodes becomes less than the given number are assigned unavailable. Obviously that the priors given on the DTs are dependent on the class boundaries and noise level in data available for training, and it is intuitively clear that the sharper class boundaries, the larger DTs should be. However in practice the use of such an intuition without a prior knowledge on favourite shape of the DTs can lead to inducing over-complicated DTs and as a result the averaging over such DTs can produce biased class posterior estimates (Chipman 1998;

Denison 2002). More over, within the standard RJ MCMC technique suggested for averaging over DTs, the required balance cannot be kept. This may happen because of over-fitting the Bayesian models (Domingos 2000). Another reason is that the RJ MCMC technique averaging over DTs assigns some moves which cannot provide a given number of data points allowed being in the splitting nodes unavailable (Chipman 1998; Denison 2002).

For the cases when the prior information of the favourite shape of DTs is unavailable, the Bayesian DT technique with a sweeping strategy has revealed a better performance (Schetinin et al. 2004). Within this strategy the prior given on the number of DT nodes is defined implicitly and dependent on the given number of data points allowed being at the DT splits. So the sweeping strategy gives more chances to induce the DTs containing a near optimal number of splitting nodes required to provide the best generalization. At the same time within this technique the number of data points allowed to be in the splitting nodes can be reasonably reduced without increasing the risk of overcomplicating the DTs.

Next we describe the standard Bayesian RJ MCMC technique and then we describe the Bayesian DT technique with the sweeping strategy.

4.3.2 The Methodology of Bayesian Averaging

In general, the predictive distribution we are interested in is written as an integral over parameters θ of the classification model

$$p(y \mid \mathbf{x}, \mathbf{D}) = \int_{\theta} p(y \mid \mathbf{x}, \theta, \mathbf{D}) p(\theta \mid \mathbf{D}) d\theta \qquad (4.17)$$

where y is the predicted class $(1, \ldots, C)$, $\mathbf{x} = (x_1, \ldots, x_m)$ is the m-dimensional input vector, and $\mathbf{D}$ denotes the given training data.

The integral (4.17) can be analytically calculated only in simple cases. In practice, part of the integrand in (4.17) which is the posterior density of θ conditioned on the data $\mathbf{D}$, $p(\theta \mid \mathbf{D})$, cannot usually be evaluated. However if values $\theta^{(1)}, \ldots, \theta^{(N)}$ are drawn from the posterior distribution $p(\theta \mid \mathbf{D})$, we can write:

$$p(y \mid \mathbf{x}, \mathbf{D}) \approx \sum_{i=1}^{N} p(y \mid \mathbf{x}, \theta^{(i)}, \mathbf{D}) p(\theta^{(i)} \mid \mathbf{D}) \qquad (4.18)$$

$$= \frac{1}{N} \sum_{i=1}^{N} p(y \mid \mathbf{x}, \theta^{(i)}, \mathbf{D}).$$

This is the basis of the MCMC technique for approximating integrals (Denison 2002). To perform the approximation, we need to generate random samples from $p(\theta \mid \mathbf{D})$ by running a Markov Chain until it has converged to a stationary distribution. After this we can draw samples from this Markov Chain and calculate the predictive posterior density (4.18).

Let us now define a classification problem presented by data $(\mathbf{x}_i, y_i)$, $i = 1, ..., n$, where n is the number of data points and $y_i \in \{1, ..., C\}$ is a categorical response. Using DTs for classification, we need to determine the probability φ_{ij} with which a datum $\mathbf{x}$ is assigned by terminal node $i = 1, ..., k$ to the jth class, where k is the number of terminal nodes in the DT. Initially we can assign a $(C - 1)$-dimensional Dirichlet prior for each terminal node so that $p(\varphi_i \mid \theta) = \mathrm{Dic}_{-1}(\varphi_i \mid \alpha)$, where $\varphi_i = (\varphi_{i1}, ..., \varphi_{iC})$, θ is the vector of DT parameters, and $\alpha = (\alpha_1, ..., \alpha_C)$ is a prior vector of constants given for all the classes.

The DT parameters are defined as $\theta = (s_i^{pos}, s_i^{var}, s_i^{rule})$, $i = 1, ..., k - 1$, where s_i^{pos}, s_i^{var} and s_i^{rule} define the *position*, *predictor* and *rule* of each splitting node, respectively. For these parameters the priors can be specified as follows. First we can define a maximal number of splitting nodes, say, $s_{max} = n - 1$, so $s_i^{pos} \in \{1, ..., s_{max}\}$. Second we draw any of the m predictors from a uniform discrete distribution $U(1, ..., m)$ and assign $s_i^{var} \in \{1, ..., m\}$. Finally the candidate value for the splitting variable $x_j = s_i^{var}$ is drawn from a uniform discrete distribution $U(x_j^{(1)}, ..., x_j^{(N)})$, where N is the total number of possible splitting rules for predictor x_j, either categorical or continuous.

Such priors allow the exploring of DTs which partition data in as many ways as possible, and therefore we can assume that each DT with the same number of terminal nodes is equally likely (Denison 2002). For this case the prior for a complete DT is described as follows:

$$p(\theta) = \left\{ \prod_{i=1}^{k-1} p(s_i^{rule} \mid s_i^{var}) p(s_i^{var}) \right\} p(\{s_i^{pos}\}_1^{k-1}). \qquad (4.19)$$

For a case when there is knowledge of the favoured structure of the DT, the above prior has been assumed as the prior probability that further splits of the terminal nodes are dependent on how many splits have already been made above them (Chipman 1998). For example, for the ith terminal node the probability of its splitting is written as:

$$p_{split}(i) = \gamma(1 + d_i)^{-\delta}, \qquad (4.20)$$

where d_i is the number of splits made above i and γ, $\delta \geq 0$ are given constants. The larger δ, the more the prior favours "bushy" trees. For $\delta = 0$ each DT with the same number of terminal nodes appears with the same prior probability.

Having set the priors on the parameters φ and θ, we can determine the marginal likelihood for the data given the classification tree. In the general case this likelihood can be written as a multinomial Dirichlet distribution (Denison 2002):

$$p(\mathbf{D} \mid \theta) = \left[\frac{\Gamma\{\alpha C\}}{\{\Gamma(\alpha)\}^C} \right]^k \prod_{i=1}^{C} \frac{\prod_j^C \Gamma(m_{ij} + \alpha_j)}{\Gamma(n_i + \sum_{j=1}^C \alpha_j)}, \tag{4.21}$$

where n_i is the number of data points falling in the ith terminal node of which m_{ij} points are of class j and Γ is a Gamma function.

4.3.3 Reversible Jumps Extension

To allow sampling DT models of variable dimensionality, the MCMC technique exploits the Reversible Jump extension (Green 1995). This extension allows the MCMC technique to sample large DTs induced from real-world data. To implement the RJ MCMC technique described in Chipman (1998) and Denison (2002) it has been suggested exploring the posterior probability by using the following types of moves.

− *Birth*. Randomly split the data points falling in one of the terminal nodes by a new splitting node with the variable and rule drawn from the corresponding priors.
− *Death*. Randomly pick a splitting node with two terminal nodes and assign it to be one terminal with the united data points.
− *Change-split*. Randomly pick a splitting node and assign it a new splitting variable and rule drawn from the corresponding priors.
− *Change-rule*. Randomly pick a splitting node and assign it a new rule drawn from a given prior.

The first two moves, *birth* and *death*, are reversible and change the dimensionality of θ as described in (Green 1995). The remaining moves provide jumps within the current dimensionality θ. Note that the *change-split* move is included to make "large" jumps which potentially increase the chance of sampling from a maximal posterior whilst the *change-rule* move does "local" jumps.

For the birth moves, the proposal ratio R is written as:

$$R = \frac{q(\theta \mid \theta')p(\theta')}{q(\theta' \mid \theta)p(\theta)}, \tag{4.22}$$

where $q(\theta \mid \theta')$ and $q(\theta' \mid \theta)$ are the proposed distributions, θ' and θ are $(k + 1)$ and k-dimensional vectors of DT parameters, respectively, and $p(\theta)$ and $p(\theta')$ are the probabilities of the DT with parameters θ and θ':

$$p(\theta) = \{\prod_{i=1}^{k-1} \frac{1}{N(s_i^{var})} \frac{1}{m} \} \frac{k}{S_k} \frac{1}{K}, \tag{4.23}$$

where $N(s_i^{var})$ is the number of possible values of s_i^{var} which can be assigned as a new splitting rule, S_k is the number of ways of constructing a DT with k splitting nodes.

For binary DTs, as given from graph theory, the number S_k is the *Catalan number*

$$S_k = \frac{1}{k+1}\binom{2k}{k}, \tag{4.24}$$

and we can see that for $k \geq 25$ this number becomes astronomically large, $S_k \geq (4.8)^{12}$.

The proposal distributions are as follows:

$$q(\theta \mid \theta') = \frac{d_{k+1}}{D_{Q'}}, \tag{4.25}$$

$$q(\theta' \mid \theta) = \frac{b_k}{k} \frac{1}{N(s_k^{var})} \frac{1}{m}, \tag{4.26}$$

where $D_{Q1} = D_Q + 1$ is the number of splitting nodes whose branches are both terminal nodes.

Then the proposal ratio for a *birth* is given by:

$$R = \frac{d_{k+1}}{b_k} \frac{k}{D_{Q1}} \frac{S_k}{S_{k+1}}. \tag{4.27}$$

The number D_{Q1} in (4.27) is dependent on the DT structure and it is clear that $D_{Q1} < k \; \forall \; k = 1, \ldots, K$. Analysing (4.27), we can also assume $d_{k+1} = b_k$. Then letting the DTs grow, i.e. $k \to K$, and considering $S_{k+1} > S_k$, we can see that the value of $R \to c$, where c is a constant lying between 0 and 1.

Alternatively, for the death moves the proposal ratio is written as:

$$R = \frac{b_k}{d_{k-1}} \frac{D_Q}{(k-1)} \frac{S_k}{S_{k-1}}, \tag{4.28}$$

and then we can see that under the assumptions considered for the birth moves, $R \geq 1$.

4.3.4 The Difficulties of Sampling Decision Trees

The RJ MCMC technique starts drawing samples from a DT consisting of one splitting node whose parameters were randomly assigned within the predefined priors. So we need to run the Markov Chain whilst it grows and its likelihood is unstable. This phase is said *burn-in* and it should be preset enough long in order to stabilize the Markov Chain. When the Markov Chain will be enough stable, we can start sampling. This phase is said *post burn-in*.

It is important to note that the DTs grow very quickly during the first burn-in samples. This happens because an increase in log likelihood value for the birth moves is much larger than that for the others. For this reason almost every new partition of data is accepted. Once a DT has grown the *change* moves are accepted with a very small probability and, as a result, the MCMC algorithm tends to get stuck at a particular DT structure instead of exploring all possible structures.

The size of DTs can rationally decrease by defining a minimal number of data points, p_{min}, allowed to be in the splitting nodes (Breiman et al. 1984; Chipman 1998; Denison 2002; Dietterich 2000; Duda 2001; Kuncheva 2004). If the number of data points in new partitions made after the birth or change moves becomes less than a given number p_{min}, such moves are assigned unavailable, and the RJ MCMC algorithm resamples such moves.

However, when the moves are assigned unavailable, this distorts the proposal probabilities p_b, p_d, and p_c given for the birth, death, and change moves, respectively. The larger the DT, the smaller the number of data points falling in the splitting nodes, and correspondingly the larger is the probability with which moves become unavailable. Resampling the unavailable moves makes the balance between the proposal probabilities biased as described in (Schetinin et al. 2004).

The disproportion in the balance between the probabilities of birth and death moves is dependent on the size of DTs averaged over samples. Clearly, at the beginning of burn-in phase the disproportion is close to zero, and to the end of the burn-in phase, when the size and form of DTs are stabilized, its value becomes maximal.

Because DTs are hierarchical structures, the changes at the nodes located at the upper levels can significantly change the location of data points at the lower levels. For this reason there is a very small probability of changing and then accepting a DT split located near a root node. Therefore the RJ MCMC algorithm collects the DTs in which the splitting nodes located far from a root node were changed. These nodes typically contain small numbers of data points. Subsequently, the value of log likelihood is not changed much, and such moves are frequently accepted. As a result, the RJ MCMC algorithm cannot explore a full posterior distribution properly.

One way to extend the search space is to restrict DT sizes during a given number of the first burn-in samples as described in Denison (2002). Indeed, under such a restriction, this strategy gives more chances of finding DTs of a smaller size which could be competitive in term of the log likelihood values with the larger DTs. The restricting strategy, however, requires setting up in an ad hoc manner the additional parameters such as the size of DTs and the number of the first burn-in samples. Sadly, in practice, it often happens that after the limitation period the DTs grow quickly again and this strategy does not improve the performance.

Alternatively to the above approach based on the explicit limitation of DT size, the search space can be extended by using a restarting strategy as Chipman et al. (1998a, b) have suggested. Clearly, both these strategies cannot guarantee that most of DTs will be sampled from a model space region with a maximal posterior. In Sect. 4.3.5 we describe our approach based on sweeping the DTs.

4.3.5 The Bayesian Averaging with a Sweeping Strategy

In this section we describe our approach to decreasing the uncertainty of classification outcomes within the Bayesian averaging over DT models. The main idea of this approach is to assign the prior probability of further splitting DT nodes to be dependent on the range of values within which the number of data points will be not less than a given number of points, p_{min}. Such a prior is explicit because at the current partition the range of such values is unknown.

Formally, the probability $P_s(i, j)$ of further splitting at the ith partition level and variable j can be written as:

$$P_s(i,j) = \frac{x_{max}^{(i,j)} - x_{min}^{(i,j)}}{x_{max}^{(1,j)} - x_{min}^{(1,j)}}, \tag{4.29}$$

where $x_{min}^{(i,j)}$ and $x_{max}^{(i,j)}$ are the minimal and maximal values of variable j at the ith partition level.

Observing (4.29), we can see that $x_{max}^{(i,j)} \leq x_{max}^{(1,j)}$ and $x_{min}^{(i,j)} \geq x_{max}^{(1,j)}$ for all the partition levels $i > 1$. On the other hand there is partition level k at which the number of data points becomes less than a given number p_{min}. Therefore, we can conclude that the prior probability of splitting P_s ranges between 0 and 1 for any variable j and the partition levels i: $1 \leq i < k$.

From (4.29), it follows that for the first level of partition, probability P_s is equal to 1.0 for any variable j. Let us now assume that the first partition split the original data set into two non-empty parts. Each of these parts contains less data points than the original data set, and consequently for the $(i = 2)$th partition either $x_{max}^{(i,j)} < x_{max}^{(1,j)}$ or $x_{min}^{(i,j)} > x_{max}^{(1,j)}$ for new splitting variable j. In any case, numerator in (4.29) decreases, and probability P_s becomes less than 1.0. We can see that each new partition makes values of numerator and consequently probability (4.29) smaller. So the probability of further splitting nodes is dependent on the level i of partitioning data set.

The earlier prior favours splitting the terminal nodes which contain a large number of data points. This is clearly a desired property of the RJ MCMC technique because it allows accelerating the convergence of Markov chain. As a result of using prior (4.29), the RJ MCMC technique of sampling DTs can explore an area of a maximal posterior in more detail.

However prior (4.29) is dependent not only on the level of partition but also on the distribution of data points in the partitions. Analyzing the data set at the ith partition, we can see that value of probability P_s is dependent on the distribution of these data. For this reason the prior (4.29) cannot be implemented explicitly without the estimates of the distribution of data points in each partition.

To make the birth and change moves within prior (4.29), the new splitting values $s_i^{rule,new}$ for the ith node and variable j are assigned as follows. For the birth and change-split moves the new value $s_i^{rule,new}$ is drawn from a uniform distribution:

$$s_i^{rule,new} \sim U(x_{min}^{1,j}, x_{max}^{1,j}). \tag{4.30}$$

The earlier prior is "uninformative" and used when no information on preferable values of s_i^{rule} is available. As we can see, the use of a uniform distribution for drawing new rule $s_i^{rule,new}$, proposed at the level $i > 1$, can cause the partitions containing less data points than p_{min}. However, within our technique such proposals can be avoided.

For the change-split moves, drawing $s_i^{\text{rule,new}}$ follows after taking new variable $s_i^{\text{var,new}}$:

$$s_i^{\text{var,new}} \sim U\{S_k\}, \tag{4.31}$$

where $S_k = \{1, \ldots, m\} \setminus s_i^{\text{var}}$ is the set of features excluding variable s_i^{var} currently used at the ith node.

For the change-rule moves, the value $s_i^{\text{rule,new}}$ is drawn from a Gaussian with a given variance σ_j:

$$s_i^{\text{rule,new}} \sim N(s_i^{\text{rule}}, \sigma_j), \tag{4.32}$$

where $j = s_i^{\text{var}}$ is the variable used at the ith node.

Because DTs have hierarchical structure, the change moves (especially change-split moves) applied to the first partition levels can heavily modify the shape of the DT, and as a result, its bottom partitions can contain less the data points than p_{min}. As mentioned earlier, within the Bayesian DT techniques (Chipman 1998; Denison 2002) such moves are assigned unavailable.

Within our approach after birth or change move there arise three possible cases. In the first case, the number of data points in each new partition is larger than p_{min}. The second case is where the number of data points in one new partition is larger than p_{min}. The third case is where the number of data points in two or more new partitions is larger than p_{min}. These three cases are processed as follows.

For the first case, no further actions are made, and the RJ MCMC algorithm runs as usual.

For the second case, the node containing unacceptable number of data points is removed from the resultant DT. If the move was of birth type, then the RJ MCMC resamples the DT. Otherwise, the algorithm performs the death move. For the last case, the RG MCMC algorithm resamples the DT.

As we can see, within our approach the terminal node, which after making the birth or change moves contains less than p_{min} data points, is removed from the DT. Clearly, removing such unacceptable nodes turns the random search in a direction in which the RJ MCMC algorithm has more chances to find a maximum of the posterior amongst shorter DTs. As in this process the unacceptable nodes are removed, we named such a strategy *sweeping*.

After change move the resultant DT can contain more than one node splitting less than p_{min} data points. However this can happen at the beginning of burn-in phase, when the DTs grow, and this unlikely happen, when the DTs have grown.

4.3.6 Performance of the Bayesian Decision Tree Technique

In this section we describe the experiments with a two-dimensional synthetic problem, in order to visualise the decision boundaries and data points. The problem was generated by a mixture of five Gaussians. The data points drawn from the first three Gaussians belong to class 1 and the data points from the remaining two Gaussians to class 2. Such a mixture of the kernels is an extended version of the Ripley data (Ripley 1994).

In our case the training data contain 250 and the test data 1,000 data points drawn from the above mixture. Because the classes overlap, the Bayes error on the test data is 9.3%. The data points of the two classes denoted by the crosses and dots are depicted in Fig. 4.9. The Bayesian decision boundary is shown here by the dashed line. Both the Bayesian and randomized DT ensemble techniques were run on these synthetic data with the pruning factor p_{min} set equal to 5.

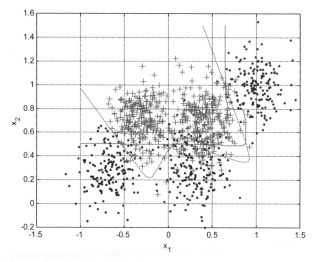

Fig. 4.9. Synthetic data points belonging to two classes. The *dashed* and *solid lines* depict the class boundaries calculated for the Bayesian rule and Bayesian DTs, respectively

The Bayesian DT technique with the sweeping strategy was run with 20,000 samples for burn-in and 10,000 for post burn-in phases. The probabilities of birth, death, change variable, and change rule were 0.1, 0.1, 0.1, and 0.7, respectively. The sample rate was set to 7. Priors on the number of nodes in DTs were set uniform. The uniform prior allows the DTs to grow by making birth moves whilst the proposed DT parameters made within the p_{min} are available.

The acceptance rates during the burn-in and post burn-in phases were equal to 0.47. The mean and variance values of DT nodes were 12.4 and 2.5, respectively.

From Fig. 4.9 we can see that the resultant class boundary depicted by the solid line is close to the optimal class boundary and the classification error on the test data is 12.4%. Thus, the Bayesian DT technique with a shrinking strategy reveals a good performance on this artificial problem.

4.3.7 The Use of the Bayesian Decision Tree Techniques for Classifying the Solar Flares

This section describes the experimental results on the comparison of the Bayesian DT techniques with the standard and sweeping strategies described in Sect. 4. The experiments were conducted on the solar flare data available for the machine learning community.

The Solar Flare Data

The solar flare data were taken from the machine learning repository (Blake and Merz 1998). Originally, these data contain three potential classes, one for the number of times a certain type of solar flare occurred in a 24 h period. Each pattern represents captured features for one active region on the Sun. The number of the patterns is 1,066. Each pattern is presented by ten features listed in Table 4.1.

Table 4.1. The solar flare data features

#	features	values
1	code for class (modified Zurich class)	{A,B,C,D,E,F,H}
2	code for largest spot size	{X,R,S,A,H,K}
3	code for spot distribution	{X,O,I,C}
4	activity	{1 = reduced, 2 = unchanged}
5	evolution	{1 = decay, 2 = no growth, 3 = growth}
6	previous 24 h flare activity code	{1 = nothing as big as an M1, 2 = one M1, 3 = more activity than one M1}
7	historically-complex	{1 = Yes, 2 = No}
8	did region become historically complex on this pass across the sun's disk	{1 = yes, 2 = no}
9	area	{1 = small, 2 = large}
10	area of the largest spot	{1 = <=5, 2 = >5}

From all the predictors listed in this table, three classes of flares produced in regions are predicted: C, M, and X classes. However, in our experiments the prediction was assigned to be a flare or non-flare with two possible outcomes.

For all the above domain problems, no prior information on the preferable DT shape and size was available. The pruning factor, or the minimal number of data point allowed being in the splits, p_{min} was given equal to 1. The proposal probabilities for the death, birth, change-split and change-rules are set to be 0.1, 0.1, 0.1, and 0.7, respectively. The numbers of burn-in and post burn-in samples were set equal 50,000 and 5,000, respectively. The sampling rate for all the domain problems was set equal to 7. All these parameters of MCMC sampling were set the same for both the standard and the proposed Bayesian techniques.

The performance of the Bayesian MCMC techniques was evaluated within the uncertainty envelope techniques described in (Fieldsend et al. 2003) within five-fold cross-validation and 2σ intervals. The average size of the induced DTs is an important characteristic of the Bayesian techniques and it was also evaluated in our experiments.

The MATLAB code of the Bayesian DT technique with MCMC sampling is shown in an Appendix I. We developed this code for experiments described in this section and can be used by the readers for exercising.

Experimental Results

Both Bayesian DT techniques with the standard (DBT1) and the sweeping (BDT2) strategies have correctly recognized 82.1% and 82.5% of the test examples, respectively. The average number of DT nodes was 17.5 and 10.1 for these strategies, respectively, see Table 4.2.

Table 4.2. The performance of the BDT1 and BDT2 on the solar flare data

strategy	number of DT nodes	perform, %	sure correct, %
BDT1	17.5 ± 1.5	82.1 ± 4.5	67.4 ± 3.3
BDT2	$\mathbf{10.1 \pm 1.6}$	82.5 ± 3.8	$\mathbf{70.2 \pm 4.2}$

Figures 4.10 and 4.11 depict samples of log likelihood and numbers of DT nodes as well as the densities of DT nodes for burn-in and post burn-in phases for the BDT1 and BDT2 strategies. From the top left plot of these figures we can see that the Markov chain very quickly converges to the stationary value of log likelihood near to -320. During post burn-in the values of log likelihood slightly oscillate around this value of log likelihood that says us that the samples of DTs are drawn from a stable Markov Chain.

As we can see from Table 4.2 both the BDT1 and the BDT2 strategies reveal the same performance on the test data. However the number of DT nodes induced by the BDT2 strategy is much less than that induced by the

BDT1 strategy. Besides, the BDT2 strategy provides more sure and correct classifications than those provided by the BDT1 strategy.

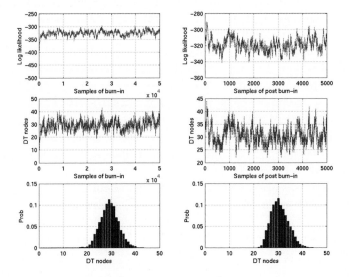

Fig. 4.10. The standard BDT1 strategy: samples of the burn-in and post burn-in

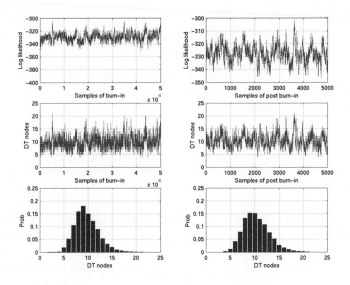

Fig. 4.11. The BDT2 strategy: samples of the burn-in and post burn-in

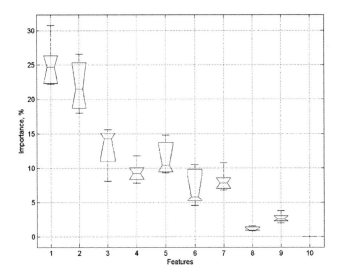

Fig. 4.12. The feature importance for the classification of the solar flares

Figure 4.12 depicts the probabilistic contribution of the ten features to the classification outcome. The most important contribution are made features x_1 (code for class) and x_2 (code for largest spot size). Much less contribution are made by features x_8 (did region become historically complex), x_9 (area) and x_{10} (area of the largest spot).

Thus we conclude that our Bayesian strategy of averaging over DTs using a sweeping strategy is able decreasing the classification uncertainty without affecting classification accuracy on the problems examined. Clearly this is a very desirable property for classifiers used to find a probabilistic explanation of classification outcomes.

4.3.8 Confident Interpretation of Bayesian Decision Tree Ensembles

This section describes our method of interpreting Bayesian DT ensembles. First we introduce the confidence in the classification outcomes of the DT ensemble which can be quantitatively estimated on the training data. Then we give a demonstrative example how a desired DT can be selected, and finally we present a DT extracted from the DT ensemble induced from the solar flare data.

The Interpretation Strategy Using the Classification Confidence

There are two approaches to interpreting of DT ensembles. The first is based on an idea of clustering DTs in two-dimensional space of DT size and DT fitness (Chipman 1998). The second approach is based on searching a DT of maximum a posterior (MAP) (Domingos 1998).

Our approach is based on the quantitative estimates of classification confidence, which can be made within the Uncertainty Envelope technique described in Fieldsend et al. (2003). The details of our technique are described as follows.

In general, the Bayesian DT strategies described in this section allow sampling the DTs induced from data independently. In such a case, we can naturally assume that the inconsistency of the classifiers on a given datum **x** is proportional to the uncertainty of the DT ensemble. Let the value of class posterior probability $P(c_j|\mathbf{x})$ calculated for class c_j be an average over the class posterior probability $P(c_j|K_i, \mathbf{x})$ given on classifier K_i:

$$P(c_j \mid \mathbf{x}) = \frac{1}{N}\sum_{i=1}^{N} P(c_j \mid K_i, \mathbf{x}), \qquad (4.33)$$

where N is the number of classifiers in the ensemble.

As classifiers $K_1, \ldots, K_N$ are independent of each other and their values $P(c_j|K_i, \mathbf{x})$ range between 0 and 1, the probability $P(c_j|\mathbf{x})$ can be approximated as follows:

$$P(c_j \mid \mathbf{x}) \approx \frac{1}{N}\sum_{i=1}^{N} I(y_i, t_i \mid \mathbf{x}), \qquad (4.34)$$

where $I(y_i, t_i)$ is the indicator function assigned to be 1 if the output y_i of the ith classifier corresponds to target t_i, and 0 if it does not.

It is important to note that the right side of (4.34) can be considered as a *consistency* of the outcomes of DT ensemble. Clearly, values of the consistency, $\gamma = \frac{1}{N}\sum_{i=1}^{N} I(y_i, t_i \mid \mathbf{x})$, lie between $1/C$ and 1.

Analyzing (4.34), we can see that if all the classifiers are degenerate, i.e. $P(c_j|K_i, \mathbf{x}) \in \{0, 1\}$, then the values of $P(c_j|\mathbf{x})$ and γ become equal. The outputs of classifiers can be equal to 0 or 1, for example, when the data points of two classes do not overlap. In other cases, the class posterior probabilities of classifiers range between 0 and 1, and the $P(c_j|\mathbf{x}) \approx \gamma$. So we can conclude that the classification confidence of an outcome is characterized by the consistency of the DT ensemble calculated on a given datum. Clearly, the values of γ are dependent on how representative the training data are, what classification scheme is used, how well the classifiers were trained within a classification scheme, how close the datum $\mathbf{x}$ is to the class boundaries, how the data are corrupted by noise, and so on.

Let us now consider a simple example of a DT ensemble consisting of $N = 1,000$ classifiers in which two classifiers give a conflicting classification on a given datum $\mathbf{x}$ to the other 998. Then consistency $\gamma = 1 - 2/1,000 = 0.998$, and we can conclude that the DT ensemble was trained well and/or the data point $\mathbf{x}$ lies far from the class boundaries. It is clear that for new datum appearing in some neighbourhood of the $\mathbf{x}$, the classification uncertainty as the probability of misclassification is expected to be $1 - \gamma = 1 - 0.998 = 0.002$. This inference is truthful for the neighbourhood within which the prior probabilities of classes remain the same. When the value of γ is close to $\gamma_{min} = 1/C$, the classification uncertainty is highest and a datum $\mathbf{x}$ can be misclassified with a probability $1 - \gamma = 1 - 1/C$.

From the earlier consideration, we can assume that there is some value of consistency γ_0 for which the classification outcome is confident, that is the probability with which a given datum $\mathbf{x}$ could be misclassified is small enough to be acceptable. Given such a value, we can now specify the uncertainty of classification outcomes in statistical terms. The classification outcome is said to be *confident and correct*, when the probability of misclassification is acceptably small and $\gamma \geq \gamma_0$. Next we describe the proposed interpretation technique.

The Interpretation Technique

The idea behind our method of interpreting the Bayesian DT ensemble is to find a single DT which covers most of the training examples classified as confident and correct. For multiple classification systems the confidence of classification outputs can be easily estimated by accounting consistency of the classification outcomes as described (Fieldsend et al. 2003).

Indeed, within a given classification scheme the outputs of the multiple classifier system depend on how well were the classifiers trained and how representative were the training data. For a given data sample, the consistency of classification outcomes depends on how close is this sample to the class boundaries. So for the ith class, the confidence in the MCS can be estimated as a ratio γ_i between the number of classifier outcomes of the ith class, N_i, and the total number of classifiers N: $\gamma_i = N_i / N$, $i = 1, ..., C$, where C is the number of classes.

Clearly the classification confidence is maximal, equal to 1.0, if all the classifiers assign a given input to the same class; otherwise the confidence is less than 1.0. The minimal value of confidence is equal to $1/C$ if the classifiers assign the input datum to the C classes in equal proportions. So for a given input the classification confidence in the MCS can be properly estimated by ratio γ.

Within the earlier framework in real-world applications, we can define a given level of the classification confidence, $\gamma_0 : 1/C \leq \gamma_0 \leq 1$, for which cost of misclassification is small enough to be accepted. Then for the given input, the outcome of the MCS is said *confident* if the ratio $\gamma \geq \gamma_0$. Clearly, on the labelled data we can distinguish between *confident and correct* outcomes and *confident but incorrect* outcomes. The last outcomes of the multiple classifier system may appear due to noise or overlapping classes in the data.

Let us now consider how the earlier estimates of classification confidence can be used to interpret the Bayesian DT ensemble. Assume the following example with five classifiers outcomes $o_1,..., o_5$ and five data points $x_1, ..., x_5$ as presented in Table 4.3. For the given datum $x_j, j = 1,..., 5$, the outcomes $o_1,..., o_5$ are written as follows $o_i = 1$, if $y_i = t_j$, otherwise $o_i = 0$, where y_i and t_j are the outcome of the $i$the classifier and the class label of datum x_j, respectively.

Table 4.3. An example of classification outcomes of the DT ensemble

datum	o_1	o_2	o_3	o_4	o_5	γ
X_1	0	0	1	1	1	3/5
X_2	0	1	1	1	1	4/5
X_3	1	0	1	1	0	3/5
X_4	1	1	1	0	1	4/5
X_5	1	1	0	1	0	3/5
sum	3	3	4	4	3	

Let us now give a confidence level $\gamma_0 = 3/5$ under which all the five samples are classified as confident and correct. Then we can see that the third and fourth DTs cove a maximal number of samples equal to four. Consequently, one of these DTs can be selected for interpreting the confident classification. Such a selection can be reasonably done with a minimal DT size criterion because such DTs provide better generalization ability.

In practice, the number of DTs in the ensemble as well as the number of the training examples can be large. Nevertheless, counting the number of confident and correct outcomes as described in the earlier example, we can find a desired DT which can be used for interpreting the confident classification. The performance of such a DT can be slightly worse than that of the Bayesian DT ensemble. Later we provide the experimental comparison of their performances. The main steps of the selection procedure are given in "The Selection Procedure".

The Selection Procedure

All what we need is to find a set of the DTs which cover a maximal number of the training samples classified as confident and correct whilst the number of misclassifications on the remaining examples keeps minimal. To find such a DT set, we can remove the conflicting examples from the training data and then select the DTs with a maximal cover of the training samples classified by the DT ensemble as confident and correct.

Thus the main steps of the earlier selection procedure are as follows:

1. Amongst a given Bayesian DT ensemble find a set of DTs, S1, which cover a maximal number of the training samples classified as confident and correct with a given confidence level γ_0.
2. Find the training samples which were misclassified by the Bayesian DT ensemble and then remove them from the training data. Denote the remaining training samples as D1.
3. Amongst the set S1 of DTs find those which provide a minimal misclassification rate on the data D1. Denote the found set of such DTs as S2.
4. Amongst the set S2 of DTs select those whose size is minimal. Denote a set of such DTs as S3. The set S3 contains the desired DTs.

The earlier procedure finds one or more DTs and puts them in the set S3. These DTs cover a maximal number of the training samples classified as confident and correct with a given confidence level γ_0. The size of these DTs is minimal and any of them can be finally selected for interpreting the confident classification.

Next we discuss the use of the earlier selection procedure and compare the performance of the resultant DTs on the real-world data.

Experimental Results

In this section we describe the experiments carried with the Bayesian DT technique described earlier and the selection procedure as described above. The goal of these experiments are to compare the performance the resultant DTs extracted from the Bayesian DT ensembles on the sunspot group with flares data as described in Sect. 4.3.7.

Figure 4.13 depicts a DT selected by the proposed selection procedure. This DT consists of four splitting nodes and provides 84.5% of correct classification on the 213 test examples. The ensemble of the Bayesian DTs provides on these test data 84.9% of correct classifications whilst the DTs consist of, on average, 18.2 splitting nodes. Clearly the selected DT provides a very good example of the DT ensemble.

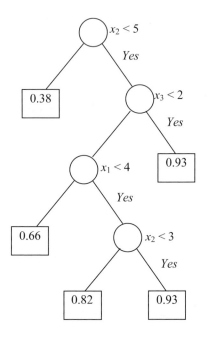

Fig. 4.13. A decision tree consisting of four nodes extracted from the DT ensemble on the solar flare data

The MAP technique has selected a DT which correctly recognized 83.1% of the test examples. This DT consists of 21 splitting nodes.

Thus, the suggested technique selects a DT which provides a better classification accuracy than that selected with the MAP technique. Meanwhile the DT obtained with the suggested technique is much shorter than that with the MAP technique.

4.3.9 Conclusions

The classification techniques based on the Bayesian model averaging described in this chapter has revealed promising results in our experiments with the probabilistic interpretation of solar flares. At the same time experts can easily interpret the resultant classification models.

Our experiments with the solar flare data have also shown the effectiveness of the procedure suggested to find a single DT covering a maximal number of sure and correct examples of training data. The resultant DT provides a maximal certainty in classification outcomes, and meanwhile it provides high performance on train data. Thus, we conclude that the proposed methodology of the Bayesian DT averaging revealed promising results in applications to astrophysical images and data.

4.3.10 Questions and Exercises

1. How does the given number of data points, p_{min}, affect the Bayesian DTs?
2. Why is the burn-in phase required?
3. How long should the Markov Chain run during the burn-in and post burn-in phases?
4. How does the value of sigma given for a Gaussian proposal affect the acceptance level?
5. What happens if value of the proposal probability for birth moves does set to 0.0?
6. Using the MATLAB program from Appendix I, provide the experimental evidences for your answers on the above questions.

Appendix MATLAB Code of the Bayesian Decision Tree Technique Running Markov Chain Monte Carlo "BDT_MCMC"

```
%
% BDT_MCMC.m
%
% Scripts and functions:
%     prop_ratio, mc_load_data,   mc_split,
%     mc_log_lik,   mc_create_move, mc_shrink, mc_prop_ratio,
%     mc_prob, mc_test, mc_rpt
%
global X Y C num mvar Pr pmin Xt pC Tp K Cf q_nom …
x_min x_max   Pp

% Set parameters:

% the maximal number of DT nodes
node_max = 50;

% the minimal number of data points in DT nodes
pmin = 3;

% the variance of Gaussian for change-rule moves
q_sig = .1;

% the variance of uniform distribution for nominal
% features
q_nom = 1;

% the number of burn-in samples
nb = 50000;
```

```
% the number of post burn-in samples
np = 5000;

% the proposal probabilities for birth, death,
% change-question, and change-rule moves
Pr = cumsum([0.2 0.2 0.2 0.4]);

% the sampling rate
sample_rate = 7;

% the level of consistency
consist_lev = 0.99;

% store the DT outputs accepted in post burn-in
hist_collect = 1;

% print each a given sample
print_sample = 500;

load prop_ratio R1 R2 node_max % see script prop_ratio

% the name of data file
fnam = 'flar';

& call function mc_load_data_cf
[C,mvar,X,Y,num,Xt,Yt,numt,Cf] = mc_load_data_cf(fnam);

Pp(1) = (Pr(2) - Pr(1))/Pr(1);
Pp(2) = Pr(1)/(Pr(2) - Pr(1));
Perf = []; % performance on the test
x_min = min(X); x_max - max(X);
K = q_sig*std(X);

% Class prior probabilities:

pC = zeros(1,C);
for i = 1:C
  pC(i) = length(find(Y == i));
end
pC = pC/num;

n_samples = nb + np;
sum_yt = zeros(C,numt); % test output
store_prb = zeros(C,numt);

if hist_collect == 1
  hist_ar = zeros(C,numt); % save the DT outputs accepted
end
Lik = zeros(1,n_samples);
Ts = zeros(1,n_samples);
ac = zeros(4,2);

% Initialize a DT:

v1 = unidrnd(mvar);
if Cf(v1) == 0 % the nominal variable
  A = x_min(v1):1:x_max(v1);
  T = struct('p',1,'v',v1,'q',A(ceil(length(A)*rand)),...
  'c',[1 2],'t',[1 1]);
```

```
else % the discrete variable
  T = struct('p',1,'v',v1,'q',unifrnd(x_min(v1),...
x_max(v1)),'c',[1 2],'t',[1 1]);
end

P = mc_split([],T,T(1).p,1:num); % call the function

lik = mc_log_lik(P);% call the function

tlen = 1;
mod1 = 1;
is = nb;
i1 = 0;

while is < n_samples
  i1 = i1 + 1;

  % Make an available move:

  while 1
    [T1,m1] = mc_create_move(T); % call the function
    N1 = mc_split([],T1,T1(1).p,1:num); % call the function

    [N2,T2] = mc_shrink(N1,T1); % call the function

    len = size(T1,2) - size(T2,2);

    switch m1
    case 1
      if len == 0
        break;
      end
    case 2
      break;
    otherwise
      if len == 0
        break;
      end
    end
  end

  lik1 = mc_log_lik(N2); % call the function

  R = mc_prop_ratio(T2,m1); % call the function

  r = exp(lik1 - lik)*R;

  if rand < r % then accept the DT
    T = T2;
    P = N2;
    lik = lik1;
    tlen = size(T2,2);
    mod1 = m1;
    accept = 1;
  else
    accept = 0;
  end
  if i1 <= nb % then burn-in phase
    Lik(i1) = lik;
```

```
      Ts(i1) = tlen;
      if accept == 1
         ac(mod1,1) = ac(mod1,1) + 1;
      end
      if mod(i1,print_sample) == 0
         fprintf('%6i %7.1f %3i\n',i1,lik,tlen)
      end
   elseif mod(i1,sample_rate) == 0 % then post burn-in phase
      is = is + 1;
      Lik(is) = lik;
      Ts(is) = tlen;
      if accept == 1
         ac(mod1,2) = ac(mod1,2) + 1;
      end
      Tp = mc_prob(size(P,1),P);% call the function
      sum_yt = mc_test(zeros(C,numt),T,T(1).p,1:numt);
      store_prb = store_prb + sum_yt;
      if hist_collect == 1
         [ym,yt] = max(sum_yt);
         for j = 1:numt
            hist_ar(yt(j),j) = hist_ar(yt(j),j) + 1;
         end
      end
      if dt_collect == 1
         is1 = is1 + 1;
         Dtp{is1,1} = T;
         Dtp{is1,2} = Tp;
      end
      if mod(is,print_sample) == 0
         fprintf('%6i %7.1f %3i\n',is - nb,lik,tlen)
      end
   end
end

mc_rpt; % call the script

Perf(ni,:) = [100-ter conf_cor uncert conf_incor, mean(Ts) std(Ts)
ac_lev(2)];
save(fnam,'Perf');


%
% prop_ratio.m
%
S = zeros(1,node_max);
S(1) = 1;
for n = 2:node_max
   if mod(n,2) == 0
      S(n) = 2*sum(S(1:n - 1));
   else
      S(n) = 2*sum(S(1:n - 1)) - S((n - 1)/2);
   end
end
R2 = S(2:node_max)./S(1:node_max - 1);
R1 = S(1:node_max - 1)./S(2:node_max);
save prop_ratio R1 R2 node_max
```

```
%
% mc_load_data
%
function [C,mvar,X,Y,num,Xt,Yt,numt,Cf] = mc_load_data_cf(smod)
load flare_data_class X Y Xt Yt

% the UCI flare data: X, Xt are the training and test data,
respectively; Y,Yt are the target vectors for training and testing,
respectively

mvar = size(X,2);
num = length(Y);
numt = length(Yt);
C = max(Y); the number of classes

% Sort features x(i) as nominal (0) or discrete (1):

discr_max = 12; % let be a criterion for discrete or nominal
Cf = zeros(1,mvar);
for i = 1:mvar
  A = 1:num;
  discr = 0;
  while isempty(A) == 0
    A1 = find(X(A,i) == X(A(1),i));
    discr = discr + 1;
    A(A1) = [];
  end
  if discr > discr_max
    Cf(i) = 1;
  end
end


%
% mc_split
%
function P = mc_split(P,T,pos,A)
global C X Y
% Find the node N in a given position:
for i = 1:size(T,2)
  if T(i).p == pos
    N = T(i);
    break
  end
end

X1 = (X(A,N.v) >= N.q);
% The left branch:
A0 = find(X1 == 0);
if N.t(1) == 0 % a splitting node
  P = mc_split(P,T,N.c(1),A(A0));
else % a terminal node
  if isempty(A0) == 0
    for i = 1:C
      P(N.c(1),i) = sum(Y(A(A0)) == i);
    end
  else
    P(N.c(1),:) = zeros(1,C); % go ahead
  end
end
```

```
% The right branch:
A1 = find(X1 == 1);
if N.t(2) == 0
  P = mc_split(P,T,N.c(2),A(A1));
else
  if isempty(A1) == 0
    for i = 1:C
      P(N.c(2),i) = sum(Y(A(A1)) == i);
    end
  else
    P(N.c(2),:) = zeros(1,C);
  end
end


%
% mc_log_lik
%
function [lik] = mc_log_lik(N2)
global C
tlen = size(N2,1);
Lik = zeros(1,tlen);
for i = 1:tlen
  lik1 = sum(gammaln(N2(i,:) + 1));
  Lik(i) = lik1 - gammaln(sum(N2(i,:)) + 1);
end
lik = sum(Lik);   % log likelihood


%
% mc_create_move
%
function [T,m1] = mc_create_move(T)
global X mvar Pr K Cf q_nom x_max x_min
tlen = size(T,2);
r1 = rand;
if r1 < Pr(1)
  m1 = 1;
elseif r1 < Pr(2)
  if tlen > 1
    m1 = 2;
  else
    m1 = 4;
  end
elseif r1 < Pr(3)
  m1 = 3;
else
  m1 = 4;
end
switch m1
case 1 % make the birth move
t1 = tlen + 1; % the number of terminals
  T1 = zeros(3,t1);
  T2 = zeros(1,tlen); % a list of node positions
  k = 0;
  for i = 1:tlen
    T2(i) = T(i).p;
    va = find(T(i).t == 1);
```

```
    for j = va
      k = k + 1;
      T1(:,k) = [i; j; T(i).c(j)];
    end
  end
  t = unidrnd(t1);      % random choice of a terminal node
  vi = T1(1,t);% an index of the above node
  va = T1(2,t);
  t3 = T(vi).c(va); % keep a replaced terminal index
  t2 = 1;      % find a skipped/new split position
  while isempty(find(T2 == t2)) == 0
    t2 = t2 + 1;
  end
  T(vi).c(va) = t2; % change the parent node
  T(vi).t(va) = 0;
  v1 = unidrnd(mvar);
  q1 = unifrnd(x_min(v1),x_max(v1));  % a new rule
  t4 = 1;      % find a skipped/new terminal node position
  while isempty(find(T1(3,:) == t4)) == 0
    t4 = t4 + 1;
  end
  N = struct('p',t2,'v',v1,'q',q1,'c',[t3 t4],'t',[1 1]);
  T(t1) = N;
  return

case 2 % make the death move
  A = [];
  for i = 1:tlen
    if sum(T(i).t) == 2
      A = [A i];
    end
  end
  t = A(unidrnd(length(A)));
  c1 = min(T(t).c);
  p1 = T(t).p;
  T(t) = [];
  tlen = tlen - 1;
  for vi = 1:tlen
    ap = find(T(vi).c == p1 & T(vi).t == 0);
    if isempty(ap) == 0
      T(vi).c(ap) = c1; % connected to a terminal node c1
      T(vi).t(ap) = 1;
      break
    end
  end
  return

case 3 % make the change-question move
  ni = unidrnd(tlen);
  vnew = 1:mvar;
  vnew(T(ni).v) = [];
  v1 = vnew(unidrnd(mvar - 1));
  T(ni).v = v1;
  if Cf(v1) == 0 % nominal
    T(ni).q = round(unifrnd(x_min(v1),x_max(v1)));
  else
    T(ni).q = unifrnd(x_min(v1),x_max(v1));
  end
  return
```

```
case 4 % make the change-rule move
  ni = unidrnd(tlen);
  v1 = T(ni).v;
  if Cf(v1) == 0  % nominal
    if rand > 0.5
      T(ni).q = T(ni).q + ceil(q_nom*rand);
    else
      T(ni).q = T(ni).q - ceil(q_nom*rand);
    end
  else % discrete
    T(ni).q = T(ni).q + K(v1)*randn;
  end
  return
end


%
% mc_shrink
%
function [N,T] = mc_shrink(N1,T1)
global num pmin
tree_len = size(T1,2);
tree_len1 = tree_len;
do_more = 1;
while do_more == 1
  del_split = zeros(tree_len1,1);
  T2 = T1;
  for i = 1:tree_len1
    No = T1(i);
    if sum(No.t) == 2
      t = [No.c(1) No.c(2)];
      ts = [sum(N1(t(1),:)) sum(N1(t(2),:))];
      if isempty(find(ts < pmin)) == 0
        del_split(i) = 1;
        p1 = No.p;
        cm = min(No.c);
        for j = 1:tree_len
          za = find(T1(j).t == 0 & T1(j).c == p1);
          if isempty(za) == 0
            T2(j).t(za) = 1;
            T2(j).c(za) = cm;
            break
          end
        end
      end
    end
  end
  A0 = find(del_split == 0);
  if isempty(A0) == 1
    A0 = 1;
  end
  T1 = T2(A0);
  tree_len1 = size(T1,2);
  if tree_len1 < tree_len
    tree_len = tree_len1;
    N1 = mc_split([],T1,T1(1).p,1:num); % call the function
  else
    do_more = 0;
```

```
      end
end

do_more = 1;
while do_more == 1
   del_split = zeros(tree_len1,1);
   T2 = T1;
   for i = 1:tree_len1
      No = T1(i);
      si = find(No.t == 0);
      if length(si) == 1
         t = No.c(3 - si);
         if sum(N1(t,:)) < pmin
            del_split(i) = 1;
            p1 = No.c(si);
            for j = 1:tree_len
               za = find(T1(j).t == 0 & T1(j).c == No.p);
               if isempty(za) == 0
                  T2(j).t(za) = 0;
                  T2(j).c(za) = p1;
                  break
               end
            end
            break
         end
      end
   end
   A0 = find(del_split == 0);
   if isempty(A0) == 1
      A0 = 1;
   end
   T1 = T2(A0);
   tree_len1 = size(T1,2);
   if tree_len1 < tree_len
      tree_len = tree_len1;
      N1 = mc_split([],T1,T1(1).p,1:num); % call the function
   else
      do_more = 0;
   end
end
N = N1;
T = T1;


%
% mc_prop_ratio
%
function R = mc_prop_ratio(T,m1)
global R1 R2 Pp node_max
lenT = size(T,2);
Dq = 0;
for i = 1:lenT
   if sum(T(i).t) == 2
      Dq = Dq + 1;
   end
end
if lenT > node_max
   lenT = node_max;
end
```

```
switch m1
case 1
  R = (lenT + 1)/(Dq + 1)*R1(lenT)*Pp(1);
case 2
  R = Dq/lenT*R2(lenT)*Pp(2);
otherwise
  R = 1;
end



%
% mc_prob
%
function Tp = mc_prob(tlen,P)
global pC C
Tp = repmat(pC,[tlen 1]);
Tn1 = sum(P')';
A1 = find(Tn1 > 0);
for i = 1:C
  Tp(A1,i) = P(A1,i)./Tn1(A1);
end



%
% mc_test
%
function P = mc_test(P,T,pos,A)
global Xt Tp C
% Find the node N for a given position:
for i = 1:size(T,2)
  if T(i).p == pos
    N = T(i);
    break
  end
end

X1 = (Xt(A,N.v) >= N.q);
% The left branch:
A0 = find(X1 == 0);
if N.t(1) == 0 % a splitting node
  P = mc_test(P,T,N.c(1),A(A0));
elseif isempty(A0) == 0   % a terminal node
  P(:,A(A0)) = repmat(Tp(N.c(1),:)',1,length(A(A0)));
end
% The right branch:
A1 = find(X1 == 1);
if N.t(2) == 0
  P = mc_test(P,T,N.c(2),A(A1));
elseif isempty(A1) == 0
  P(:,A(A1)) = repmat(Tp(N.c(2),:)',1,length(A(A1)));
end
return



%
% mc_rpt
%
ac_lev = [sum(ac(:,1))/nb sum(ac(:,2))/np];
```

```
fprintf('Data "%s" %3i/%5i/%5i/%2i/%4.2f\n',...
fnam,mvar,[nb np]/1000,sample_rate,q_sig)
fprintf(' pmin = %i,',pmin)
fprintf(' accept: %5.3f %5.3f\n',ac_lev)
end1 = size(Ts,2);
dt_mean  = mean(Ts(nb+1:end1));
dt_sigma = std(Ts(nb+1:end1));
fprintf('DT mean and sigma: %5.2f %5.2f\n',dt_mean,dt_sigma)
[dum,mp] = max(store_prb);
ter = 100*mean(mp' ~= Yt);
fprintf('Test error = %5.2f%%\n\n',ter)

subplot(3,2,1)
plot(Lik(1:nb)), grid on
xlabel('Samples of burn-in'), ylabel('Log likelihood')
subplot(3,2,3)
plot(Ts(1:nb)), grid on
ylabel('DT nodes'), xlabel('Samples of burn-in')
subplot(3,2,5)
[a,b] = hist(Ts(1:nb),1:max(Ts(1:nb)));
bar(b,a/nb);
grid on
xlabel('DT nodes'), ylabel('Prob')
subplot(3,2,2)
plot(Lik(nb+1:end1)), grid on
xlabel('Samples of post burn-in'), ylabel('Log likelihood')
subplot(3,2,4)
plot(Ts(nb+1:end1)), grid on
ylabel('DT nodes'), xlabel('Samples of post burn-in')
subplot(3,2,6)
[a,b] = hist(Ts(nb+1:end1),1:max(Ts(nb+1:end1)));
bar(b,a/np);
grid on
xlabel('DT nodes'), ylabel('Prob')

if hist_collect == 1
  fprintf('1: %5.2f%% %5.2f%%\n',100 - ter,ter)
  hist_ar_norm = hist_ar/(np*rs);
  Yq = zeros(numt,1);
  for c1 = 1:C
    A = find(hist_ar_norm(c1,:) >= consist_lev);
    Yq(A) = c1;
  end
  conf_cor = 100*sum(Yq == Yt)/numt;
  A = find(Yq > 0);
  conf_incor = 100*sum(Yq(A) ~= Yt(A))/numt;
  uncert = 100*sum(Yq == 0)/numt;
  fprintf('2:%5.2f%%%5.2f%%%5.2f%%\n',conf_cor,…uncert,conf_incor)
end
```

angular templates. The mixture coefficients thus depend on the frequency responses of the measuring instruments and, through the emission spectra, on the radiative properties of the source processes. The instrumental noise affecting the data is normally very strong. Sometimes, owing to the different observation times allowed for different points in the sky, the noise power also depends on the particular pixel. Given this information, the problem is to reconstruct the maps of the individual source processes. The source emission spectra are normally known more or less approximately. Assuming that they are perfectly known leads to the so-called *non-blind* separation procedures. Instead, letting them be estimated by the algorithm as additional unknowns leads to *blind* or *semi-blind* separation procedures, depending on the possible introduction of additional knowledge.

In this section, we present the model we assumed to account for our observed data, with the necessary approximations to be made. Blind and semi-blind approaches to separation will be treated in Sect. 4.5. As is always the case, the different approximated models significantly affect the results of the separation procedures. Some of the difficulties arising with these approximations will be stated here, whereas their significance will be detailed in the next two sections. In Sect. 4.4.2, we specify the data model we assumed to face the separation problem for the *Planck* mission. In Sect. 4.4.3, the separation problem is formally introduced. In Sect. 4.4.4, the possibility of finding a priori models for the source emission spectra and intensity distributions is envisaged, based on possibly known physical properties of some of the radiating sources. As will be shown elsewhere, this can lead to build semi-blind separation procedures. Finally, in Sect. 4.4.5, a statistical model for the instrumental noise affecting the *Planck* data is given. Some brief remarks conclude the section.

4.4.2 A Linear Instantaneous Mixture Model for Astrophysical Data

Let us suppose we have M radiometric sensors pointing at the sky through a telescope, with bearing (ξ, η), and working on M different frequency bands. Let ν be the frequency. As usual (Hobson et al. 1998; Baccigalupi et al. 2000), we assume that each radiation process $\tilde{s}_j(\xi, \eta, \nu)$ from the microwave sky can be represented by the product of a frequency-independent angular pattern $s_j(\xi, \eta)$ and a bearing-independent frequency spectrum $F_j(\nu)$:

$$\tilde{s}_j(\xi, \eta, \nu) = s_j(\xi, \eta) F_j(\nu) \qquad (4.35)$$

Assuming that N distinct astrophysical sources are observable at any fixed frequency, the total radiation received at that frequency is the sum of

N signals (*processes*, or *components*) of the type (4.35), where subscript j is a process index. Normally, the optical aperture of the telescope depends on frequency. This means that the maps formed over the different channels will have different angular resolution. In general, the telescope beam is a frequency-dependent, space-variant smearing kernel. Instead, if we assume that this is a convolutive kernel, the total signal observed by the ith sensor through the telescope at bearing (ξ, η) will be

$$\hat{x}_i(\xi,\eta) = \iiint h_i(\xi - \psi, \eta - \zeta) b_i(v) \sum_{j=1}^{N} s_j(\psi,\zeta) F_j(v) d\psi d\zeta dv \tag{4.36}$$

$$\tilde{x}_i(\xi,\eta) * h_i(\xi,\eta), \qquad i = 1,2,...M$$

where $h_i(\xi,\eta)$ is the telescope beam function (assumed constant over the passband of the ith instrument channel), $b_i(v)$ is the frequency response of the ith sensor, the asterisk denotes 2D linear convolution, and

$$\tilde{x}_i(\xi,\eta) = \sum_{j=1}^{N} A_{ij} s_j(\xi,\eta), \qquad i = 1,2,...M \tag{4.37}$$

with

$$A_{ij} = \int b_i(v) F_j(v) dv \qquad i = 1,2,...M; \quad j = 1,2,...N \tag{4.38}$$

If the source spectra are approximately constant within the passbands of the different channels then (4.38) can be rewritten as:

$$A_{ij} = F_j(v_i) \int b_i(v) dv \tag{4.39}$$

The coefficient A_{ij} is thus proportional to the spectrum of the jth source at the centre-frequency v_i of the ith channel.

Let us rewrite (4.36):

$$\hat{x}_i(\xi,\eta) = \sum_{j=1}^{N} A_{ij} s_j(\xi,\eta) * h_i(\xi,\eta) = \sum_{j=1}^{N} A_{ij} \bar{s}_{ji}(\xi,\eta) \tag{4.40}$$

The observed functions are thus given by combinations of the $M \cdot N$ "modified" source functions $\bar{s}_{ji}(\xi,\eta)$. Note that, if the convolutive kernels are equal over all the instrument channels, (4.40) can be rewritten as:

$$\hat{x}_i(\xi,\eta) = \sum_{j=1}^{N} A_{ij} s_j(\xi,\eta) * h(\xi,\eta) = \sum_{j=1}^{N} A_{ij} \hat{s}_j(\xi,\eta) \tag{4.41}$$

and, again, we have M data functions expressed as combinations of N source functions. In the case of channel-independent telescope beams, then, the observed data model is the linear instantaneous mixture model (4.41), where any pixel of a data map is only obtained by combining the values of the same pixel in the source maps. When the M kernels $h_i(\xi,\eta)$ are perfectly known, we can always give the same angular resolution to all the data maps (Salerno et al. 2000), thus obtaining a data model of the type (4.41) or (4.37), which are formally analogous. In this case, the total radiation measured over the ith channel is the quantity specified in (4.37) added with the instrumental noise.

$$x_i(\xi,\eta) = \tilde{x}_i(\xi,\eta) + n_i(\xi,\eta) = \sum_{j=1}^{N} A_{ij} s_j(\xi,\eta) + n_i(\xi,\eta), \qquad (4.42)$$

$$i = 1,2,...M$$

where $n_i(\xi,\eta)$ is the instrumental noise of the ith sensor at location (ξ,η). As we specify later, the noise power is generally different in different map locations. For any pixel considered, $A_{ij} s_j$ is the contribution of the jth source process to the ith channel measurement. In view of (4.39) and (4.42), the M sensed signals can be represented in vector form:

$$\mathbf{x}(\xi,\eta) = \mathbf{A}\mathbf{s}(\xi,\eta) + \mathbf{n}(\xi,\eta) \qquad (4.43)$$

where $\mathbf{x} = \{x_i, i = 1, ... M\}$ is the M-vector of the observations, i being a channel index, $\mathbf{A}$ is a location-independent $M \times N$ matrix, whose entries A_{ij} have the form (4.39), $\mathbf{s} = \{s_j, j = 1, ... N\}$ is the N-vector of the individual source processes and $\mathbf{n} = \{n_i, i = 1, ... M\}$ is the M-vector of the instrumental noise. Hereafter, matrix $\mathbf{A}$ will be referred to as the *mixing matrix*. The ith row of $\mathbf{A}$ contains the relative strengths of all the source maps on the ith channel. The jth column of $\mathbf{A}$ is proportional to the frequency emission spectrum of the jth source process.

4.4.3 The Source Separation Problem

Let us assume to have a set $\mathbf{x}(\xi,\eta)$ of radioastronomical data maps that can be thought of as generated following the model (4.43) or (4.40). Our astrophysical interest is to derive the set $\mathbf{s}(\xi,\eta)$ of the individual source maps. This is the formal definition of the astrophysical source separation problem. The ways to address this problem differ from one another depending on the information we have available. Besides the data set $\mathbf{x}$, we normally have some information on the statistical noise distribution and some knowledge on the mixing matrix. Also, the telescope beams $h_i(\xi,\eta)$ are normally known. Note from (4.39) that the elements of the mixing matrix are proportional to the source spectra through coefficients that only depend on known instrumental frequency responses. The source spectra, however, are not known for many astrophysical processes. This means that in most cases the mixing matrix is only approximately known.

One approach to separation is to consider **A** to be perfectly known. This is called the *non-blind* approach. Something about this strategy can be found in Chap. 5, but we do not treat it in detail, since we are only interested in the solution of the separation problem when the mixing matrix cannot be considered known. If it is assumed to be totally unknown, the problem is to obtain **A** and $\mathbf{s}(\xi, \eta)$ from the observations $\mathbf{x}(\xi, \eta)$ alone. This is called a *blind source separation* problem (Haykin 2000; Cichocki and Amari 2002), and can be solved by several, mainly statistical, approaches. Note that, when **A** is totally unknown, any scaling applied to the source maps still gives a valid solution, provided that a reciprocal scaling is applied to the related column of **A**. This is called the *scaling ambiguity* of the blind separation problem. When a partial knowledge on the mixing matrix can be exploited, the solution strategy is denoted as *semi-blind*. A state of the art in blind and semi-blind separation is outlined in Sect. 4.5. Since statistical processing always takes advantage from the availability of probabilistic models, in this chapter we specify some of the possibilities to introduce prior knowledge in our problem in the form of probabilistic models for the source processes and the instrumental noise.

As is seen, several approximations have been used to reach the linear instantaneous formulation (4.43). All such "model errors" can affect the performances of the different separation strategies.

4.4.4 Source Models Parametrizing the Mixing Matrix

As already said, (4.39) is related to known instrumental features and to the emission spectra of the individual source processes. These are not always known, but we often have some knowledge in the form of parametric relationships. In the microwave and millimeter-wave ranges, the dominant radiations are the CMB, the galactic dust, the free–free emission and the synchrotron emission (see De Zotti et al. (1999), Tegmark et al. (2000)). Another significant signal comes from the extragalactic radio sources.

The emission spectrum of the CMB is perfectly known, being a blackbody radiation. In terms of antenna temperature, it is:

$$F_{CMB}(\nu) = \frac{\tilde{\nu}^2 \exp(\tilde{\nu})}{\left[\exp(\tilde{\nu}) - 1\right]^2} \tag{4.44}$$

where $\tilde{\nu}$ is the frequency in GHz divided by 56.8. Thus, from (4.39) and (4.44), it can be seen that the column of **A** related to CMB is known.

For the thermal emission of the galactic dust, we have

$$F_{dust}(v) \propto \frac{\overline{v}^{\beta_{dust}+1}}{\exp(\overline{v})-1} \tag{4.45}$$

where $\overline{v} = hv/kT_{dust}$, and where h is the Planck constant, k is the Boltzmann constant and T_{dust} is the physical dust temperature. If we assume this temperature constant, the frequency law (4.45), that is, the column of $\mathbf{A}$ related to dust emission, only depends on a scale factor and the spectral index β_{dust}:

The free–free emission spectrum can be considered known, following the law

$$F_{ff}(v) \propto (v/10)^{-2}(v/40)^{-0.14} \tag{4.46}$$

For the synchrotron radiation, we have the following power law:

$$F_{syn}(v) \propto v^{-\beta_{syn}} \tag{4.47}$$

Thus, the column of $\mathbf{A}$ related to synchrotron only depends on a scale factor and the spectral index β_{syn}.

From the formulas earlier, we derive that the columns of $\mathbf{A}$ related to CMB and free–free are perfectly known, and that each element of the other columns has the form:

$$A_{ij} = c_j g_j(v_i; \beta_j) \tag{4.48}$$

where c_j is independent of frequency, g_j is a known function of frequency and of the spectral index β_j, and v_i is the centre frequency of the ith measurement channel. Bearing in mind that the blind source separation problem has a solution up to scaling, we can assume a modified matrix whose generic element is

$$\tilde{A}_{ij} = \frac{A_{ij}}{A_{1j}} = \frac{g_j(v_i; \beta_j)}{g_j(v_1; \beta_j)} \tag{4.49}$$

Note that the elements in the first row of the modified matrix are equal to 1, provided that, with no loss of generality, all the elements of the original first row are non-zero. The properties of matrix $\mathbf{A}$ (and of this scaled version of it) are explicitly exploited in some of the approaches we investigated for the astrophysical separation problem.

As far as the extragalactic radio sources are concerned, in most of the experimentation reported in Chap. 5, we assumed that these have been removed from the data by one of the specific techniques proposed in the literature (Tenorio et al. 1999; Cayón et al. 2000; Vielva et al. 2001). In fact, these techniques cannot remove totally the point sources, but they do remove the brightest ones (the most important for us, since they affect more than others the study of the CMB, see Tegmark et al. (2000)).

A Mixture-of-Gaussians Probabilistic Source Model

Additional information on the source signals can derive from their intensity distributions. As we will see in the forthcoming sections, blind separation approaches mostly neglect the source distributions in estimating the mixing matrix. However, there is a specific approach (the independent factor analysis) that tries to make a parametric estimation of the source distributions, proposing a Mixture-of-Gaussians (MoG) model for them. To check the opportunity of adopting such a model in our case, we tried to derive the actual probability densities from source maps that are as realistic as possible (see Baccigalupi et al. (2000)). In Fig. 4.14, we report synthetic maps of CMB, galactic dust and synchrotron on a square sky patch with a 15° aperture, together with the related histograms. The standard inflationary cosmological theory tells us that the CMB anisotropies are Gaussian distributed. The galactic dust emission is expected to be significant in the high-frequency region of our measurement band. The related antenna temperature values were extrapolated from available measurements at frequencies outside our range. It is clear from the histogram that the amplitude distribution of galactic dust is non-gaussian. The curve actually is multimodal and asymmetric. The galactic synchro-tron radiation has a negative-power-law spectrum, and its influence on the total measured field is significant in the low-frequency measurement channels. The map showed has been extrapolated from already available observations. We looked into the amplitude distributions of both dust and synchrotron, to see whether the MoG model can fit them through a reasonable number of parameters. The MoG parametric model to fit a generic density $p_s(s)$ is the following:

$$p_s(s) = \sum_i w_i \frac{1}{\sqrt{2\pi}\sigma_i} \exp(-\frac{(s-\mu_i)^2}{2\sigma_i^2}) \qquad (4.50)$$

where the parameters to be estimated are the weights w_i, whose sum over i must be equal to one, the mean values μ_i, and the standard deviations σ_i. For estimating these parameters, we used an expectation-maximization (EM) algorithm (Dempster et al. 1977). The results are shown in Fig. 4.14, plotted with the dashed curves. As can be seen, the fits are very close to the histograms. It is interesting to note that we used only three Gaussian components in the mixture for the dust, and four for the synchrotron, which is far less than we expected. We repeated our experiments on 15 different sky patches of the same size, and in all cases we have seen that the Gaussian mixture model provides very good fits using less than five

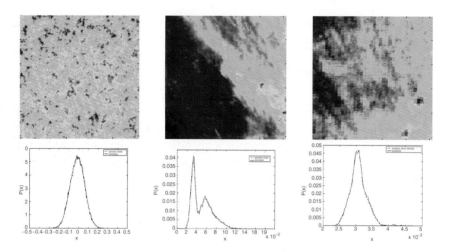

Fig. 4.14. CMB anisotropy, galactic dust and synchrotron radiation maps on a $15° \times 15°$ sky patch, with related histograms (*solid curves*) and Gaussian mixture model fits (*dashed curves*)

components (Kuruoglu et al. 2003). However, since the dust and synchrotron emissions only come from our galaxy, their distribution vary sensibly with the galactic latitude, and the mixture parameters are different for each different sky patch.

4.4.5 Noise Distribution

Many separation algorithms do not consider noise in their data models, thus the noise realization **n** does not even appear in (4.43). Some approaches assume white Gaussian noise with known, uniform, covariance matrix. In some cases, the elements of the covariance matrix can be estimated along with the other unknowns. White Gaussian noise is a reasonable assumption, but the noise variance can depend on the particular location in the sky. In our problem, indeed, the noise is Gaussian but space-varying, since the value of each pixel in the data maps is obtained by integrating the related measurements over the observation time, and the scanning strategy of the sensors does not guarantee the same observation time for all the pixels. Of course, the longer is the integration time the smaller is the noise in the measurement. The noise variance is thus different in different sky regions. In any case, once the scanning strategy has been fixed, the observation time per pixel is known very well. Thus, given the instrument sensitivity, the noise variance is also known. In Fig. 4.15, we show a typical noise map and the related standard deviation map. The pattern due to non-uniform scanning is clearly appreciable.

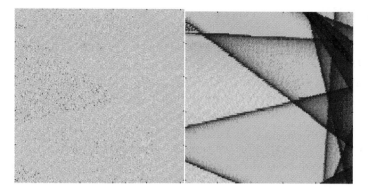

Fig. 4.15. Instrumental noise from uneven sky coverage. Left: typical realization. Right: space-varying RMS map

Some of the separation methods that we will introduce in the next chapter are able to take noise variability into account; some others assume uniform noise in the data model but show some robustness against moderate non-stationarity.

4.4.6 Conclusion

In this section, we introduced the problem of source separation applied to the detection of the different diffuse astrophysical sources in multichannel radioastronomic survey maps. From a generic mixture model, we have shown the approximations needed to build a noisy linear instantaneous data model that has proved to be useful for our application. The difference between the blind and non-blind approaches has been introduced, and some suggestions have been given to exploit available knowledge on both the sources and the noise.

4.5 Blind and Semi-blind Source Separation

Emanuele Salerno, Luigi Bedini, Ercan E. Kuruoglu and Anna Tonazzini

Abstract

After a brief state-of-the-art description, we introduce the fundamental concepts of independent component analysis, which is the reference technique we used to solve the blind- or semi-blind source separation problem for astrophysical images. The specific aspects of this problem are then presented, and the results we obtained with some specific techniques are described.

4.5.1 Introduction

In this section, we present a very brief state of the art in blind and semi-blind source separation of astrophysical maps. The basic concepts of the independent component analysis approach to blind source separation for linear instantaneous mixtures are introduced in general, with reference to the most recent literature. Then, a number of issues that are specific to the astrophysical application are reviewed, pointing out their peculiarities with respect to the general case, with the possibilities of exploiting semi-blind approaches and of generalizing the data model and the separation approach, thus permitting to override the strict assumptions of independent component analysis.

After the first presentation, we report our experience with blind and semi-blind astrophysical source separation methods. We briefly introduce our attempts with totally blind approaches, to treat more diffusely our strategies to avoid the related drawbacks. In particular, we developed an IFA strategy to include non-stationary noise and knowledge on the source priors in our model. This approach is further extended by a Bayesian approach based on a Markov random field map model that allows spatial autocorrelation to be accounted for. The introduction of significant cross-correlations between different sources led us to move away from the strict ICA strategy. We developed a method that allows us to neglect higher-order statistics by relying on the spatial autocorrelation, and to include non-stationary noise in the model. This semi-blind method enables us to estimate the mixing matrix, the source priors and the unknown source cross-correlations. Finally, the most complete data model is included in our implementation of particle filtering for source separation. This model can include non-stationary sources and mixing operator, non-stationary noise, non-instantaneous and even non-linear mixing.

A first approach to the problem of astrophysical source separation could be to assume the mixing coefficients in (4.37) to be perfectly known. The problem can thus be solved by a number of linear systems, in a spatial or a transformed domain. This approach has been followed by Hobson et al. (1998) and Stolyarov et al. (2002, 2005), who used the maximum entropy criterion to regularize the linear systems to be solved, also assuming the source power spectra to be known, and by Bouchet et al. (1999), who adopted the multi-frequency Wiener filter separation strategy, assuming Gaussian distributions for the source signals. We denote this type of approaches, which we do not treat in detail here, as *non-blind* methods. As in (4.38), the mixing coefficients depend on the emission spectra of the individual sources at the working frequencies adopted, and these are not known for all the sources. This means that assuming known mixing coefficients is rather unrealistic and could over constrain the problem, thus leading to unphysical solutions. This drawback has been recognized, and attempts are being made to relax the strict assumption of a perfectly known data model (Barreiro et al. 2004).

A totally different strategy to deal with this problem is to consider the mixing coefficients as unknown or only partially known (blind source separation, BSS). Different strategies can be adopted to address this problem, mostly relying on the independent component analysis (ICA) principle (Comon 1994). The application of the classical ICA approach only requires mutual independence and non-gaussianity of the sources. The separation approaches that only rely on this information are referred to as *totally blind*. Conversely, the approaches that are able to take other pieces of information into account are denoted as *semi-blind* approaches. Since we became involved in the astrophysical source separation problem (Baccigalupi et al. 2000), we adopted blind or semi-blind approaches, and obtained promising results with different strategies. In this section, we summarize the main literature in blind and semi-blind separation, with particular reference to the astrophysical application.

We first proposed an ICA approach for the problem of totally blind astrophysical source separation (Baccigalupi et al. 2000) relying on the supposed statistical independence between the CMB and the foreground radiations. We also assumed that the foreground maps, in their turn, can be considered approximately independent of each other. In this way, we were able to try ICA strategies to separate the sources from the observed maps. The scale and permutation indeterminacies typical of ICA are not critical to our application, since other physical constraints can help recovering the proper magnitudes of the signals at the relevant frequencies (Maino et al. 2002).

To avoid the possible drawbacks of ICA, we tried to specialize our separation techniques in order to include the different pieces of prior information we have in this particular problem.

Each approach is based on a different set of assumptions. Herein, we summarize and comment our experience in astrophysical source separation algorithms, with particular reference to the data that will be made available by the *Planck* mission. Some of the algorithms we experimented have been taken from the existing literature and adapted to our specific problem. Some others are somewhat novel approaches to source separation that, besides astrophysics, could also be applied to other imaging problems.

This section is organized as follows. We first present the general ICA concepts, as developed during the last decade, and then take into account the non-idealities that characterize the astrophysical application, from both the points of view of problem-specific knowledge and of the justification of some of the approximations made to apply ICA procedures to astrophysical maps. We then summarize our first proposal to apply totally blind separation techniques to CMB data. This proposal has triggered a certain interest in both information processing and astrophysics researchers, thus starting a new stream of papers in the international literature. In successive sections, we report our proposals to overcome the drawbacks related to the poor flexibility of ICA with respect to prior information and noise treatment. Some of the results are also shown and compared to the ICA results. The Markov random-field approach, moreover, is capable of accounting for the possible spatial autocorrelation of the source maps. Section 4.5.5 outlines our proposal to treat the problem of non-negligible cross-correlations between different sources, which, strictly speaking, cannot be addressed by ICA approaches. Also in this case, some results are shown, and the advantages and drawbacks of this very efficient technique are highlighted. Another section describes the application of particle filtering to astrophysical source separation. This is the most general approach we devised so far, and is able to treat many aspects of non-stationarity in the problem, as well as possible non-linearities. The two concluding sections contain our views on future trends in astrophysical source separation and some remarks on the results obtained.

4.5.2 Independent Component Analysis Concepts

Problems of blind source separation first arose in audio processing and in telecommunications. A typical example is the so-called *cocktail party problem*, where the individual voices or different sounds in a room must be separated from the mixed signals sensed at a number of different locations, without knowing the features of the propagation paths. BSS thus consists in estimating both the individual source signals and the mixing operator (a constant mixing matrix in the linear instantaneous case) from the observed mixed signals. Strictly speaking, this problem has two aspects that can be treated both separately and simultaneously: the first is estimation of the mixing matrix $\mathbf{A}$ (*model learning* or *system identification*); the second is estimation of the source set $\mathbf{s}$ (*source separation* or *estimation*). One classical approach to BSS (Comon 1994; Bell and Sejnowski 1995; Pope and Bogner 1996; Amari and Cichocki 1998) is based on the ICA concept. Assuming that all the source processes are statistically independent, ICA techniques aim at transforming the observed data $\mathbf{x} = \mathbf{As}$ by means of a linear transformation $\mathbf{W}$ so as to obtain an output vector $\mathbf{y} = \mathbf{Wx}$ whose elements are mutually independent. The ICA principle states that, if the independent sources (except at most one) are non-gaussian and the output signals y_i are mutually independent, then the latter reproduce the original sources up to scaling and permutation. The extra condition of non-gaussianity is essential to avoid ambiguous solutions in the case of stationary processes. If the source processes are non-stationary or auto-correlated, this condition can be overlooked (see for example Belouchrani et al. (1997)).

A matrix $\mathbf{W}$ that produces independent outputs can be estimated by several strategies. Perhaps the most immediate approach relies on the evaluation of higher-order cumulants (Comon 1994). Indeed, ICA can be seen as an extension of the classical principal component analysis (or PCA: for an exhaustive overview, see Cichocki and Amari (2002)), which obtains uncorrelated signals through the diagonalization of the data covariance matrix, namely, by only relying on second-order statistics. The ICA solution is a higher-order refinement to the PCA solution, as pointed out by De Lathauwer (1997), who includes ICA in the framework of BSS by tensor algebra strategies.

Enforcing independence by optimizing suitable penalty functions is conceptually different from constraining the higher-order statistics. However, it can be proved that this approach implicitly sets to zero all the cross-central-moments of orders greater than one. The minimum mutual information approach (Bell and Sejnowski 1995) gets independent outputs by minimizing the mutual information between them, finding maximum likelihood solution if the actual source priors are available. The optimization is accomplished by a feed-forward neural network implementing a stochastic gradient descent. A similar solution is proposed by Amari and Cichocki (1998), with the difference that the convergence of their uniform gradient algorithm do not depend on the conditioning of the mixing operator. The infomax method (Lee et al. 1999) exploits entropy as a measure of independence, whilst other methods rely on the minimization of higher-order-statistics contrast functions (Cardoso 1998, 1999). The strict relationships among the various methods have been pointed out by Amari and Cichocki (1998), among others. Knuth (1998), Lee and Press (1998) and Mohammad-Djafari (2001) propose Bayesian estimation as a unifying framework for BSS, within which all the proposed methods can be viewed as special cases. All these approaches have shown good performances in those applications where the preliminary assumptions are justified. As far as computational complexity issues are concerned, a very efficient algorithm has been proposed, called *FastICA* and relying on minimizing the negentropy function (Hyvärinen and Oja 1997).

4.5.3 Application-Specific Issues

Independence

The basic assumptions of ICA for stationary signals are the mutual independence between the elements of the source vector and the non-gaussianity of all but at most one of them. Indeed, the first hypothesis we made to apply stationary ICA to astrophysical maps was that all the sources are mutually independent. As far as non-gaussianity is concerned, we have seen that the only source expected to be Gaussian is the CMB.

Physically, we can justify the independence of CMB from all the fore-grounds, but not the independence between different foregrounds. Nevertheless, our simulated experiments with ICA have given satisfactory results, also showing some robustness against non-independence. Of course, when the cross-correlations between source pairs are large, the ICA separation strategies fail, since this fundamental deviation from the ideal data model prevents the ICA principle from being successfully exploited. In these cases, the problem should be faced by relying on dependent component analysis (Barros 2000; Bach and Jordan 2002).

Since significant cross-correlations are expected between the foreground maps, we were led to consider dependent component analysis to solve our problem. In treating this aspect, we were also able to exploit additional information, when available. Our algorithms are capable of estimating the source power spectra and statistical distributions, by taking into account possible information on the mixing parameters and on the structure of the source covariance matrices (Bedini et al. 2003, 2005).

Spatial and Temporal ICA

In a general BSS problem, we have a vector data set originated from a temporal sequence of sensed spatial patterns or a multispectral acquisition of a spatially distributed radiation (the latter is the case in our astrophysical problem). This data set is assumed to be the superposition of a number of underlying factors, or sources, characterized by specific statistical properties in both the temporal (frequential) and the spatial domains. Enforcing mutual source independence in the spatial domain leads to *spatial ICA*, which maximizes the independence between any pair of sources at fixed time instants leaving unconstrained the statistical properties of the related temporal (frequential) "sequences". Conversely, *temporal ICA* assumes independence in the time (frequency) domain, and does not put any constraint on the related spatial patterns. These two approaches are perfectly equivalent, as far as the data model is concerned. Also in this case, the problems arise when the independence assumptions are not perfectly verified in reality. This can imply that a perfect independence enforced in one domain induces unphysical solutions in the other domain. This difficulty can be overcome by the so-called *spatiotemporal ICA* (Stone et al. 2002), which finds a tradeoff between the requirements introduced in both the temporal and the spatial domains. This aspect is not relevant to our separation problem, since we are able to make a deterministic assumption on our frequential sequences (see (4.35)), thus adopting a pure spatial separation approach. Our assumption, however, comes from more or less rough approximations, thus, although we put no emphasis on the distinction between spatial and temporal ICA, a future refinement of the separation methods should also take this aspect into account.

Noise

Although noise is ubiquitous, most ICA algorithms have been developed for noiseless data (Cardoso 1999; Hyvärinen and Oja 2000). This does not mean that these algorithms cannot be used in practice, since they all are somewhat robust against noise. For even moderate noise, however, some specific measure may be required. One particular formulation of *FastICA* includes additive Gaussian noise in the data model (Hyvärinen and Oja 1997; Hyvärinen 1999a, b). This approach introduces contrast functions based on Gaussian moments, which are not sensitive to Gaussian stationary noise. Moulines et al. (1997) and Attias (1999) proposed an independent factor analysis method (IFA), assuming Gaussian mixture source models. This approach can handle noise as well, since the data model explicitly accounts for stationary Gaussian noise, whose covariance matrix can also be estimated from the data.

To manage noisy mixtures, we first adopted the version of *FastICA* that is insensitive to a certain amount of uniform Gaussian noise (Maino et al. 2002; Baccigalupi et al. 2004). As mentioned, however, our noise can depend on location. We then implemented a particular version of IFA (Kuruoglu et al. 2003 and Sect. 4.5.5 later) that has been shown to be more robust against noise than *FastICA* and is capable to account for non-stationary noise. Moreover, the underlying data model includes an additional layer with respect to the ICA model, which allows us to both estimate the source marginals parametrically and introduce specific constraints in the relevant parameters. This makes the IFA approach more flexible than ICA, and not necessarily totally blind. The possibilities of ICA as a totally blind separation approach, indeed, are clearly an advantage in the cases where the only reliable information is non-gaussian and mutual source (possibly approximate) independence. On the other hand, poor flexibility is a disadvantage when additional knowledge is available. In our case, as shown in the last section, we do possess some information about the relationships among the mixing coefficients, the source and the noise marginals, and all this knowledge should be exploited.

Convolutive Mixtures

Another obstacle to practical applications is the instantaneousness of the model. Even assuming that the data are generated by linearly mixing the source signals, the mixtures are generally not instantaneous, and thus the mixing operator is not simply a matrix–vector product, since any element of the mixing matrix is actually a smearing kernel (convolutive, as in (4.36), for a time- or space-invariant model). As a consequence, the data value at any location does not only depend on the source functions at the same location, but on all the values in a neighbourhood. This feature is apparent when dealing with signals in the time domain (e.g. audio signals), where the spectral properties and the temporal delays of the individual channels are generally different even in a stationary environment (Haykin 1994, 2000; Torkkola 1996; Cichocki and Amari 2002). In this case, an additional ambiguity besides scaling and permutation is that the separation can be obtained up to a common convolution kernel. In the spatial domain, e.g. in the 2D case of registered multispectral images, each spectral component can be affected by a specific defocusing, thus implying a necessity of both separation and deconvolution. When the defocusing features are the same for all the channels, as in (4.41), the problem becomes again one of separation from an instantaneous mixture.

So far (Tonazzini and Gerace 2005), we dealt with the problem of simultaneous separation and deconvolution only for simplified settings (e.g. 2×2 mixtures and small-size, low-pass FIR kernels). Moreover, we assumed the knowledge of a specific autocorrelation model for each source. Our problem with *Planck* data will require further attention to this issue. In fact, our radiometric maps will be obtained by looking at the sky through a telescope, whose aperture is sensibly channel-dependent, especially at low frequencies. This means that, strictly speaking, the mixture model to be adopted should not be instantaneous, since the data vector elements at each pixel are influenced by different neighbourhoods in the source maps. The peculiarity of our problem, however, is that the telescope radiation patterns at the different measurement frequencies are known very well. Thus, if we assume that the mixing matrix is constant all over the sky, we can estimate it blindly after degrading the resolutions of the data maps as described in Salerno et al. (2000). The subsequent estimation of the individual sources can then be made by means of non-blind techniques exploiting the available resolutions as much as possible. A further approach can consist in performing both model learning and source separation in a transformed domain, as has been done for non-blind separation by Hobson et al. (1998) in the Fourier space, and by Stolyarov et al. (2002, 2005) in the spherical harmonic space.

Non-Stationary Data and Models

The possible non-stationarity of the data, usually caused by time- or location-dependent noise, signals, or mixing coefficients, cannot be treated by either ICA or IFA, both being based on sample statistics.

Non-stationarity is especially relevant in the astrophysical application when all-sky maps are treated. In our particular case, noise is a Gaussian white process with space-dependent covariance. This introduces a first level of non-stationarity in our data, which can be faced by performing the source separation over small sky patches, or adopting our version of IFA or the second-order approach described in the next chapter. Some of the foreground sources in our problem lie in our galaxy. This means that their statistical features are likely to be different at different galactic coordinates. At least, they are much stronger on the galactic plane than at high galactic latitudes. This is another source of non-stationarity. Furthermore, the mixing matrix can depend on the sky region. As an example, the frequency spectrum of the galactic dust thermal emission (4.45) depends on the physical dust temperature, which is not uniform in the sky. Again, these non-stationarities can be dealt with by applying the separation strategies over distinct sky portions of suitable sizes. As an alternative, the particle filtering approach described in the next chapter (Cichocki and Amari 2002; Costagli et al. 2004) can handle any kind of non-stationary data much better than ICA. The problem in our case is its high computational complexity.

Prior Knowledge

The ICA-based separation algorithms normally do not permit to exploit any additional knowledge besides independence and non-gaussianity. Some additional knowledge, however, is commonly available in real-world cases. Semi-blind algorithms able to exploit this information are thus needed. Any possible problem-specific knowledge should be taken into account case by case. Moreover, even if no specific knowledge is available, real-world signals normally show some kind of structure. In principle, being a step beyond IFA, our Bayesian approach (Tonazzini et al. 2003, 2006; Kuruoglu et al. 2004) allows us to exploit all the available information, including possible dependencies between different sources. Particularly suited for 2D problems, this approach has a great flexibility provided that all the information available can be formalized in a source prior, whether factorizable or not.

The least we can expect from a real signal is a certain amount of autocorrelation. For 1D signals, this amounts to assume a certain degree of smoothness in the sources. In 2D cases, this assumption is much stronger, since the two-dimensional structure of an image can be constrained in many ways, also accounting for the structure of a possible non-smooth subset. Other relevant knowledge often comes from the autocorrelation properties of the individual sources, and can be particularly useful for the robust separation of noisy mixtures, especially in 2D cases. When used as a constraint, autocorrelation is known to be a good regularizer to many ill-posed problems. Whilst most ICA algorithms do not account for possible correlations inside the individual sources, some techniques have been proposed that exploit this information (Tong et al. 1991). Our specific contribution is described in the Sect. 4.5.6, and relies on Markov-random-field source models (Tonazzini et al. 2003; Kuruoglu et al. 2004).

Other non-specific prior information could consist in partial or total knowledge of the individual source marginals. This can be taken into account by the IFA algorithms. Our IFA procedure allows us to constrain the statistical features of the noise and the sources, and also some of the mixing coefficients. Each of these extra constraints normally reduces the dimensionality of the search space, thus alleviating the computational complexity of our method, which, however, remains high.

When approaching particular problems, many other pieces of prior knowledge can be available, such as particular features of the noise variance maps, known correlations between different sources, specific constraints for the mixing operators. A class of problems that are unsolvable if no additional knowledge can be exploited arise when the data model is non-linear (Cichocki and Amari 2002). Indeed, in these cases, finding a solution implies assuming a specific non-linearity in the model, from which it is possible to efficiently find a solution for a specific application. Very few real-world phenomena actually follow linear laws, and approximate linear models are not always suitable to solve the related problems. A number of applications requiring separation of sources from non-linear mixtures justify a continued attention in this field. In principle, the data model underlying the particle filtering approach we proposed for astrophysical map separation can include non-linear mixtures.

4.5.4 Totally Blind Approaches

Solving (4.43) for **A** and **s** with no other information is an under-determined problem for many reasons, thus, some extra assumptions must be made. The ICA strategy assumes mutual independence and non-gaussianity of the source signals (see Hyvärinen et al. (2001)). This strategy was also our first attempt to astrophysical source separation, in spite of the poor flexibility to the inclusion of prior information and the noise sensitivity shown by most algorithms proposed in the literature. Indeed, we chose to explore the performance of this totally blind approach and neglected all the prior information we might have but mutual independence. Also, we tried to assess the robustness of ICA against noise in our specific application.

The ICA strategy to separation tries to obtain mutually independent signals by enforcing a factorizable joint probability density function.

Nothing more can be done if the actual marginal source distributions are not known: one can only choose different functions that are able to improve the results for different kinds of statistical distributions (sub-Gaussian or super-Gaussian, see also Bell and Sejnowski (1995)). In practice, a linear transformation is applied to the observed data, such that the output components are as independent as possible. To check independence, a suitable measure should be chosen. Let $\mathbf{u}$ be the output vector and $\mathbf{W}$ the matrix associated to the linear transformation $\mathbf{u} = \mathbf{Wx}$. The Kullback-Leibler divergence $R(\mathbf{W})$ between the actual probability density of $\mathbf{u}$ and a factorizable joint density can be used as an independence measure:

$$R(\mathbf{W}) = \int p_\mathbf{u}(\mathbf{u}) \log \frac{p_\mathbf{u}(\mathbf{u})}{q(\mathbf{u})} d\mathbf{u} \qquad (4.51)$$

where $p_\mathbf{u}(\mathbf{u})$ is the probability density of the output vector and $q(\mathbf{u})$ the product of N functions each depending on a single component of $\mathbf{u}$. If $q(\mathbf{u})$ were the true joint source density, the minimizer of $R(\mathbf{W})$ multiplied by $\mathbf{x}$ at each pixel would be the maximum likelihood estimate of the source vector $\mathbf{s}$. When the mixing matrix $\mathbf{A}$ is square and non-singular, matrix $\mathbf{W}$ is an estimate of its inverse.

The first blind technique we proposed to solve the separation problem in astrophysics (Baccigalupi et al. 2000) was based on this approach, and allowed model learning and source estimation to be performed simultaneously. The separating strategy we used is based on a feed-forward neural network whose synaptic weight matrix at convergence is $\mathbf{W}$ and the neuron activation functions are derived from $q(\mathbf{u})$. The online weight update rule (the *learning algorithm*) was derived from a uniform gradient search (Amari and Cichocki 1998). We applied this algorithm to maps in a $15° \times 15°$ sky region, artificially built by using simulated emissions from CMB, extragalactic radio sources, galactic synchrotron and thermal dust at 30, 44, 70 and 100 Ghz: the four lowest frequencies of the *Planck* sensors. Under idealized conditions (resolution higher than the one expected for Planck and noiseless data), promising results were obtained in terms of CMB estimation accuracy (1% level), recovery of extragalactic radio sources, and estimation of the frequency dependencies of the strongest components. The weaker galactic components (synchrotron and dust), instead, were recovered with a 10% rms error, and poor frequency scaling estimates.

Intuitively, but this has also been proved formally, independence can also be enforced by maximizing non-gaussianity. Indeed, by the central limit theorem, a sum of Gaussian random variables is "more Gaussian" than any of its components. Hence, if we manipulate a number of combinations of non-gaussian variables to maximize their non-gaussianity, we are able to get the original sources. This is the strategy that has been followed by Hyvärinen and Oja (1997) to develop their *FastICA* algorithm, which sets negentropy as a measure of non-gaussianity. Given a random vector **y**, the negentropy function, J, is defined as follows:

$$J(\mathbf{y}) = H(\mathbf{y}_{gauss}) - H(\mathbf{y}) \qquad (4.52)$$

where H is entropy, and $\mathbf{y}_{gauss}$ is a Gaussian vector whose covariance matrix is the same as the covariance matrix of **y**. The FastICA algorithm approximately maximizes negentropy by a very fast fixed point batch algorithm. Maximizing negentropy, or an approximation to it, has been proved to be equivalent to minimize the Kullback-Leibler divergence (4.51). A very nice information page on FastICA can be found in a Web page created by Aapo Hyvärinen (http://www.cs.helsinki.fi/u/ahyvarin/whatisica.shtml), containing tutorials, demonstrations, and a FastICA software package free for noncommercial use. A "noisy" version of FastICA has also been designed (Hyvärinen 1998, 1999b), which builds penalty functions that are insensitive to uniform additive Gaussian noise, thus being more robust to noisy data, provided that the noise covariance matrix is known. We investigated the performance of FastICA and noisy FastICA for astrophysical source separation (Maino ct al. 2002). The experimental conditions we established were more realistic than in the previous case. The simulated observations covered the whole sky, with angular resolution and mean instrumental noise levels at the same orders of the *Planck* specifications. Yet, the telescope beam widths were assumed frequency-independent, and the noise stationary. The components considered were CMB, galactic synchrotron, thermal dust and free–free emissions, and the observed vectors were drawn from different combinations of the *Planck* frequency channels ranging from 30 to 353 GHz. We observed a considerable improvement over the above online neural algorithm, in terms of both computational speed and accuracy. CMB and thermal emissions were accurately reconstructed, and synchrotron was also recovered on all scales where it has significant power. The frequency scalings were recovered with precision better than 1%. The robustness of noisy FastICA, however, was not found sufficient for the expected noise levels.

4.5.5 Independent Factor Analysis

Although an exact knowledge of the source densities has been demonstrated unnecessary for separation, the data model assumed by ICA lacks relevant information, which can be useful, for example, in very noisy cases. The possibility of estimating the source densities along with the mixing parameters is the main feature of the independent factor analysis (IFA) method (Moulines et al. 1997; Attias 1999). This approach permits to treat noisy mixtures properly and also to estimate the covariance matrix of stationary Gaussian noise. The source distributions are modelled as MoGs, as in (4.45), whose parameters are estimated jointly with the mixing matrix. This method is still totally blind, but, as we already stated, the MoG model is well suited for our particular application. Thus, in principle, IFA offers an attractive alternative to noisy FastICA. The learning and the separation steps are performed sequentially as in Attias (1999). In the learning step, the mixing matrix, the noise covariance and the MoG parameters are estimated via an EM algorithm. In the separation step, the sources are estimated by using the densities obtained in the first step, via maximum likelihood or MAP estimation. We modified the algorithm proposed by Attias to allow space-varying noise to be treated, and compared the performance of our version with the one of the EM learning proposed by Attias.

In the learning step, each source s_i is given a probability density in the form (4.50), where index i is replaced by index q_i, which is now the index of the generic Gaussian factor for the ith source. The corresponding parameters are denoted by μ_{i,q_i}, σ_{i,q_i} and w_{i,q_i} (that is, the mean, the standard deviation and the weight, respectively, of the q_ith Gaussian factor describing source s_i). Indices q_i run from 1 to the maximum allowed number of factors for the ith source, n_i, which, in general, is not the same for all the sources. To have a normalized probability density function for each source, the sum of the w_{i,q_i} over q_i must be equal to one. The joint distribution of all the sources is thus:

$$p(\mathbf{s} \mid \theta) = \prod_{i=1}^{N} p(s_i \mid \theta_i) = \sum_{\mathbf{q}} w_{\mathbf{q}} G(\mathbf{s} - \mu_{\mathbf{q}}, \mathbf{V}_{\mathbf{q}}) \tag{4.53}$$

where:

$$\theta_i = \{w_{i,q_i}, \mu_{i,q_i}, \sigma_{i,q_i}\}, \tag{4.54}$$

$$\mathbf{q} = (q_1, ..., q_N)$$

$$w_{\mathbf{q}} = \prod_{i=1}^{N} w_{i,q_i} = w_{1,q_1} \times ... \times w_{N,q_N}$$

$$\mu_{\mathbf{q}} = (\mu_{1,q_1}, ..., \mu_{N,q_N})$$

$$\mathbf{V}_{\mathbf{q}} = diag(\sigma_{1,q_1}, ..., \sigma_{N,q_N})$$

and G represents the N-variate Gaussian density with mean vector $\mu_{\mathbf{q}}$ and covariance matrix $\mathbf{V}_{\mathbf{q}}$.

The conditional probability of a particular source vector $\mathbf{s}$ given $\mathbf{q}$ is:

$$p(\mathbf{s}\,|\,\mathbf{q}) = G\,(\mathbf{s} - \mu_{\mathbf{q}}, \mathbf{V}_{\mathbf{q}}) \tag{4.55}$$

The observations are the simple linear mixtures of the sources summed with noise, as in (4.42).

The probability density function of the Gaussian, zero-mean noise is $p(\mathbf{n}) = G\,(\mathbf{n}, \Lambda)$. Denoting the collective parameters as $W = (\mathbf{A}, \Lambda, \theta)$, the resulting data probability density function is:

$$p(\mathbf{x}|W) = \int p(\mathbf{x}|\mathbf{s})p(\mathbf{s})d\mathbf{s} = \int G(\mathbf{x} - \mathbf{As}, \Lambda)\prod_{i=1}^{N} p(s_i|\theta_i)d\mathbf{s} \tag{4.56}$$

and, owing to the simple Gaussian factors, can be evaluated analytically for any parameter vector W. To estimate all the parameters in W, Attias (1999) minimizes the Kullback-Leibler divergence between the above density and the density $p(\mathbf{x})$ calculated from the available data maps. An EM optimization algorithm has several advantages in this case, since the relevant densities are linear combinations of Gaussians, and the hidden variables can be averaged out analytically in the E-step. This leads to explicit update rules for all the parameters, provided that the noise is stationary, that is, Λ does not depend on location. To model the astrophysical map separation problem more closely, we dropped the noise stationarity hypothesis and renounced to estimate the noise variances, assuming a known space-dependent noise covariance matrix. However, the EM approach showed several inconveniences. First, as is known, EM convergence is assured only locally, and thus we often obtained different results for different initial guesses. Secondly, prior information on source emission spectra, such as parametric laws of the type (4.40)–(4.42), is not easy to be taken into account.

We developed an algorithm based on simulated annealing (see Laarhoven and Aarts (1987)), where the source parameters are updated pixel by pixel to consider the different noise variances (Kuruoglu et al. 2003). The resulting computational complexity is sensibly increased, but our approach reaches the same solution for all the starting points, and permits to constrain the unknown parameters within known ranges. Moreover, mixing matrices specified parametrically as in (4.44) can be treated very easily, with the additional advantage of reducing the total number of unknowns.

The elapsed times of our first implementation are in any case very high. For this reason, our simulated experiments only considered mixtures of two sources (CMB and galactic dust, or CMB and synchrotron). The noise was assumed space-variant, with an RMS map of the type shown in Fig. 4.15, and the coefficients of the actual mixing matrix $\mathbf{A}_o$ were derived from typical values expected for the spectral indices β_{syn} and β_{dust}. With the EM algorithm, when all the parameters are assumed unknown, the mixing matrices were not estimated correctly. By fixing all the source parameters to the values found for our simulated maps and only leaving the mixing matrix unknown, the result of the EM learning algorithm was considered satisfactory for non-stationary noise, with an average SNR of 30 dB, provided that a good initial estimate had been chosen. Figure 4.16 shows the result of the ML separation of CMB and dust, on the basis of the mixing matrix learned and the source distributions assumed. However, as was predictable, the performance deteriorated when the SNR decreased. This suggested us that EM learning is not efficient enough for the IFA estimation when the source distributions are left unknown.

Let us suppose now to have a one-parameter emission spectrum for each foreground source, as in (4.45) and (4.47). By exploiting this information, we can rewrite the mixing matrix in the form (4.49); if we have N sources and M measurement channels, the number of unknowns that specify $\mathbf{A}$ are now reduced from NM to just $N-2$, considering that the columns related to CMB and free–free are perfectly known from (4.44) and (4.46). In the case considered earlier, for example, a single spectral index is sufficient to specify the entire 2×2 mixing matrix. This formulation cannot be introduced

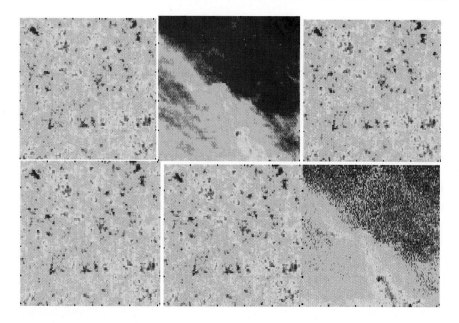

Fig. 4.16. ML-IFA estimation of CMB and dust radiations from their mixtures. Average SNR = 30dB. Top: original source maps; middle: mixture maps; bottom: source estimates

the EM scheme proposed by Attias. Simulated annealing, instead, only needs that the objective function can be calculated for any parameter vector W. For this purpose, there is no difference between the cases where W is specified by the spectral indices or by the mixing coefficients. Simulated annealing is thus naturally suited to solve the learning problem formulated in this way. Our simulations were performed in the same conditions as for the EM case, but with randomly chosen starting points. When all the model parameters were considered unknown, we did not obtain good results, with either stationary or non-stationary noise. Instead, experiments with fixed model parameters showed a superior performance if compared with the EM experiments, even with significantly lower SNRs. For example, the results of SA learning in the same case described earlier, but with an average SNR of about 14 dB, were comparable to the ones obtained by EM. In particular, the CMB coefficients were always estimated very well, whereas the dust coefficients were affected by a significant error, due to both the much higher noise power and the very weak dust emission for the frequencies and the sky region considered.

With the mixing matrix expressed as in (4.49) and all the other parameters left unknown, simulated annealing always converged to the correct mixing matrix, source means and weight coefficients. The variances of the source components, especially when associated to small weights, however, were not estimated so well. As an example, compare the two results shown in Fig. 4.17 for the synchrotron map. These are obtained from the same mixtures, with different noise realizations. The plot shown in the top panel represents a case where all the MoG parameters were estimated satisfactorily. The bottom panel, instead, shows a case where the variances of the Gaussian factors are not estimated correctly. This bad estimation, however, was shown not to have a great influence on separation: in Fig. 4.18, top row, we show the original CMB and synchrotron maps. These were mixed and added with the usual non-stationary noise realization, to give the observed maps shown in the second row. The sources estimated using the badly estimated parameters are shown in the third row. For comparison, we also show (bottom row) the source estimates obtained by *FastICA*.

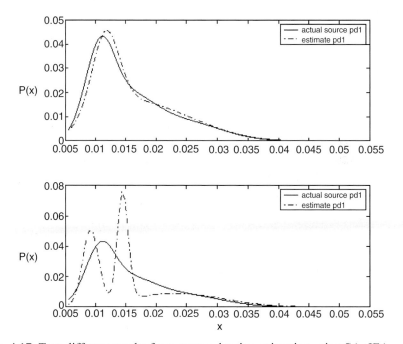

Fig. 4.17. Two different results from source density estimation using SA+IFA

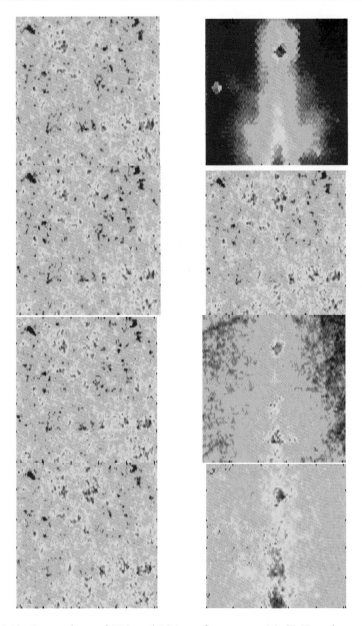

Fig. 4.18. Comparison of IFA and ICA performances. (**a**) CMB and synchrotron sources; (**b**) mixtures with 18 dB SNR; (**c**) IFA+SA+ML estimates; (**d**) ICA estimates

It is apparent that ICA yields a much more noisy result than simulated annealing IFA (SA-IFA). This suggests us that IFA is more successful than

ICA in recovering weak sources, albeit at a much higher computational cost. Note also the non-stationary noise pattern in the ICA-estimated synchrotron map.

4.5.6 Bayesian Source Separation Using Markov Random Fields

As explained earlier, in Baccigalupi et al. (2000), we used a self-organizing neural network ICA approach in a very idealized setting; in Maino et al. (2002), we used *FastICA* and its noisy version, which showed some success in separating artificial mixtures in the presence of stationary noise. These techniques are totally blind, except for the independence assumption. In Kuruoglu et al. (2003), we improved our results, even with non-stationary noise, by adopting a generic model for the sources and a priori knowledge about the mixing operator. Other authors have also addressed the problem of (more or less blind) separation of astrophysical components. In Cardoso et al. (2002), separation is performed by matching the spectra of the estimated data maps to the spectra of the observed data. In Snoussi et al. (2001), an EM algorithm is employed in the spectral domain, making use of some generic priors for the sources. These approaches assume stationary noise and signals, and suffer the common drawbacks of EM, i.e. local optimality and computational complexity. Moreover, these techniques ignore the local spatial information present in the images. The source maps, indeed, carry important information in their spatial structure (see Barros and Cichocki (2001)). In particular, foreground sources show significant local correlation. The spatial structure, in turn, is different from one source to another. Tonazzini et al. (2003) and Kuruoglu et al. (2004) developed a technique to model the local correlation structure in the images and achieve separation using Markov random fields (MRF) and Bayesian estimation. This technique was shown to be robust against noise, and has been applied with some success to astrophysical images as well.

Again, the data maps $\mathbf{x}$ are modelled as simple linear instantaneous noisy mixtures of the source maps $\mathbf{s}$, as in (4.35)–(4.43). The problem of estimating the mixing matrix $\mathbf{A}$ and the source samples $\mathbf{s}$ can be stated in a Bayesian setup as a maximum a posteriori estimation problem:

$$\left(\hat{\mathbf{s}},\hat{\mathbf{A}}\right)= \arg\max_{\mathbf{s},\mathbf{A}} P(\mathbf{s},\mathbf{A}|\,\mathbf{x}) = \arg\max_{\mathbf{s},\mathbf{A}} P(\mathbf{x}|\mathbf{s},\mathbf{A})P(\mathbf{s})P(\mathbf{A}) \quad (4.57)$$

where, from the independence assumption:

$$P(\mathbf{s}) = \prod_k P_k(s_k)$$

(4.58)

Each marginal source prior can be written by means of the Gibbs/MRF formalism:

$$P_k(s_k) = \frac{1}{Z_k} \exp\{-U_k(s_k)\}$$

(4.59)

where Z_k is the partition function and $U_k(s_k)$ is the prior energy in the form of a sum of potential functions over the set of cliques of interacting locations. The number of different cliques, as well as their shape, is related to the order of the neighbourhood system adopted for the MRF. This, in turn, depends on the extent of correlation among the pixels, whilst the functional form of the potentials determines the correlation strength. In the specific case at hand, we define $U_k(s_k)$ as:

$$U_k(s_k) = \alpha_k \sum_{i,j} f_k(s_k(i,j)) + \beta_k \sum_{i,j} \sum_{(m,n) \in N_{i,j}} \phi_k(s_k(i,j) - s_k(m,n))$$

(4.60)

where α_k and β_k are two positive weights, the so-called regularization parameters, and $N_{i,j}$ is the first-order neighbourhood for location (i,j). Functions f_k and ϕ_k are to be chosen, respectively, according to the expected probability density of each $s_k(i,j)$, and to the expected correlation between adjacent source samples.

We chose the priors on the basis of the typical features of our maps. In particular, the CMB distribution is known to be Gaussian, therefore for f_{CMB} in (4.60) we adopted a quadratic error function $\|\cdot\|^2$, which is the logarithm of the Gaussian kernel. The main goal of this prior is to regularize the solution against receiver noise. This is particularly important for sky patches located on the galactic plane, where the CMB is weaker than the galactic foregrounds and the presence of noise hampers its extraction from the total signal. In the second part of the regularizer, function ϕ_k models the neighbouring pixel interactions. The solution features to be preserved are smoothness in uniform regions and edges, that is, the lines where smoothness breaks down. Edges are crucial in many imaging and vision tasks, since they carry important information. In order for a regularizer to be edge-preserving, it must increase less than quadratically. Among this type of regularizers, we experimented in particular the one suggested by Shulman and Herve (1989), which has the form:

$$\phi(\xi) = \begin{cases} \xi^2 & \text{if } |\xi| \le \Delta \\ (2\Delta |\xi| - \Delta^2) & \text{if } |\xi| > \Delta \end{cases} \qquad (4.61)$$

where parameter Δ is a threshold above which the stabilizer becomes linear. In view of (4.60), this means that the image derivatives are penalized quadratically when their values are not greater than Δ, and only linearly otherwise. The regularizer is thus adaptive to discontinuities, that is, it is able to preserve sharp edges. This is likely to regularize correctly the estimated synchrotron and galactic dust maps, which are characterized by steep intensity changes separating relatively smooth areas. Note that this regularizer is convex. In our case, provided that the prior density for $\mathbf{A}$ is convex, this implies that the log of the functional to be optimized is also convex in both variables, thus enabling us to devise an efficient optimization strategy.

As is usual when the joint maximization with respect to two groups of variables is concerned, we proposed to solve the problem in (4.57) by means of alternating maximization. This results in iteratively alternating steps of estimation of $\mathbf{A}$ and $\mathbf{s}$, respectively:

$$\mathbf{A}^{(k)} = \arg\max_{\mathbf{A}} P(\mathbf{x} | \mathbf{s}^{(k-1)}, \mathbf{A}) P(\mathbf{A}) \qquad (4.62)$$

$$\mathbf{s}^{(k)} = \arg\max_{\mathbf{s}} P(\mathbf{x} | \mathbf{s}, \mathbf{A}^{(k)}) P(\mathbf{s}) \qquad (4.63)$$

When very general assumptions are made about the distributions involved, an implementation of the earlier scheme ensuring convergence without excessive complexity is based on a simulated annealing estimation of $\mathbf{A}$, following (4.62). The overall scheme is modified by keeping $\mathbf{s}$ constant for each constant temperature cycle, and only updating $\mathbf{s}$, by (4.63), at the end of the cycle. In our case, significant simplifications in the alternating maximization scheme are allowed. Indeed, we assumed no prior information for the mixing matrix, so that $P(\mathbf{A})$ is the uniform distribution and, as observed earlier, the problem is convex in both $\mathbf{s}$ and $\mathbf{A}$. An alternating optimization scheme can thus be implemented with no need for stochastic relaxation. A gradient ascent can be used to solve (4.63), whilst solving (4.62) yields a closed-form update for $\mathbf{A}$ (Kuruoglu et al. 2004).

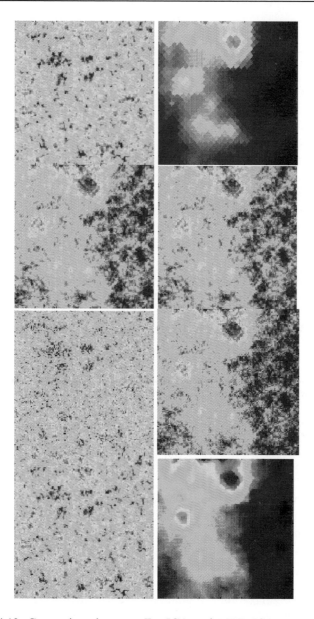

Fig. 4.19. Comparison between FastICA and MRF+ICA source estimates from mixtures with 20 dB SNR. Top: original source maps. Second row: mixture maps. Third row: FastICA estimates. Bottom: MRF+ICA estimates

In Fig. 4.19, we compare the results obtained by this method and the corresponding FastICA results. Although the algorithm is valid for mixtures of more components, this experiment was conducted on mixtures of two sources only (CMB and synchrotron). The assumed data maps simulate the *Planck* 100 and 70 GHz channels. We used the Shulman-Herve regularizer for both the sources, though with different regularization parameters α_k and β_k. The superiority of our technique over FastICA is apparent in the figure. From a quantitative point of view, the signal to interference ratio (SIR) given in Table 4.4 also shows a very significant difference (about 10 dB for synchrotron) between the performances of MRF+ICA and FastICA.

Table 4.4. Comparison of SIR performances of MRF+ICA and FastICA

SIR (dB)	MRF+ICA	FastICA
CMB	9.23	−7.07
synchrotron	16.52	5.99

4.5.7 A Semi-Blind Second-Order Approach

In principle, the problem of estimating all the model parameters and source signals cannot be solved by just using second-order statistics, since these are only able to enforce uncorrelation. However, this has been done in special cases, where additional hypotheses on the spatial correlations within the individual signals are assumed (Tong et al. 1991; Belouchrani et al. 1997; Delabrouille et al. 2002; Yeredor 2002; Patanchon et al. 2003). Another possibility of solving the separation problem on the basis of second-order statistics alone is when the relevant constraints are such that the total number of parameters to be estimated can be reduced. This can be done in our case by exploiting relationships of type (4.44)–(4.47). Indeed, when each source spectrum is either known or can be described by using a single parameter, the number of the unknowns is drastically reduced. We showed that, under these hypotheses, a very fast model learning algorithm can be applied, based on matching some calculated data covariance matrices to the corresponding matrices estimated from the observed data (Bedini et al. 2005).

This strategy also offers additional advantages, such as the possibility of estimating the source probability density functions and relaxing the strict assumption of uncorrelation between the source signals. The latter is an interesting issue, since, whilst all the contributions on blind astrophysical source separation have been assuming statistically independent sources, it is well known that this is not always true. For example, in sub-mm observations, the galactic foregrounds are not completely independent. Accounting for covariances at different shifts, the mixing and the source covariance matrices at these shifts can be estimated, thus performing a dependent component analysis.

We give here a brief description of this approach, with emphasis on model learning. Once the model has been learned, a number of standard reconstruction procedures are available to separate the individual sources. Working on second-order statistics alone is advantageous with respect to robustness against noise, since their estimation is more immune from erratic data than estimation of higher-order statistics.

Let us consider the source and noise signals in (4.43) as realizations of two stationary vector processes. The covariance matrices of the source process at shifts (τ, ψ) are

$$C_s(\tau,\psi) = \left\langle [s(\xi,\eta) - \mu_s][s(\xi + \tau,\eta + \psi) - \mu_s]^T \right\rangle \qquad (4.64)$$

where μ_s is the mean vector of process s. The noise processes in all the channels are assumed white, zero-mean, and mutually uncorrelated. Thus, the noise covariance matrices are diagonal, and zero at any non-zero shift pair (τ, ψ):

$$C_n(0,0) = \left\langle nn^T \right\rangle = diag(\sigma_1^2,...,\sigma_M^2); \qquad (4.65)$$

$$C_n(\tau,\psi) = 0. \quad \forall(\tau,\psi) \neq (0,0)$$

If the noise variances at all the channels are known, then matrix C_n is known. If the sources are known to be independent, or at least uncorrelated, then the unknown matrix C_s is known to be diagonal. Otherwise, if we know that any two sources are significantly correlated, then we must allow the corresponding entries in C_s to be non-zero.

Exploiting (4.43), the covariance matrix of the observed data can be written as:

$$C_x(\tau,\psi) = \left\langle [x(\xi,\eta) - \mu_x][x(\xi + \tau,\eta + \psi) - \mu_x]^T \right\rangle = \qquad (4.66)$$

$$= AC_s(\tau,\psi)A^T + C_n(\tau,\psi)$$

Equation (4.66) suggests us that matrices A and C_s can be estimated, provided that a sufficient number of measurement channels is available and matrix C_x is non-zero for a sufficient number of shift pairs. This is equivalent to say that, even if the mutual independence assumption cannot be made, separation is still possible by using second-order statistics, provided that the sources have significant spatial structures (i.e. spatial correlations). Let us define the matrices

$$\mathbf{H}(\tau,\psi) = \mathbf{A}\mathbf{C}_s(\tau,\psi)\mathbf{A}^T = \mathbf{C}_x(\tau,\psi) - \mathbf{C}_n(\tau,\psi) \qquad (4.67)$$

In estimating matrices $\mathbf{A}$ and $\mathbf{C}_s$, the total number of unknowns is the number of parameters specifying matrix $\mathbf{A}$ plus the number of non-zero elements in matrices $\mathbf{C}_s$ at each shift pair considered. Once matrices $\mathbf{H}$ (τ, ψ) are given, the total number of useful relationships made available by (4.67) is the number of their distinct elements. Each of these equations will be non-linear in the problem unknowns. We do not possess matrices $\mathbf{H}$ as defined by (4.66) and (4.67), but we assumed matrices $\mathbf{C}_n$ to be known, and we can estimate $\mathbf{C}_x(\tau,\psi)$ from the data:

$$\hat{\mathbf{C}}_x(\tau,\psi) = \frac{1}{N_p} \sum_{\xi,\eta} \left[\mathbf{x}(\xi,\eta) - \mu_x \right] \left[\mathbf{x}(\xi+\tau,\eta+\psi) - \mu_x \right]^T \qquad (4.68)$$

where the summation is extended over all the N_p data samples. Estimates $\hat{\mathbf{H}}(\tau,\psi)$ of the matrices in (4.67) can thus be obtained.

By minimizing the distance between the two members of each equality in (4.67) as functions of the unknown parameters, we can estimate matrix $\mathbf{A}$ and matrices $\mathbf{C}_s$. In formulas, if β is the vector of the parameters specifying matrix $\mathbf{A}$ (4.44)–(4.49) and $\sigma(\tau,\psi)$ are the vectors of all the non-zero elements in the source covariance matrices, our estimation takes the form:

$$(\beta,\sigma(\tau,\psi)) = \arg\min_{\beta,\sigma(\tau,\psi)} \sum_{\tau,\psi} \| \mathbf{A}(\beta)\mathbf{C}_s(\sigma(\tau,\psi))\mathbf{A}^T(\beta) - \hat{\mathbf{H}}(\tau,\psi) \| \qquad (4.69)$$

where the Frobenius norm was adopted. Our present strategy to find the minimizer is an annealing scheme on β. At each update of vector β, matrices $\mathbf{C}_s$ can be first evaluated analytically; their entries corresponding to uncorrelated sources are then forced to zero. The penalty function in (4.69) can now be calculated and used to update β.

Our method can be considered a *semi-blind dependent component analysis source separation* method. In fact, it is not totally blind, as we must assume to know which sources are uncorrelated.

By exploiting the results of model learning, there are several strategies to estimate the individual source maps and their probability density. For example, let us assume to have an estimate of $\mathbf{A}$. Let $\mathbf{B}$ be its Moore–Penrose generalized inverse. In our case we have $M \geq N$, thus, as is known, $\mathbf{B} = (\mathbf{A}^T\mathbf{A})^{-1}\mathbf{A}^T$. Let us denote each of the N rows of $\mathbf{B}$ as an M-vector $\mathbf{b}_i$,

$i = 1,..., N$, and consider the generic element y_i of the N-vector $\mathbf{Bx} = \mathbf{s} + \mathbf{Bn}$:

$$y_i := \mathbf{b}_i^T \cdot \mathbf{x} = s_i + \mathbf{b}_i^T \cdot \mathbf{n} := s_i + n_{ti} \qquad (4.70)$$

The probability density function of y_i, $p(y_i)$, can be estimated from the histogram of the set $\mathbf{b}_i \cdot \mathbf{x}(\xi, \eta)$, for all the available pairs (ξ, η), whilst the probability density function of n_{ti}, $p(n_{ti})$, is a Gaussian, whose parameters can be easily derived from $\mathbf{C}_n$ and $\mathbf{b}_i$. The pdf of y_i is the convolution between $p(s_i)$ and $p(n_{ti})$:

$$p(y_i) = p(s_i) * p(n_{ti}) \qquad (4.71)$$

From this relationship, $p(s_i)$ can be estimated by deconvolution. As is well known, deconvolution is normally an ill-posed problem and, as such, it lacks a stable solution. In our case, we can regularize it by the obvious positivity constraint and the normalization condition for pdfs.

Vector $\mathbf{Bx}$ gives a rough estimate of the source processes, though corrupted by amplified noise. To reduce the influence of noise, a simple strategy could be a suitable filtering. More sophisticated Bayesian strategies could exploit the availability of the source distributions to regularize the solution. In the case examined here, the source distributions should be estimated from the data, since they are normally unknown, save for CMB, which should be Gaussian. This strategy has been attempted, as already shown, when using the IFA approach. The source densities are modelled as mixtures of Gaussians, and the related parameters are estimated by means of a very expensive algorithm. Minimization in (4.69) followed by deconvolution of (4.71) is not very expensive computationally, thus, if included in the IFA strategy, it can reduce the overall cost of the learning-separation process.

The strategy described earlier is effective even with noise levels above the *Planck* specifications, and with foreground radiation stronger than CMB. The latter condition is likely to occur near the galactic plane, where the emitting regions along the line of sight are much stronger than the ones found at high galactic latitudes. None of the previously proposed techniques has reached such a result so far. It is to remark that both model learning and separation failed at high galactic latitudes, if the same levels of noise are introduced. In those regions, indeed, the only dominant signal is the CMB, and the foregrounds are often well below the noise level. In any case, however, CMB is estimated very well with all the assigned SNRs.

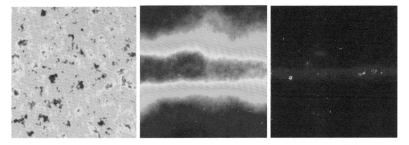

Fig. 4.20. Source maps from a 15°×15° patch centred at 0° galactic latitude and 20° galactic longitude, at 100 GHz. Their size is 512×512, and thus each pixel is about 1.7 arcmin wide. Left: CMB; centre: synchrotron; right: thermal dust

We report here just a few examples from our extensive experimentation. We assumed $N = 3$ source signals, CMB, galactic synchrotron and thermal dust emissions, and $M = 4$ measurement channels, centred at 30, 44, 70 and 100 GHz. In this case, we only used the zero-shift covariance matrix, and exploited the fact that the 4×3 mixing matrix can be specified by only two parameters.

Figure 4.20 shows the three source maps extracted from the simulated database. The standard deviations of the source processes at 100 GHz were 0.084 mK for CMB, 0.043 mK for synchrotron, and 0.670 mK for dust, in antenna temperature. We assigned the sources s_1 to CMB, s_2 to synchrotron and s_3 to dust, and the signals x_1, x_2, x_3 and x_4 to the observations at 100 GHz, 70 GHz, 44 GHz and 30 GHz, respectively. The true mixing matrix, $\mathbf{A_O}$, has been derived from (4.44), (4.45) and (4.47), with spectral indices $\beta_{syn} = 2.9$ and $\beta_{dust} = 1.8$.

We adopted a noise covariance matrix for which the noise standard deviation at 100 GHz is the same as the CMB standard deviation. The other values have been established by scaling the standard deviation at 100 GHz in accordance with the expected *Planck* sensitivity at each channel. In Fig. 4.21, we show our noisy data at 100 GHz and 44 GHz, respectively. The case examined does not fit the assumptions of mutual independence, since the original source covariance matrix at zero shift is:

$$\mathbf{C_s}(0,0) = \begin{bmatrix} 0.00716 & 0.00003 & -0.00112 \\ 0.00003 & 0.00183 & 0.01778 \\ -0.00112 & 0.01778 & 0.44990 \end{bmatrix} \quad (4.72)$$

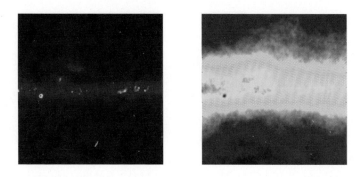

Fig. 4.21. Data maps containing CMB, synchrotron and dust emissions, as well as instrumental noise, at 100 and 44 GHz

showing a significant correlation (62%) between dust and synchrotron, and negligible correlations between CMB and synchrotron and CMB and dust (0.9% and 1.9%, respectively). The correlation coefficient between any two sources is defined as $r_{ij} = c_{ij} / \sqrt{c_{ii} c_{jj}}$, where c_{ij} are the elements of matrix $\mathbf{C_s}$. For the data described earlier, we ran our learning algorithm for 500 different noise realizations; for each run, 5,000 iterations of our minimization procedure were performed. The unknown parameters were the spectral indices β_{syn} and β_{dust} and the covariances σ_{11}, σ_{22}, σ_{33} and σ_{23}, setting to zero the values of σ_{12} and σ_{13}, representing the correlations between the CMB and the foregrounds. The typical elapsed times per run were a few tens of seconds on a 2 GHz CPU computer, with a Matlab interpreted code. In a typical case, we estimated β_{syn} =2.8904 and β_{dust} =1.8146. As a quality index for the estimation of $\mathbf{A}$, we adopted the matrix $\mathbf{Q} = (\mathbf{A}^T \mathbf{C_n}^{-1} \mathbf{A})^{-1} (\mathbf{A}^T \mathbf{C_n}^{-1} \mathbf{A_o})$, which, in the case of perfect model learning, should be the 3×3 identity matrix $\mathbf{I}$, so that the Frobenius norm of matrix $\mathbf{Q} - \mathbf{I}$ should be zero. In this case, it is 0.0669. The estimated source covariance matrix to be compared with the matrix in (4.72), is:

$$\mathbf{C_s}(0,0) = \begin{bmatrix} 0.00717 & 0. & 0. \\ 0. & 0.00188 & 0.01802 \\ 0. & 0.01802 & 0.44803 \end{bmatrix} \qquad (4.73)$$

The accordance between the estimated and the original matrices is apparent.

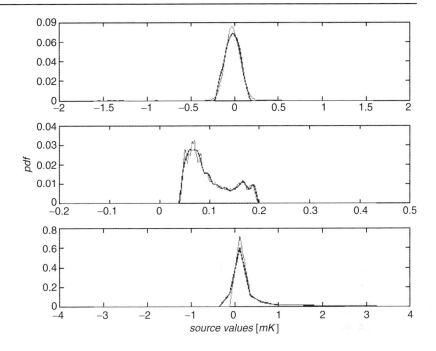

Fig. 4.22. Real (blue) and estimated (red) source density functions. Top: CMB; middle: synchrotron; bottom: dust

The reconstructed probability density functions of the source processes, estimated from (4.70) and (4.71), are shown in Fig. 4.22. The reconstructed densities are often indistinguishable from the original functions. In any case, the ranges of the individual processes have been identified very sharply. We simply separated the sources by multiplying the data matrix by the pseudoinverse of the estimated mixing matrix, obtaining visually very good results, as shown in Figure 4.23, left column. Part of the residual noise was then reduced by Wiener filtering the reconstructed maps (Fig. 4.23, right column). To evaluate more quantitatively the results of the whole learning-separation procedure, we compared the power spectrum of the CMB map with the one of the reconstructed map (Fig. 4.24). As mentioned, the reconstructed spectrum can be corrected for the known theoretical spectrum of the noise component n_{tl}, obtained as in (4.70). As can be seen from the figure, the reconstructed spectrum is very similar to the original within a multipole $l = 2,000$.

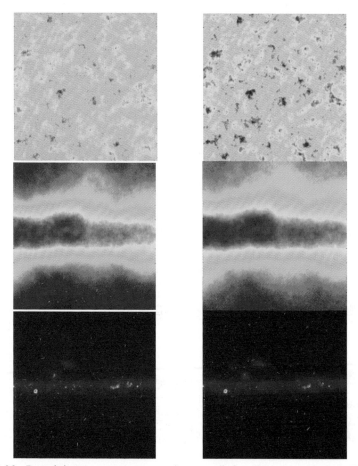

Fig. 4.23. Pseudoinverse reconstructed maps (left column), and their Wiener-filtered versions (right column). Top: CMB; middle: synchrotron; bottom: dust

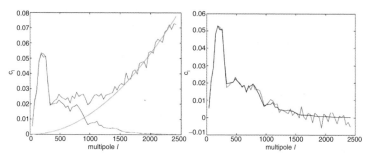

Fig. 4.24. Left: real (blue) and estimated (red) power spectra. The green line is the theoretical power spectrum of the noise component n_{tl} in (4.70), evaluated for the noise covariance and the Moore–Penrose pseudoinverse of the estimated mixing matrix. Right: real (blue) and estimated (red) CMB power spectrum, corrected for the theoretical noise

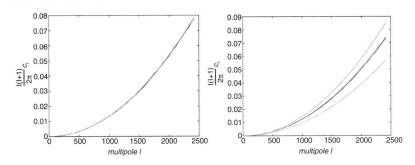

Fig. 4.25. Left: averaged power spectrum of 100 CMB residual maps on the $(0°, 20°)$ sky patch. The curve is almost indistinguishable from the theoretical noise spectrum and the dispersion band. Right: averaged power spectrum of 100 CMB residual maps on the $(15°, 20°)$ sky patch (blue). The red curve represents the theoretical noise spectrum, and the green curves delimit the dispersion band of the individual spectra

Averaging the learned results over the 500 runs, we have found mean values 2.8894 for β_{syn} and 1.8146 for β_{dust}, with standard deviations of 0.0015 and 0.0019, respectively. We compared all the reconstructed sources with the original ones, and computed the averaged power spectrum of the residual maps. The result for 100 runs with different noise realizations is shown in Fig. 4.25, left, where the averaged residual spectrum almost coincides with the theoretical noise spectrum.

We repeated this analysis for several patches and different noise levels. Apart from the already mentioned failure of the algorithm at high galactic latitudes, we have found that, in all the patches taken into account, the average estimated parameters are the same as the ones reported earlier. As an example, the analysis carried out on a patch centred at a galactic latitude of $15°$ and longitude $20°$, with the noise level specified earlier, gave average values 2.8897 and 1.8157 for β_{syn} and β_{dust}, respectively, almost identical to the previously shown values. However, since the noise covariance is the same, but the signal power is smaller than before, the signal-to-noise ratio is in this case significantly smaller than before. This fact induces much more dispersion than in the previous case. In fact, we have standard deviations of 0.0094 for β_{syn} and 0.1959 for β_{dust}, about 9 and 100 times, respectively, larger than the values found on the previous patch. In Fig. 4.25, right, we show the averaged residual spectrum for this case.

Obviously, the noise on the estimated source maps increases as the noise on the data maps increases. Wiener filtering can be used to improve the appearance of the final maps. Conversely, spectrum estimation has been found fairly insensitive to system noise, and this is a good point in favour of our algorithm.

4.5.8 Particle Filtering

Although the methods presented earlier are somewhat robust against non-stationary data, all the data models we used assume that the mixing operator, the source signals and the noise are independent of the position in the sky. These assumptions are not always justified. In particular, as mentioned, the noise is non-stationary, the source radiations are also non-stationary unless they are considered in small patches, and the mixing operator may depend on the position. Taking these non-stationarities into account would be useful to extend the possibilities of blind or semi-blind separation strategies.

In this section, we describe a relatively new technique, *particle filtering*, which overcomes the difficulties mentioned earlier. This technique can also account for both the inter-pixel dependence in the source maps and any prior information available on the source marginals. Particle filtering is an extension of Kalman filtering to non-linear systems and non-Gaussian signals, implemented through numerical Bayesian techniques. It has recently found applications mainly in tracking problems, and its theory is well covered in Doucet et al. (2001).

In our work, the basic particle filter has been adapted and implemented in order to improve its performances to our purpose, using the prior information available about the statistical properties of the sources. This technique is very promising if compared to batch methods, as its sequential structure permits a more appropriate analysis of non-stationary signals and noise.

The general filtering problem for non-stationary sources can be expressed by the following *state* and *observation* equations:

$$\mathbf{s}_t = f_t(\mathbf{s}_{t-1}, \mathbf{v}_t) \tag{4.74}$$

$$\mathbf{x}_t = h_t(\mathbf{s}_t, \mathbf{n}_t) \tag{4.75}$$

The state equation (4.74) describes the evolution of the state vector $\mathbf{s}_t$ at step t, f_t being a possibly non-linear function, $\mathbf{s}_{t-1}$ the state vector at the preceding step, and $\mathbf{v}_t$ the so-called *dynamic noise process*. The observation equation (4.75) describes the evolution of the data $\mathbf{x}_t$ at step t through a possibly non-linear function h_t, given the current state $\mathbf{s}_t$ and the *observation noise* realization $\mathbf{n}_t$ at step t. In the case of astrophysical source map separation, the data model (4.74)–(4.75) tells us that the observed data are given by a generally non-linear and non-stationary mixing of the source signals, and each source sample depends probabilistically on the neighbouring samples. In our 2D case, parameter t represents a pixel and, whilst (4.74) and (4.75) can always be read as if t were a 2D index, the practical 2D implementation of this technique is not a trivial extension of the 1D case. When f_t and h_t are linear and $\mathbf{v}_t$ and $\mathbf{n}_t$ are Gaussian, estimating $\mathbf{s}$ from $\mathbf{x}$ reduces to Kalman filtering.

Particle filtering is a technique that solves the general filtering problem by using numerical Bayesian (sequential Monte Carlo) techniques. Although known since the late 1960s (Handschin and Mayne 1969), this technique has been overlooked until the early 1990s because of the lack of computational power. More recently, it has been introduced in source separation (Ahmed et al. 2000; Andrieu and Godsill 2000; Everson and Roberts 2000). In particular, Everson and Roberts considered a linear instantaneous mixing in which the sources and the noise are stationary whilst the mixing matrix is non-stationary. They assumed generalized Gaussian models for the sources. Andrieu and Godsill, instead, considered the problem of convolutional mixing, and adopted a time-varying autoregressive model for the sources, assumed to be Gaussian. The mixing was also assumed to evolve following a time-varying autoregressive Gaussian process. At present, we assume non-stationary sources and noise, and stationary mixing operator. Yet, we assume to have random source model parameters to fully exploit the potentials of Bayesian modelling. As in the case of independent factor analysis, we model the source priors as mixtures of Gaussians.

The basis of particle filtering is the representation of continuous pdfs with discrete points (*particles*). In our case, the source posteriors will be approximated by:

$$p(\mathbf{s}_{0:t} \mid \mathbf{x}_{1:t}) \approx p_K(d\mathbf{s}_{0:t} \mid \mathbf{x}_{1:t}) = \frac{1}{K}\sum_{i=1}^{K}\delta_{\mathbf{s}_{0:t}^{(i)}}(d\mathbf{s}_{0:t}) \tag{4.76}$$

where $\delta_{\mathbf{s}_{0:t}^{(i)}}$ denotes the delta-Dirac mass and K is the number of points where the continuous pdf is discretized. In this case, a minimum-mean-square-error (MMSE) estimate of a function of interest $I(f_t)$ can be obtained as:

$$I_{MMSE_K}(f_t) = \int f_t(\mathbf{s}_{o:t}) p_K(d\mathbf{s}_{0:t} \mid \mathbf{x}_{1:t}) = \sum_{i=1}^{K} f_t(\mathbf{s}_{0:t}^{(i)}) \tag{4.77}$$

Unfortunately, it is usually impossible to sample from the posterior distribution. A classical solution is to use *importance sampling*, by replacing the true posterior distribution by an *importance function*,

$\pi(\mathbf{s}_{0:t} \mid \mathbf{x}_{1:t})$, which is easier to sample from. Provided that the support of $\pi(\mathbf{s}_{0:t} \mid \mathbf{x}_{1:t})$ includes the support of $p(\mathbf{s}_{0:t} \mid \mathbf{x}_{1:t})$, we get the identity

$$I(f_t) = \frac{\int f_t(\mathbf{s}_{0:t})\zeta(\mathbf{s}_{0:t})\pi(\mathbf{s}_{0:t} \mid \mathbf{x}_{1:t})d\mathbf{s}_{o:t}}{\int \zeta(\mathbf{s}_{0:t})\pi(\mathbf{s}_{0:t} \mid \mathbf{x}_{1:t})d\mathbf{s}_{o:t}} \tag{4.78}$$

where $\zeta(\mathbf{s}_{0:t}) = p(\mathbf{s}_{0:t} \mid \mathbf{x}_{1:t})/\pi(\mathbf{s}_{0:t} \mid \mathbf{x}_{1:t})$ is known as the *importance weight*. Consequently, it is possible to obtain a Monte Carlo estimate of $I(f_t)$ by using K particles $\{\mathbf{s}_{0:t}^{(i)}, \quad i = 1,...,K\}$ sampled from $\pi(\mathbf{s}_{0:t} \mid \mathbf{x}_{1:t})$:

$$\overline{I}_K(f_t) = \frac{\frac{1}{K}\sum_{i=1}^{K} f_t(\mathbf{s}_{0:t}^{(i)})\zeta(\mathbf{s}_{0:t}^{(i)})}{\frac{1}{K}\sum_{i=1}^{K}\zeta(\mathbf{s}_{0:t}^{(i)})} = \sum_{i=1}^{N} f_t(\mathbf{s}_{0:t}^{(i)})\overline{\zeta}_t^{(i)} \tag{4.79}$$

where the *normalized importance weights* $\overline{\zeta}_t^{(i)}$ are given by: $\overline{\zeta}_t^{(i)} = \zeta(\mathbf{s}_{0:t}^{(i)})/\sum_{i=1}^{K}\zeta(\mathbf{s}_{0:t}^{(i)})$. This integration method can be interpreted as a sampling method, where the posterior distribution is approximated by:

$$\overline{P}_K(d\mathbf{s}_{0:t} \mid \mathbf{x}_{1:t}) = \sum_{i=1}^{K}\overline{\zeta}_t^{(i)}\delta_{\mathbf{s}_{0:t}^{(i)}}(d\mathbf{s}_{0:t}) \tag{4.80}$$

When the importance function is restricted to the general form:

$$\pi(\mathbf{s}_{0:t} \mid \mathbf{x}_{1:t}) = \pi(\mathbf{s}_{0:t-1} \mid \mathbf{x}_{1:t-1})\pi(\mathbf{s}_t \mid \mathbf{s}_{0:t-1}, \mathbf{x}_{1:t}) = \tag{4.81}$$

$$= \pi(\mathbf{s}_0)\prod_{k=1}^{t}\pi(\mathbf{s}_k \mid \mathbf{s}_{0:k-1}, \mathbf{x}_{1:k})$$

the importance weights and, hence, the posterior can be evaluated recursively. Our source priors are finite mixtures of Gaussians, as in (4.53). The difference with respect to the case treated in Sect. 4.5.3 is that to model non-stationary processes we need to let the parameters defined in (4.54) depend on t. Let $\mathbf{q}_{1:N,t}$ be the index vector defined in (4.54) for the sample of index t. It is possible to describe its discrete probability distribution by assuming that the state indicators $q_{i,t}$ have identical and independent distributions. If we want to introduce a correlation between the samples of a particular source, we can consider the first-order Markov model, where the state vector evolves as a homogeneous Markov chain for $t > 1$:

$$p(\mathbf{q}_{1:N,t} = \mathbf{q}_l \mid \mathbf{q}_{1:N,t-1} = \mathbf{q}_j) = \tag{4.82}$$

$$= \prod_{i=1}^{N} p(q_{i,t} = [\mathbf{q}_l]_i \mid q_{i,t-1} = [\mathbf{q}_j]_i) = \prod_{i=1}^{N} \tau_{jl}^{(i)}$$

where $\tau_{jl}^{(i)}$ is an element of the $n_i \times n_i$ real-valued *transition matrix*, $\mathbf{T}^{(i)}$, for the states of the ith source. The state transition can thus be parametrized by a set of N transition matrices $\{\mathbf{T}^{(i)}, i = 1, ..., N\}$.

Given the observations $\mathbf{x}_t$ (assuming that the number of sources N, the number of Gaussian components n_i for each source, and the number of sensors M are known), we would like to estimate all the following unknown parameters, grouped together:

$$\theta_{0:t} = \left[\mathbf{s}_{1:N,0:t}, \mathbf{q}_{1:N,0:t}, \{w_{i,q_i,0:t}\}, \{\mu_{i,q_i,0:t}\}, \{\sigma_{i,q_i,0:t}^2\}, \{\mathbf{T}_{0:t}^{(i)}\} \right] \tag{4.83}$$

where all the symbols have already been defined, here or in (4.54). To reduce the size of the parameter set to be estimated, we find the values of the mixing matrix subsequently, by means of the Rao-Blackwellisation technique (Casella and Robert 1999). As said, it is not easy to sample directly from the optimal importance distribution: this is the reason why a sub-optimal method has been employed, taking the importance distribution at step t to be the prior distribution of the sources to be estimated. The mixture of Gaussians model allows the prior distribution to be factorized into several easy-to-sample distributions related to the model parameters. Details can be found in Costagli et al. (2004).

This algorithm has been tested on two 64×64 maps obtained by mixing CMB and synchrotron, and corrupting them by realistic space-varying noise processes, with 10 dB average SNR. The indices of the Gaussian components have been assumed Dirichlet distributed, the variances are drawn from an inverted-Gamma distribution, and the means from a Gaussian centred at the value of the previous particle. As usual, the CMB map has been assumed Gaussian, whilst the synchrotron distribution has been approximated by a mixture of three Gaussian components. At each step, the algorithm generated 1,000 particles for each parameter of interest.

Since we are not aware of any other work that considers separation of non-stationary sources under non-stationary noise environment, we compared our results with the ones obtained by FastICA from the same data set. As seen in Fig. 4.26 whereas FastICA fails to recover CMB and

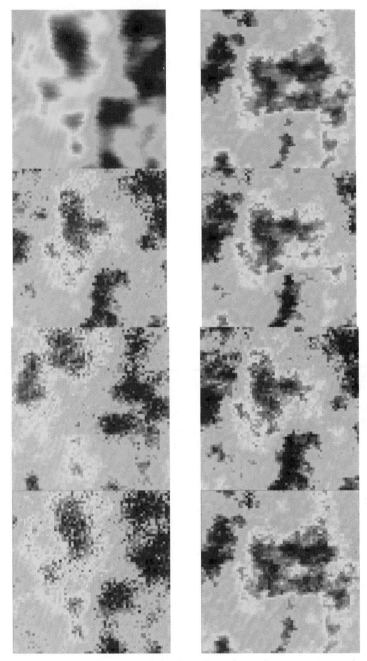

Fig. 4.26. From top to bottom: original CMB and synchrotron maps; mixtures at 100 and 30 GHz; FastICA estimates; particle filter estimates

gives a synchrotron estimate with high interference, particle filtering recovers the original maps, although corrupted by residual noise. The signal-to-interference ratios shown in Table 4.5 quantify this performance. A fundamental step in the implementation of particle filtering is the choice of the importance function. So far, we have set it to the prior distribution for analytical ease: this choice is far from being optimal, since it does not allow us to exploit any information about the observed data. Our present research takes this aspect into consideration.

Table 4.5. SIR values for FastICA and particle filtering

SIR (dB)	FastICA	Particle Filtering
CMB	1.14	10.46
synchrotron	1.62	19.16

4.5.9 Future Trends

The use of blind and semi-blind techniques for astrophysical source separation is now widely proposed in the literature. Indeed, using rigid data models for this problem has proved not to be suitable for the levels of instrumental noise often at hand. In this sense, even non-blind separation methods have been provided with means to relax both the rigidity of the model and the stationarity of the data. Many of the techniques proposed are now able to treat significant noise levels and non-idealities in the data.

A number of problems still remain to be faced. Some of them are of specific interest to astrophysics, but can also be useful for source separation at large. Separating dependent sources is one of these problems, only being treated in the very recent literature. Our second-order algorithm, specifically designed for astrophysical separation, could also lead to more general approaches, especially when prior information is available.

Another problem of general interest is the separation of non-stationary sources. Non-stationarity can be intrinsic in the source process or the mixing mechanism, or be generated by a non-stationary noise process. Particle filtering has been proposed very recently for two-dimensional source separation, and it seems to be the most promising approach to solve these problems, as well as to be capable of treating non-stationary mixing operators. Developing practically applicable procedures, however, still requires an ability to manage a very high computational complexity.

A problem that has interest for both our specific problem and in general is the separation of convolutive data. The linear instantaneous data model of (4.43), indeed, is no more valid in that case, operator **A** being still linear, but not instantaneous. Separating the convolutive mixtures generated by such an operator, even though up to a common residual kernel, is a problem that has only received practically useful solutions in special cases. In our case, the problem of separating convolutive mixtures arises when the telescope beamwidth depends on frequency, and is thus different for the different measurement channels. Now, unlike the general convolutive problem stated earlier, all the entries in each row of matrix **A** are proportional to a single convolution kernel, which, in turn, is likely to be known, both in advance, since it is a specification of the telescope design, and *a posteriori*, since it must be measured exactly over each channel during the instrument calibration and alignment. Although this is a simplified version of the general convolutive problem, it has no satisfactory solution yet. A deconvolution prior to separation, exploiting the knowledge of the kernels, has not demonstrated to be feasible so far. We proposed to degrade the resolution of all the data maps to the worst value, so as to be able to evaluate the constant coefficients of the mixing matrix. Map separation at the maximum feasible resolution could then be achieved by non-blind approaches, working in the original or in some transformed space. The ability of these methods of reconstructing the source maps at full resolution even in the presence of significant noise, however, is questionable, since they rely on either too generic (the maximum entropy principle), or too specific assumptions, unlikely to be verified (exact knowledge of the source angular power spectra). Some help in this sense could come from the second-order statistical approaches. In any case, the problem is still open to research.

4.5.10 Conclusion

We have presented and discussed several methods for component separation in astrophysical images. Some of them are straightforward adaptations of existing algorithms to the specific problem at hand. Some others are novel approaches or original extensions of previous methods. In particular, we developed strategies for handling non-stationary noise and for introducing available a priori information into the problem. Possible information regards the autocorrelation and cross-correlation properties of the source maps and the relationships among the mixing coefficients. Knowledge of the variance maps of possibly non-stationary noise can also be exploited in some cases. Introducing this prior information led us to devise efficient algorithms.

Our research started by testing the performances of existing techniques, such as the ones based on mutual independence alone. The need to include a suitable source model and to treat non-stationary noise suggested us to adopt more flexible strategies, such as independent factor analysis and particle filtering, whose computational complexity, however, still prevents us from providing a complete experimental assessment. On the other hand, we also addressed the possibility of accounting for the source spatial structure and the correlation between pairs of sources. This led us to formulate our Markov-random-field and our second-order approaches, which belong to the emerging field of dependent component analysis. Even though it has only been designed to take stationary noise into account, our second-order approach led us to a fast procedure that is also robust against moderately non-stationary noise.

In implementing all the algorithms, we bore in mind the knowledge coming from the physics of our particular problem. Moreover, all the experimental assessment has been made against quite realistic data. This should always be made in facing practical problems, and has given good results in our astrophysical case. In our opinion, this is one of the reasons why blind and semi-blind techniques are now being considered so often in astrophysical imaging.

4.6 Intelligent Segmentation of Ultrasound Images Using Cellular Neural Networks

Damjan Zazula and Boris Cigale

Abstract

Ultrasound imaging is one of the fastest growing application fields of image processing. Ultrasound-based medical examinations mean a well-established, wide-spread, harmless and cost-effective diagnosing approach. Modern ultrasonic devices have a variety of sophisticated built-in image processing options. However, precise detection of different organ and tissue structures and, in particular, their growth still needs additional, mainly off-line processing with highly specialised algorithms.

It has been shown the detection of image objects can be completed by adequately trained cellular neural networks (CNNs). In this section, we describe how an intelligent machine vision system is feasible based on

CNNs. Their learning capabilities are discussed and exemplified with ovarian ultrasound images. It becomes evident the task of ultrasound image analysis is too complex for a single CNN. Therefore, we introduce a grid of interlinked CNNs that are individually trained for separate steps of a segmentation process. This novel approach is illustrated by the detection of ovarian follicles and statistically validated. The proposed algorithms are revealed in the Matlab code.

The readers are also encouraged to tackle the problems appended to this section. With basic knowledge of computer usage and programming, they will gain hands-on experience on the concepts of CNNs and their application to image recognition.

4.6.1 Introduction

A steady growth of computer performance and decrease of their costs have influenced all the computer-assisted applications and services. One of the most rapidly developing fields encompasses medical imaging with a variety of imaging facilities and technologies (Bronzino 1995). Ultrasound, for example, is relatively inexpensive, so its use has spread widely although the ultrasound image interpretation is not a straightforwards and easy task.

We have to extract process and recognise a vast variety of visual structures from the captured images. The first attempt is supposed to reduce the complexity. An image must be segmented into smaller constituent components, such as edges and regions. The task is far from being plain even if the environment conditions are stable. Whenever the conditions change or the observed scene behaves with a certain dynamics, the processing methods must resort to adaptive solutions (Perry et al. 2002). The human brain has attained a significantly high level of adaptive processing. It also possesses a rather efficient ability of learning. Both the adaptability and learning can be hold up as an example for the developers of the computer recognition and classification algorithms.

In this section, we are going to explain how the structure and functions of the human central neural system have been mimicked in order to design flexible and intelligent machine vision systems that are capable of performing the demanding tasks of image processing and pattern recognition (Bow 2002). We are focusing on medical images and derive new approaches which are in particular suitable for the images obtained by ultrasonic diagnostic devices. Our basic development vehicle is propelled by so called cellular neural networks (CNN) whose fundaments, meaning and utilization in image processing are revealed in Sect. 4.6.2. Implementation of the CNN computer simulators is discussed in Sect. 4.6.3. Their learning ability and the corresponding approaches are explained in Sect. 4.6.4. Complexity of the ultrasound image segmentation requires more versatility and performance as one CNN can bid. Therefore, we suggest a grid of interlinked CNNs in Sect. 4.6.5 and exemplify the approach described in previous subsections by the detection of follicles in real ultrasound images of women's ovaries. Section 4.6.6 discusses the features of CNNs and their application to ultrasound image processing, and concludes the section.

We want to make the topics comprehensible and as much educational as possible. Readers may tackle the instructional problems which accompany the chapter for their revision and comprehension. Some of the assignments need a special computer environment for running the CNNs. Our solutions suggest MATLAB compatible routines revealed in Sects. 4.6.3 and 4.6.4.

4.6.2 Basics of Cellular Neural Networks

Visual information conveyed by the observed images to the viewer depends on the inhomogeneity in images. Variations among different images are great, yet only a small number of image feature types are needed to characterize them. These can be attributed to smooth regions, textures and edges (Perry et al. 2002). Thus, any live or artificial image processing system must primarily recognize these features.

Human brains master even much more than this; it provides the abilities of perceptual interpretation, reasoning and learning (Bow 2002). The processing power needed originates in billions of interconnected biological neurons. They create neural networks which act as a very complex non-linear system.

In the last three decades, the idea of copying this structure matured in a variety of classes of artificial neural networks (ANNs). Technically speaking, ANNs belong to the category of distributed systems, consisting of many identical processing elements, i.e. the neurons, which perform a rather simple but non-linear function. Only the ensemble of all neurons, the network, features an interesting behaviour (Hänggi and Moschzty 2000). ANNs must be trained first to adopt the desired behaviour.

In general, any number of the neurons' outputs can be fed back into any number of the inputs. A fully connected network where all the outputs connect back to all the inputs is known as the Hopfield network (Hänggi and Moschzty 2000). Such global complexity impedes the integration of this kind of ANNs in VLSI technology.

In 1988, Chua and Yang introduced a special network structure which resembles the continuous-time Hopfield ANN, but it implements strictly local connections. The proposal was influenced by the neural networks on the one hand (Hopfield 1982) and by the cellular automata on the other hand (Hänggi and Moschzty 2000). The new construct was called cellular neural networks (CNNs). The basic building elements incorporate characteristics of essentially simple dynamic systems. Structurally, they represent equivalents of the ANN's neurons, but with the CNNs their name is cells (Chua and Yang 1988).

Every cell behaves accordingly to its inputs, internal states and feedbacks. The cell's output is a non-linear function of the cell's internal state. Its inputs and feedbacks from the outputs enter the cell weighted by the parameters which actually define the cell's behaviour. Cells are locally coupled to their neighbours, whereas various topologies have been adopted, e.g. a square or a hexagonal grid.

Not only that such a construct suits VLSI implementation by design, but its topology can also be tailored to the needs of the image processing tasks. Talking about local image operators (Russ 2002), every image pixel must be processed in the same way, whilst its new value depends on its old value and the values of neighbouring pixels. By assuming a CNN cell responsible for an image pixel, topology of a planar CNN appears (Fig. 4.27).

Referring to square topology of 2-D planar CNNs, denote the neighbourhood of cell $C(i,j)$ by $N_r(i,j)$, where index r stands for the neighbourhood radius measured as chessboard distance (Russ 2002). Figure 4.28 depicts the cell neighbourhoods for $r = 1$ and $r = 2$, which are also known under the names "3×3 neighbourhood" and "5×5 neighbourhood", respectively.

The simplest CNN cell is supposed to exhibit first-order dynamics, which is described by:

$$\frac{d}{dt}x_t(i,j) = -x_t(i,j) + \sum_{k,l \in N_r(i,j)} a(k,l)y_t(i+k,j+l) + \sum_{k,l \in N_r(i,j)} b(k,l)u(i+k,j+l) \quad (4.84)$$

where indices (i,j) determine the cell position in the network, $x_t(i,j)$ stands for its internal state in time t, $u(i,j)$ for, assumingly, time-independent inputs and $y_t(i,j)$ for the outputs, whilst $a(k,l)$ and $b(k,l)$ are the so called feedback and control parameters comprised in template masks **A** and **B**, and I is the bias term. Our notation omits the option of having the cell-position dependent parameters, because this is not common in image processing. Throughout this chapter we are also going to assume templates **A** and **B** the same and stationary for all the CNN cells.

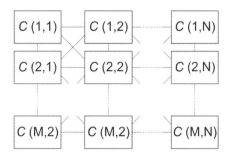

Fig. 4.27. Square topology of a 2-D planar CNN

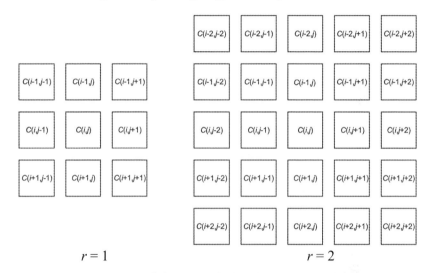

$r = 1$ $r = 2$

Fig. 4.28. Two examples of the cell neighbourhoods: 3×3 neighbourhood (left) and 5×5 neighbourhood (right)

Equation (4.84) determines the temporal evolution of each cell. The output $y_t(i,j)$ relates to the cell's state $x_t(i,j)$ by a non-linear function $f(\cdot)$. Function $f(\cdot)$ can be considered the sign (signum) function, saturation function, sigmoidal function or other. Usually, the saturation or signum functions are supposed as depicted in Fig. 4.29 and as follows:

$$\text{Saturation function: } f(x) = \frac{1}{2}\left(|x+1| - |x-1|\right) \qquad \text{Signum function: } f(x) = \frac{x}{|x|} \qquad (4.85)$$

The presented cell structure may be modelled as in Fig. 4.30 (Crounse and Chua 1995): boldface letters designate vectors and matrices, whereas the dot over **x** stands for time derivatives. Inputs **u** are convolved by the control template **B** and outputs **y** by the feedback template **A** (correlation sums form (4.84) can be rewritten as convolution sums).

The feedbacks in the cell model from Fig. 4.30 cause a non-linear temporal evolution which may not converge. It was shown in Crounse and

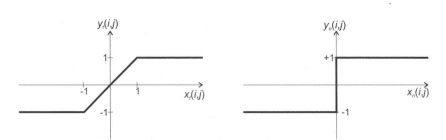

Fig. 4.29. The saturation (left) and signum (right) output functions

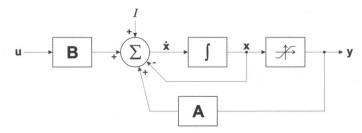

Fig. 4.30. Block diagram of a CNN cell structure (boldface letters denote vectors and matrices, whilst the dot over **x** stands for time derivatives)

Chua (1995) that a stable equilibrium is always achieved if the template **A** is symmetric, i.e. $\forall k,l \in N_r(i,j): a(k,l)=a(-k,-l)$. This is an important finding for all practical solutions.

Another interesting fact was pointed out in Crounse and Chua (1995): if $a(0,0)>1$, then all outputs in steady state are either $+1$ or -1 (owing to the saturation-function non-linearity).

If the cell feedback is suppressed ($\forall k,l \in N_r(i,j): a(k,l)=0$) for all cells $C(i,j)$, temporal evolution of system (4.84) vanishes and the stable state appears immediately after the first iteration. This actually means the cells behave as FIR filters whose characteristic is determined by the template **B**.

Following the suggestions in Harrer and Nossek (1992), we can compute the CNNs using discrete-time approach. In contrast to the continuous-time variant, the discrete-time CNN (DTCNN) is clocked in such a way that the time passes through successive iterations in discrete steps. Also the CNN output values are restricted to -1 and $+1$. The internal states depend only on the inputs and output feedback, whilst their time-derivatives are omitted:

$$x_n(i,j) = \sum_{k,l \in N_r(i,j)} a(k,l) y_n(i+k,j+l) + \sum_{k,l \in N_r(i,j)} b(k,l) u(i+k,j+l) + I$$

$$y_{n+1}(i,j) = \operatorname{sgn}\left[x_n(i,j)\right]$$

(4.86)

where n stands for the iteration step (replaces the time from (4.84)).

In the following section, we are going to enlighten the implementation of CNNs in image processing. We will explain how the inputs and initial states must be arranged for a CNN, how the desired operations mirror in the template coefficients, and how the output images are obtained. The explanation will be supported by a few simple examples.

How to Implement a CNN for Image Processing

As we have mentioned, the natural way of applying CNNs to image processing is to make individual CNN cells responsible for single image pixel neighbourhoods. Therefore, the size and geometry of the CNN is equal to the size of image $\mathbf{P}$, whereas pixel $p(i,j)$ and its r-neighbourhood correspond to the cell $C(i,j)$. This cell produces an output value which is assigned to the corresponding pixel, $y_t(i,j)$, of an interim output image $\mathbf{Y}_t$.

The steady states of the CNN cell outputs swing either to $+1$ or to -1. Consequently, the resulting output image pixels would be of the same values. On the other hand, the acquired pixel values in real images occupy integer values between 0 and 2^{n-1}, where n stands for the number of bits determining the image contrast resolution. The lowest value, 0, designates black colour, the highest, 2^n-1, white colour. To unify the coding of input and output images and to adapt it for processing by a CNN, the best way is to normalize the pixel values into an interval of $[0,1]$. After this normalization, 0 stands for black, 1 for white.

Suppose the image to be processed by the CNN has been normalized. Now, it can be used for the CNN input image, so that

$$u(i, j) = 1 - 2p(i, j) \tag{4.87}$$

if $u(i,j)$ is the input image pixel at co-ordinates (i,j) and $p(i,j)$ corresponds to the (i,j) pixel value in the normalized image. We anticipated the black pixels are fed into the CNN as values $+1$, the white pixels as -1.

Additionally, the question of the initial CNN cell status, $x_0(i,j)$, remains open. For the time being, we can just admit that in the algorithms which follow in Sect. 4.6.5 we set $x_0(i,j) = u(i,j)$ for all image pixels, thus obtaining an initial state image $\mathbf{X}_0$.

Similarly, after the CNN operation reaches the steady state the resulting image, $\mathbf{Y}_4$, can be transformed into a normalized output image, $\mathbf{S}$, accordingly:

$$s(i, j) = \frac{1 - y_\infty(i, j)}{2} \tag{4.88}$$

In practical implementations, the outcomes of (4.88) must be rounded to the closest integer (either 0 or 1) to compensate the calculation inaccuracies.

Let us illustrate the described connections among images and the CNN by Fig. 4.31.

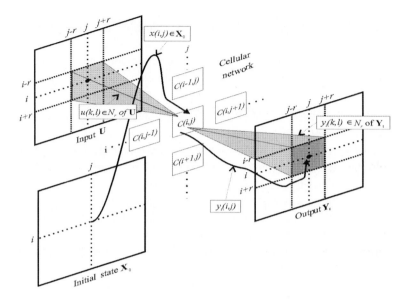

Fig. 4.31. Schematic presentation of the CNN application to image processing: new output values are computed by the CNN cells using the input neighbourhoods, initial states and previous output neighbourhoods

We have now got to the point where an appropriate CNN could already be deployed to solve our problems. However, we emphasised that only the templates **A** and **B** would conceive a certain CNN's behaviour. To comprehend their influence, we will elaborate a few basic examples in the sequel, whilst some more can be found among the problems at the end of chapter.

Example 4.6.1:

Begin with a trivial situation; assume the DTCNN with no feedback (all coefficients in the template **A** equal 0), $\mathbf{A} = \begin{bmatrix} 0,0,0 \\ 0,0,0 \\ 0,0,0 \end{bmatrix}$, whilst the template

B of size 3×3 has the following contents: $\mathbf{B} = \begin{bmatrix} 0,0,0 \\ 0,1,0 \\ 0,0,0 \end{bmatrix}$.

The 1-neighbourhood is not even necessary here; nevertheless it is shown the same as in other examples. Now, (4.86) shrinks to:

$$x(i, j) = u(i, j) + I \qquad (4.89)$$

The DTCNN outputs depend on the states $x(i,j)$ through the signum function. So, to obtain black output pixels, +1, the bias term must yield:

$$I > -u(i, j) \tag{4.90}$$

In other words, if I is fixed to a certain value between -1 and $+1$, all input pixels having their grey-level below $-I$ will appear white at the output, all with grey-levels above $-I$ black. Hence, a binary output image would be obtained.

A simple conclusion from this example says the bias I plays a role of a threshold, and **B** initialized as shown here causes binarization of images.

Example 4.6.2:

Suppose binary input images and all feedback coefficients in the template **A** equal 0 again. The obtained DTCNN acts as a local image operator controlled by the template **B**. Choose **B** according to the Laplace 3×3 operator (1-neighbourhood):

$$\mathbf{B} = \begin{bmatrix} 0, & 1, & 0 \\ 1, & -4, & 1 \\ 0, & 1, & 0 \end{bmatrix},$$

and reformulate (4.86):

$$x(i, j) = -4u(i, j) + u(i-1, j) + u(i, j-1) + u(i, j+1) + u(i+1, j) + I \tag{4.91}$$

where iterations, i.e. the time dependence, are not needed any more owing to the constant nature of all the right-hand side terms in (4.91).

Let us pay some additional attention to the bias term I. The condition for obtaining a black output pixel (the CNN's output +1) reads $x(i,j) > 0$. From this condition and referring to (4.91), bias I can be expressed as follows:

$$I > 4u(i, j) - u(i-1, j) - u(i, j-1) - u(i, j+1) - u(i+1, j) \tag{4.92}$$

Suppose we deal with a black input pixel $u(i,j) = +1$ and w neighbouring pixels at $(i-1,j)$, $(i,j-1)$, $(i,j+1)$ and $(i+1,j)$ are white, $w \in [0,4]$. Then, the output will attain black only if

$$I > 2w \tag{4.93}$$

By setting $1 < I < 2$, the constructed CNN would detect edges, actually the borders between black and white (notice that detecting an edge, cell $C(i,j)$ must cover a black pixel). All homogeneous (black or white) regions would appear black, whereas the black input regions gain a white lining.

Example 4.6.3:

This example is going to demonstrate how a CNN can be programmed to behave like a cellular automaton. Cellular automata perform rather simple tasks assigned to the individual automaton's cells with a primary intention to spread the results through a network of cells in an iterative manner.

Suppose a plain task: the output image should have every column, where a binary input image contains at least one object anywhere in this column, marked by a black pixel in the bottom row. The solution is conceptually clear: we have to scan the input columns from the bottom up and the first black pixel found turns the corresponding bottom pixel in the output image black. But dealing with cellular automata or CNNs, such approach fails because we can only either create the output image by converting the contents of the input image without feedback ($\mathbf{A} = \mathbf{0}$), which eliminates iterative operation, or do an iterative computation where also the information from the output image is fed back and influences the outcomes in next steps. Neither of these possibilities can cope with our initial conceptual solution. So we are forced to find another algorithm which will transform given image of objects into the desired output marks iteratively.

Actually, we can start with the original image by scanning the columns down from the top. Whenever a black pixel comes in sight, we shift it one row towards the image bottom and erase its previous position.

This approach suggests that we just have to initialize the CNN output image, $\mathbf{Y}_0$, by those image contents we want to process. No inputs to the CNN are requested, so that $\mathbf{B} = \mathbf{0}$. The bias I secures correct outputs when the feedback equals 0, therefore $0 < I < 2$. The CNN's initial states are taken 0 as well: $\mathbf{X}_0 = \mathbf{0}$. Following the Example 4.6.2 in Harrer and Nossek (1992), the template $\mathbf{A}$ can be composed as:

$$\mathbf{A} = \begin{bmatrix} 0, & 2, & 0 \\ 0, & 1, & 0 \\ 0, & -1, & 0 \end{bmatrix}$$

which gives the following cell computation:

$$x_n(i, j) = 2y_n(i-1, j) + y_n(i, j) - y_n(i+1, j) + I$$
$$y_{n+1}(i, j) = \operatorname{sgn}[x_n(i, j)] \tag{4.94}$$

You can learn more about this solution if you consider Problem 4.6.5 at the end of this section.

4.6.3 CNN Simulation

Because of the regular structure and local connectivity the CNNs are appealing for VLSI implementation. Actually many attempts are reported in the literature (Chua et al. 1998) and even commercial products exist

(AnaLogic Web page). However, instead of experimenting with hardware a more usual way is to recourse to the computer simulations of CNNs first (Loncar et al. 2000).

In the sequel, we are going to show a CNN simulation can be rather straightforwards if we make use of a powerful tool such as MATLAB. Entire network of cells must be simulated which means computations with entire CNN input, state and output matrices, i.e. images. The control and feedback templates, **A** and **B**, process the images in locally limited convolutions. MATLAB supports such operations by the IMFILTER function. The main loop of the CNN simulator advances the time in small increments, Δt, if we deal with a continuous-time solution, and just takes care of iterations if a discrete-time solution is implemented.

The continuous-time implementation deploys (4.84) and estimates the new state value for each cell in time $t+\Delta t$, $x_{t+\Delta t}(i,j)$, using the previous state value in time t, $x_t(i,j)$, and the information on the state time-derivative, $\dfrac{dx_t(i,j)}{dt}$. Because (4.84) is an ordinary differential equation, the Euler method can be applied for numerical integration:

$$x_{t+\Delta t}(i,j) \cong x_t(i,j) + \Delta t \frac{dx_t(i,j)}{dt} \qquad (4.95)$$

where the initial status conditions, $x_0(i,j)$, are taken from the initial image matrix $\mathbf{X}_0$.

By substituting (4.95) into (4.84), the next-step state value of cell $C(i,j)$ yields:

$$x_{t+\Delta t}(i,j) = (1-\Delta t)x_t(i,j) + \Delta t \left\{ \sum_{k,l \in N_r(i,j)} a(k,l)y_t(i+k,j+l) + \sum_{k,l \in N_r(i,j)} b(k,l)u(i+k,j+l) + I \right\} \qquad (4.96)$$

In most practical cases the choice of $\Delta t = 0.1$ satisfies the needed accuracy and leads to a convergent solution (Chua and Roska 2001).

A simulator of the linear CNN written in MATLAB is depicted in Fig. 4.32. It requires Image Processing Toolbox installed within MATLAB.

The discrete-time CNN simulator can be easily derived from the continuous-time version shown in Fig. 4.32. We just have to bear in mind that the internal states are computed in successive iterations directly from the inputs and feedback. No time dependence and no time derivatives are involved. Develop your own DTCNN simulator by solving the assignment of Problem 4.6.7 at the end of this section. It is worthwhile to recall that the CNN simulators basically work with greyscale images, where the pixel grey-levels are expected between -1 (white pixels) and $+1$ (black pixels). The same range and meaning is valid for the CNN outputs. Note also the duration of a CNN simulation depends on the number of iterations, which, however, must be decided high enough to reach the steady state.

```
function output=linearCNN(input,state,
                    Atem,Btem,I,dt,steps,bin,bout);
%
% linearCNN - This function simulates the continuous-time
% CNN operation.
%   The meaning of parameters:
%   input - the input image
%   state - the initial values of internal states
%   Atem  - the feedback template
%   Btem  - the control template
%   I     - the bias term
%   dt    - time increment
%   steps - the number of simultaion steps
%   bin   - values outside the input matrix.
%            Could be value (e.g. -1) or
%            'symmetric', 'replicate' or 'circular'.
%            Look at MATLAB help for imfilter.
%            Parameter is optional. If not defined, 0 is used!
%   bout  - values outside the output matrix.
%            Could be value (e.g. -1) or
%            'symmetric', 'replicate' or 'circular'.
%            Look at MATLAB help for imfilter.
%            Parameter is optional. If not defined, 0 is used!
%
%The input image and the internal state matrix must be of the same
%dimensions. The output image adopts these dimensions as well.
% Syntax:
% output=linearCNN(input,state,Atem,Btem,I,dt,steps,bi,bst);
%
% Written by Boris Cigale in July 2005.
%
% Last two parameters are optional. Check if exist,
% and define them if not!
if (not(exist('bin')))
  bin=0;
end
if (not(exist('bout')))
  bout=0;
end

% The constant part of the CNN computation depending on the inputs
% and bias term is calculated only once (entire images, i.e. all
% CNN cells are processed here).
binput=imfilter(input,Btem,bin)+I;

% The output values are clipped according to the saturation
% nonlinear function. To apply signum function, the subsequent
% output computations must be replaced by: output=sign(state);
output=0.5*(abs(state+1)-abs(state-1));

% CNN operates in the preselected number of steps.
% Different termination conditions may be introduced by
% replacing the following FOR-loop with a WHILE-loop.
for n=1:steps,
  % Calculation of the state time-derivatives.
  dstate=-state+imfilter(output,Atem,bout)+binput;
  % Estimation of new states.
  state=state+dt*dstate;
  % Computation of new outputs.
```

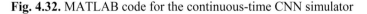

Fig. 4.32. MATLAB code for the continuous-time CNN simulator

Templates

It has been stressed that the behaviour of the CNN is completely defined with the templates **A**, **B** and bias *I*. In our case all templates will be time invariant. Beside that, the input and the initial internal states of the CNN are crucial for its behaviour. Usually, the internal states are initialized equal to the inputs; nevertheless, other possibilities exist (e.g. by initializing all the states to 0). The choice depends on the problem considered and the solution introduced.

It is also very important how we treat the boundary conditions, i.e. the cells at the image boundaries. The most frequent solution is that the virtual input and output pixels outside the image boundaries are assumed zero-valued. Nevertheless, we have seen in the examples of "How to Implement a CNN for Image Processing" some different boundary conditions were necessary. In our example later, all virtual pixels will have value -1.

When applying the CNN simulator to a concrete problem, we already have to have the templates and bias values available. Either we deploy a training phase with one of the algorithms described in Sect. 4.6.4, or we convert one of many elaborated image processing solutions with given CNN templates and bias. Such standard templates can be found in Yang (2002), for example. We will use a few of them in our illustrative examples. When an operation is selected, e.g. optimal edge detection, both templates and bias are given along with the initial values of the input, state and output matrices.

To exemplify the implementation of the predetermined CNN parameters, we are going to make a simple computation on an ultrasound image.

Example 4.6.4:

Ultrasound image processing and recognition need rather sophisticated approaches to be successful. The main problems encountered are speckle noise, weak contrast and reverberations (Yoo 2004). Nevertheless, some even quite simple operations can turn out well in the image preprocessing stages.

In general, the segmentation is one of the first steps in digital image processing (Castleman 1996). Image thresholding is the simplest method where the pixels in the image are classified into two categories: those having the observed feature (usually greyness) below a threshold, and those with the feature equal to or exceeding the threshold. The CNNs are quite capable in thresholding the images, as we have already seen in Example 4.6.1 ("How to Implement a CNN for Image Processing"). Consider here another example of thresholding using the CNN parameters as suggested by Roska et al. (1998) depicted in Fig. 4.33.

$A = \begin{bmatrix} 0 & 0 & 0 \\ 0 & 2 & 0 \\ 0 & 0 & 0 \end{bmatrix}$ $B = \begin{bmatrix} 0 & 0 & 0 \\ 0 & 0 & 0 \\ 0 & 0 & 0 \end{bmatrix}$
 I = threshold, Δt = 1; steps = 20;
 $\mathbf{X_0}$ = Observed image; $\mathbf{U} = \mathbf{0}$;
	`A=[0 0 0;0 2 0;0 0 0];`
	 `B=0;`
	 `image_out=linearCNN(zeros(...`
	 `  size(image_in)),...`
	 `  image_in,A,B,I,1,20);`

Fig. 4.33. CNN for image thresholding: threshold templates (left) and MATLAB implementation (right)

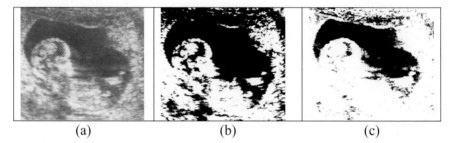

(a)	(b)	(c)

Fig. 4.34. An example of ultrasound recording of uterus with human foetus (**a**), and this image binarized by CNN using the bias term I = 0 (**b**), and I = -0.4 (**c**)

The MATLAB implementation uses our continuous-time CNN simulator as depicted in Fig. 4.32.

The ultrasound image we are going to process is shown in Fig. 4.34a. A human uterus and foetus tissues are more echogeneous, whilst the water in uterus absorbs ultrasound and appears black. After applying a certain threshold, such image would be binarized. The objects which remain visible in the binarized image depend on the threshold value selected. Grey-levels above the threshold become black in the output image, those below the threshold become white. As we can see in Fig. 4.34, middle and right, inechogeneous water remains visible because of its black colour, more echogeneous tissues, however, begin disappearing when thresholds tend towards the value of black (recall the threshold value equals $-I$).

Study also Problem 4.6.8 to comprehend thresholding of colour images. Problem 4.6.9 suggests templates for optimized edge detection to be applied to the image created in Problem 4.6.8.

Combining Several CNNs

Some simpler problems, as we have shown, can be solved using just one CNN, but the majority of real-life tasks cannot be accomplished by a single CNN. Therefore, solutions that use several interconnected CNNs or one CNN applied several times have been reported (ter Brugge et al. 1998; Cigale and Zazula 2004). CNNs can be utilised consecutively, with the output of one CNN being the input into another. On the other hand, CNNs can also be organised in parallel, so that they share the same inputs. In both cases we can consider one CNN as a computer instruction. Like in a programming language, several CNNs can be structured into more complex functions.

We are going to describe the implementation of several CNNs using the flow chart. Other presentations also exist, such as Alpha language (Chua and Roska 2001).

Illustrate the operation of three consecutive CNNs by a modification of Example 4.6.4.

Example 4.6.5:

Figure 4.34, right, shows a big part of the uterus interior in black, whilst the foetus and some other tissues are excluded. To be able to outline the uterus only, we have to get rid of all darker regions which do not belong to the uterus. We can remove them by the following procedure which assumes that the uterus is the largest area in the segmented image. Firstly, all the regions are eroded by morphological operator for erosion. The CNN templates for erosion are depicted in Fig. 4.35a, whilst the resulting image is depicted in Fig. 4.36b. Erosion is performed ten times, until all smaller darker regions disappear. Then the CNN operator which extracts selected regions is applied. The input in the CNN is the binarized image from Fig. 4.36a and the initial state of the CNN is a marker of objects to remain. In this case the image from Fig. 4.36b is used as marker. The result of the CNN operation is depicted in Fig. 4.36c, whilst the templates used are revealed in Fig. 4.35b.

$A = \begin{bmatrix} 0 & 0 & 0 \\ 0 & 0 & 0 \\ 0 & 0 & 0 \end{bmatrix}$ $B = \begin{bmatrix} 1 & 1 & 1 \\ 1 & 1 & 1 \\ 1 & 1 & 1 \end{bmatrix}$ $I = -8, \Delta t = 1;$ steps $= 1; \mathbf{X}_0 = \mathbf{0};$ $\mathbf{U} =$ Binarized input image	$A = \begin{bmatrix} 0.5 & 0.5 & 0.5 \\ 0.5 & 4 & 0.5 \\ 0.5 & 0.5 & 0.5 \end{bmatrix}$ $B = \begin{bmatrix} 0 & 0 & 0 \\ 0 & 4 & 0 \\ 0 & 0 & 0 \end{bmatrix}$ $I = -3, \Delta t = 1;$ steps $= 500;$ $\mathbf{X}_0 =$ Image with markers; $\mathbf{U} =$ Observed image
(a)	(b)

Fig. 4.35. CNN parameters needed in operations depicted in Fig. 4.36: **(a)** erosion templates, **(b)** template for the selected object extraction

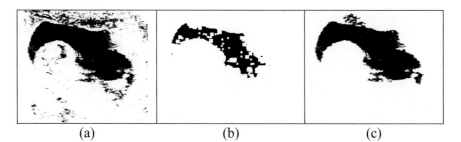

(a)   (b)   (c)

Fig. 4.36. An example of the ultrasound image segmentation using multiple CNNs: (**a**) binarized image obtained by $I = -0.4$ from the first CNN (Example 4.6.4); (**b**) resulting uterus image after erosion performed by the second CNN in ten iterations; (**c**) the object of interest from image (a) is marked by image (b), whereas the selected objects extraction CNN operator is applied

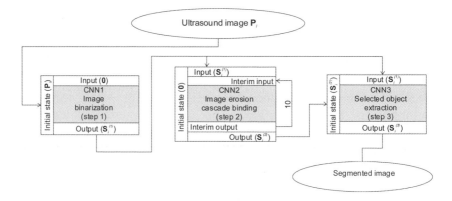

Fig. 4.37. Block diagram of grid with three CNNs solving the problem from Example 4.6.5: the loop with CNN2 indicates iterative execution

A grid of multiple CNNs was needed in Example 4.6.5. Figure 4.37 depicts the operation of this grid in a block diagram. The loop in the diagram designates the number of CNN2 iterations.

4.6.4 Training of CNNs

When we intend to use CNNs for more complex image processing tasks the question arises how to set the template elements. If the explicit solutions have not been published for the problem envisaged, the template elements can also be found by using one of the known training approaches. Whilst many such learning algorithms exist, it should be stressed that none of them is superior to others in all the cases. Thus, the choice of learning algorithm is based on the speed of convergence on a given problem.

The images available for training must be divided into two groups: learning set required by learning algorithm, and testing set which is used for verification of the resulting templates. To avoid biased results, the learning set should not include images which belong to the testing set. The learning set must be carefully chosen to represent a wide range of possible images, because the learning procedure actually carries out a search for global minimum of multivariable non-linear function, also called the cost function.

For every image $\mathbf{P}_i$ used in learning process we should have a corresponding manually annotated binary image $\mathbf{P}_i^{(m)}$. Hence, the desired solution is known, so the quality of the final CNN output could be quantified with a cost function. The original images $\mathbf{P}_i$ should be annotated by experts in the field and the objects in image are usually presented with white contours on the black background.

We can treat learning algorithm as an optimisation problem of minimising the cost function. In this section we are going to consider two optimisation algorithms: genetic algorithm and simulated annealing.

Cost Function

Cost function actually measures the similarity of two images. One of the most used cost functions when dealing with images is Euclidian distance, d_e, between a CNN output image $\mathbf{S}_i$ and the corresponding manually annotated image $\mathbf{P}_i^{(m)}$, both of the same size $M \times N$:

$$d_e = \sqrt{\sum_{x=1}^{M}\sum_{y=1}^{N}\left(p_i^{(m)}(x,y) - s_i(x,y)\right)^2} \qquad (4.97)$$

where $p_i^{(m)}(x,y)$ and $s_i(x,y)$ stand for the pixels with co-ordinates x and y in the compared images containing the manual annotations and the CNN output, respectively.

Euclidian distance as a cost function proves inappropriate when the image objects lie close to each other but it is important that the solution keep them separated. The cost does not change much if just one pixel appears which joins two objects. Utilising Euclidian distance, the trained templates are inclined to merge the object.

In Cigale and Zazula (2004), the authors suggested a modified cost function which is sensitive to narrow gabs between objects. This function compares a set of annotated objects $\mathbf{F}_i^{(m)} = \{f_{1,i}^{(m)}, f_{2,i}^{(m)}, ..., f_{L,i}^{(m)}\}$ of image

$\mathbf{P}_i^{(m)}$ to the regions $\mathbf{R}_i^{(o)} = \{r_{1,i}^{(o)}, r_{2,i}^{(o)}, ..., r_{K,i}^{(o)}\}$ recognised in the CNN output image $\mathbf{S}_i$.

Each region $r_{k,i}^{(o)}$ is checked against all annotated objects $\mathbf{F}_i^{(m)}$. The recognized object $f_{l,i}^{(m)}$ is supposed to be the one that shares the biggest area with region $r_{k,i}^{(o)}$. To verify the similarity between $r_{k,i}^{(o)}$ and $f_{l,i}^{(m)}$, a more sophisticated metrics has been proposed by Potočnik and Zazula (2002).

Potočnik and Zazula (2002). This metrics measures sensitivity and specificity of a recognition algorithm using two ratios, $\rho^{(1)}$ and $\rho^{(2)}$, between the intersection of the compared regions and the whole area of the regions. $\rho^{(1)}$ stands for the ratio between the areas of intersection and annotated object, whilst $\rho^{(2)}$ for the ratio between the areas of intersection and segmented region. The regions totally overlap if both ratios, $\rho^{(1)}$ and $\rho^{(2)}$, are equal to 1. In general, the closer the values of $\rho^{(1)}$ and $\rho^{(2)}$ to 1, the better is the matching of the regions being compared. This also holds for the product of $\rho^{(1)}\rho^{(2)}$, which is taken as a unique index of merit in our new cost function.

After all the segmented regions are compared to all the annotated objects in an image, the matching of each pair is obtained by:

$$m_i(k,l) = \rho_i^{(1)}(k,l)\rho_i^{(2)}(k,l) \tag{4.98}$$

where indices k and l stand for the kth segmented region and lth annotated object, respectively, whilst i denotes the ith image. These values are considered the elements of matrix $\mathbf{M}_i$ whose rows correspond to the segmented regions in image $\mathbf{S}_i$ and columns to the annotated objects in $\mathbf{P}_i^{(m)}$.

Selection of the best fitting pairs is based on scanning the matrix $\mathbf{M}_i$ for maximum $m_i(k,l)$ value, i.e. finding the region $r_{k,i}^{(o)}$ which best recognizes the object $f_{l,i}^{(m)}$. The found value contributes to the cost, whilst row k and column l in the matrix $\mathbf{M}_i$ are omitted from further iterative comparisons. The procedure is repeated until all the objects are detected or the maximum value in the matrix is zero (there is no further matching at all).

The result of these comparisons, i.e. the cost, corresponds to the sum of the $m_i(k,l)$ values for the recognized objects. In order to limit the value in the interval [0, 1], the average $m_i(k,l)$ is calculated by dividing the obtained cost by the number of recognized objects, O:

$$\varepsilon_i = \frac{1}{O}\sum_{k,l} m_i(k,l) ; \, k,l \in \{\max \arg (\mathbf{M}_i)\}, N \leq L. \tag{4.99}$$

Some undetected annotated objects or some superfluous regions can exist. The number of such regions is taken into consideration using two factors which multiply the cost as calculated by (4.99). The first factor is a ratio between the number of recognized objects, O, and the number of all regions in $\mathbf{R}_i^{(o)}$, K, the second factor is a ratio between the number of recognized objects and the number of all objects in $\mathbf{F}_i^{(m)}$, L.

The algorithm returns the cost value, ε_i, always between 0 and 1. The higher the value the better is the solution. Such behaviour is typical for fitness function, used in genetic algorithm, which is inversely proportional to the cost function value. The MATLAB code of the fitness computation is depicted in Fig. 4.38.

```
function fit = czfitness(R, F);
%
% This function implements fitness function
% described in (Cigale, 2004).
% Input:
%    F - binary image with annotated objects
%    R - binary image with recognised regions
% Function returns:
%    fit - fitness in [0,1]. Higher values mean
%          bigger similarity of input images.
%
% Written by B. Cigale, August 2005.
%
% Syntax: fit = czfitness(R, F);
%

[Ri,K]=bwlabel(R);
% If there are to many objects or no objects at all
if (K>50)||(K==0)
  fit=0; return;
end
% Annotated objects are labelled.
[Fi,L]=bwlabel(F);
Mi=zeros(K,L); % Matrix Mi is filled with zeros.
% Detect objects from Fi that intersect with the regions from Ri:
intrscF=Fi.*(Ri>0); fi=unique(intrscF);
% First detected region is background. Delete it.
fi(1)=[];
% Detect objects from Ri that intersect with the regions from Fi
intrscR=Ri.*(Fi>0); ri=unique(intrscR);
ri(1)=[];
% Calculate the area for objects in Fi
% (This information is needed later.)
for f=1:length(fi),
  area_f(f)=sum(Fi(:)==fi(f));
end
```

(Fig. 4.38 Contd.)

```
% For every significant region from Ri:
for r=1:length(ri),
  R=Ri==ri(r);        % isolate region on image
  area_r=sum(R(:));   % calculate area of region
  % For every significant region from Ri:
  for f=1:length(fi),
    % calculate area of the intersection
    % between ri(r) and fi(f).
    intrsc=(Fi==fi(f)).*(R);
    intrsc_area=sum(intrsc(:));
    if (intrsc_area>0)  % If area greater than 0:
      % calculate ratio rho1
      rho1=intrsc_area/area_f(f);
      % calculate ratio rho2
      rho2=intrsc_area/area_r;
      mi=rho1*rho2;
      % Product of rho1 and rho2 goes to matrix Mi.
      Mi(ri(r),fi(f))=mi;
    end
  end
end
% Begin the fitness calculation:
epsilon=0;
N=0;
% While not all the objects detected:
while any(Mi(:)),
  % find the maximum value in Mi,
  % i.e. the best fitting pair:
  [v_max, kl]=max(Mi(:));
  % calculate which region best recognises the
  % object:
  [k,l]=ind2sub(size(Mi),kl);
  % Region k and object l are now omitted from
  % further comparison.
  Mi(k,:)=0;
  Mi(:,l)=0;
  % The number of recognised objects is increased:
  N=N+1;
  % Add value to the fitness
  epsilon=epsilon+v_max; %
end
% The fitness is normalised and multiplied by the
% ratios of the number of recognised objects
% against all objects and of the number of
% recognised regions against all regions.
fit=(epsilon*N)/(K*L);
```

Fig. 4.38. The MATLAB code of fitness function

Genetic Algorithm

Genetic algorithm (GA) is a stochastic optimisation algorithm inspired by evolutionary biology, therefore the terminology is taken from the same field. The idea is to have a population of chromosomes, where each chromosome encodes a possible solution of the problem to be solved. The GA evolves those chromosomes iteratively in order to improve the encoded solution. The evolution is based on the mechanisms of natural selection and genetics in the form of selection, mating and mutation (Haupt and Haupt 2004).

Traditionally, GA treats the variables of the function to be optimised as binary strings of 0s and 1s. If the function has V variables given by v_1, $v_2, ..., v_V$, then the chromosome is written as a V-element row vector $c = [v_1, v_2, ..., v_V]$, where each value is coded in binary form with a predefined number of bits. If, for example, variables v_1, v_2, and v_3 appeared with values 5, 7, and 4, the chromosome c would be coded as $c = [101111100]$, supposing that each variable occupies three bits.

The GA which uses the earlier mentioned coding is named the binary genetic algorithm. To arrange the variables' codes in one long binary sting, a coding and a decoding function must exist. Note that such binary representation is more suitable for integer-valued parameters and less for the floating-point parameters.

Binary GA is usually outperformed by the continuous, or real-valued, generic algorithm, where the chromosomes comprise the floating-point parameters. The V variables given by v_1, $v_2, ..., v_V$ are written as an array of V elements, so that $c = [v_1, v_2, ..., v_V]$, where each value is coded as a floating-point number. If, for example, variables v_1, v_2, and v_3 appeared with values 5.2, 7.4, and 4.2, the corresponding chromosome c would be coded as $c = [5.2, 7.4, 4.2]$.

The difference of the internal structure of chromosomes in binary and continuous GA necessitates different approaches to the mating and mutation operators, whereas all other steps in deriving the algorithms can be the same.

The chromosomes of one iteration are supposed to form one generation. All the chromosomes in each generation are evaluated by the fitness function which is inversely proportional to cost function – higher values mean a better solution.

In order to proceed towards a converging solution, the so called *natural selection* is deployed which get rid of less successful chromosomes. Only the fittest chromosomes survive to mate and, thus, participate in creation of the new generation. There are two possibilities to implement this strategy. The first one sorts all chromosomes in a generation according to their fitness. Only a predefined number of chromosomes survive, the rest are deleted. There is no rule how many chromosomes should survive, but 50% is the most suggested value (Haupt and Haupt 2004). Another strategy of natural selection utilises thresholding. In this approach only the chromosomes whose fitness surpasses a given threshold survive. If the fitness of all the chromosomes remains below this threshold, a totally new generation is generated.

```
new_generation=[generation(ind(1),:); generation(ind(2),:)];
% The rest of the generation is generated in standard way:
% selection, mating and mutation.
for f=3:(number_c/2)+1,
    % Find the index of the first chromosome whose cumulative
    % probability is greater than or equal to a random number.
    % This is the first parent.
    i=find(rand<=cdf);
    parent1=generation(ind(i(1)),:);
    % Find second parent.
    i=find(rand<=cdf);
    parent2=generation(ind(i(1)),:);
    % One point crossover approach is implemented:
    % calculation of crossover point.
    cross=round(rand*(length_c-1));
    % Generation of offspring:
    offspring1=[parent1(1:cross) parent2(cross+1:length_c)];
    offspring2=[parent2(1:cross) parent1(cross+1:length_c)];
    % Mutation - offspring 1:
    if (rand<mutaton_probability)||...
        (strcmp(offspring1,parent1))||(strcmp(offspring1,parent2))
        % In case of mutation a random bit is altered.
        p=round(rand*(length_c-1))+1;
        offspring1(p)='1'-offspring1(p)+'0';
    end
    % Mutation - offspring 2:
    if (rand<mutaton_probability)||...
        (strcmp(offspring2,parent1))||(strcmp(offspring2,parent2))
        p=round(rand*(length_c-1))+1;
        offspring2(p)='1'-offspring2(p)+'0';
    end
    % Add both offspring to the new generation:
    new_generation=[new_generation; offspring1; offspring2];
end
% Previous generation is replaced by the new one.
generation=new_generation;
end
```

Fig. 4.40. MATLAB code of binary GA

Template Training Using Continuous GA

In contrast to binary GAs, the continuous GA templates need no coding/decoding because of the floating-point variables used. However, it is recommendable to limit the template values to approximate range of [–5, 5], for the same reason as when using binary GAs.

Figure 4.41 reveals which parts of the binary GA MATLAB implementation must be modified in order to create the continuous GA algorithm. Both templates **A** and **B** are assumed of size 3×3. Together with the bias term we have 19 variables. The first generation is generated randomly and all variable values lie between –5.12 and 5.11. In each generation 50% of the fittest chromosomes survive. The best two chromosomes always go to the next generation. Cost weighted random

```
% Function receives a different name:
function [chromosome,fit_list] = continuous_GA(fcost,cr);

% The chromosome length is different to the one with binary GA:
length_c=19; % Number of variables in a chromosome.
% First generation must be generated with floating-point numbers:
% First generation is generated randomly:
generation=10.23*rand(number_c,length_c)-5.12;

% Continuous fitness function must be called as follows:
  % Calculate the fitness. EXTERN function which
  % returns the fitness for each chromosome.
  fit=fitness(generation,fcost,'cont');

% Instead of one-point-crossover approach implemented in binary
% GA, continuous GA implements the following (after
% parent2=generation(ind(i(1)),:); in Fig. 4.40):

    % Make a random string of 0s and 1s to decide which variable
    % goes where. When 1, the variable goes from parent 1 to
    % offspring 1, otherwise the variable goes from parent 1
    % to offspring 2.
    swap=round(rand(1,length_c));
    % Generation of offspring:
    offspring1=swap.*parent1+(1-swap).*parent2;
    offspring2=(1-swap).*parent1+swap.*parent2;
    % Mutation - offspring 1:
    if rand<mutaton_probability
      % In case of mutation the random
      % variable is replaced by new value.
      p=round(rand*(length_c-1))+1;
      offspring1(p)=10.23*rand-5.12;
    end
    % Mutation - offspring 2:
    if rand<mutaton_probability
      p=round(rand*(length_c-1))+1;
      offspring1(p)=10.23*rand-5.12;
    end

% Continues by "Add of both offspring to the new generation,"
% as in Fig. 4.40.
```

Fig. 4.41. The modifications from binary to continuous GA

pairing is implemented to select chromosomes to mate. A randomly generated binary string decides which one of the two offspring receives which variable from the parents. The probability of mutation is 10% for each offspring. In case of mutation, the value of a randomly selected variable in chromosome is changed.

CNN Training Example

The following paragraphs explain how to train a CNN to detect image objects whose learning set is available. The explanation is backed by an example of the detection of ovarian follicles. Actually, to train a CNN it means its execution on the images comprising the learning set in an iterative manner. The template values are adjusted after each iteration to obtain a better fitness. The fitness is evaluated according to the referential annotations in the learning set.

In this section, we use only one 1-neighbourhood CNN. The time step Δt is set to 0.3 and 30 steps are performed in each simulation. The virtual pixels in input and output images are set to -1. The initial state is equal to the input image. The values in the input image are scaled according to (4.87). The referential image is assumed to identify the annotated objects as black regions on the white background. Dealing with ovarian detection, these regions correspond to manually annotated follicles.

In the learning set we have two subsets: ovarian ultrasound images and manually annotated positions of follicles in each image. Figure 4.42a depicts one of those ultrasound images. Manually annotated positions of follicles are depicted in Fig. 4.42b.

Suppose the CNN training is done both with the binary and continuous GA, and in both cases Euclidian distance is used as fitness function. The resulting detection capabilities of the CNNs trained this way are illustrated by Fig. 4.42c for binary and by Fig. 4.42d for continuous GA.

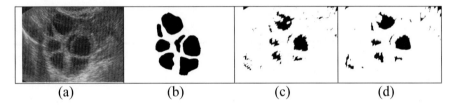

(a)   (b)   (c)   (d)

Fig. 4.42. An example of CNN training on ovarian follicles: (**a**) original image; (**b**) annotated positions of follicles; (**c**) the detection result after the CNN trained by binary GA; (**d**) the detection result after the CNN trained by continuous GA

The code of generic fitness function is depicted in Fig. 4.43, whilst Fig. 4.44 shows the variant with Euclidian distance. The generic fitness function is called from the GA code (Figs. 4.40 and 4.41) with an adequate name of the distance-measuring function to be respected.

```
function fit=fitness(generation, fcost, type);
%
%The general fitness function.
%Inputs:
%  generation - the matrix with chromosomes
%  fcost      - the name of fitness function (i.e. 'efitness')
%  type       - the type of GA. Could be 'binary' or 'cont'.
%Output:
%  fit        - fitness
%Global variables:
%L_IMAGES     - Input images in learning set
%L_INITIAL    - Images for initial state
%L_MASKS      - Expected results
%L_NO_IMAGES  - The number of images in learning set
%
%Written by B. Cigale, August 2005.
%
%Syntax: fit=fitness(generation, fcost, type);
%
global L_IMAGES;
global L_MASKS;
global L_INITIAL;
global L_NO_IMAGES;

fcall=[fcost '(output,mask)'];
[r,c]=size(generation);
for f=1:r,
  if strcmp(type,'binary')
    [A,B,I]=decoding(generation(f,:),[3 3],[3 3]);
  else
    I=generation(f,1);
    A=generation(f,2:10);
    B=generation(f,11:19);
  end

  cost=0;
  for g=1:L_NO_IMAGES,
    im=squeeze(L_IMAGES(g,:,:));
    in=squeeze(L_INITIAL(g,:,:));
    output=linearCNN(im,in, A, B, I,0.3, 30);
    output=output>0;
    mask=squeeze(L_MASKS(g,:,:));
    cost=cost+eval(fcall);
  end
  fit(f)=cost/L_NO_IMAGES
end
```

Fig. 4.43. The generic form of fitness function

The fitness is expected to change and improve (increase) through the generations. Figure 4.45 illustrates how the fitness is evolved in our experiment with ovarian follicles. Note the obtained fitness is normalised, i.e. all values are in the interval [0, 1]. We generated 30 generations, each with 30 chromosomes.

```
function fit = efitness(R, F);
%
% This function implements Euclidian fitness function.
% Input:
%      F - binary image with annotated objects
%      R - binary image with recognised regions
% Function returns:
%      fit - fitness between [0,1]. Higher values mean
%            bigger similarity of input images.
%
% Written by B. Cigale, August 2005.
%
% Syntax: fit = efitness(R, F);
%
s=size(R);
po=prod(s);
d=sum(abs(R(:)-F(:)));
fit=1-d/po;
```

Fig. 4.44. The variant of fitness function with Euclidian distance

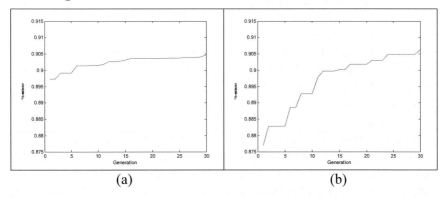

(a)   (b)

Fig. 4.45. Increase of the fitness of the best chromosome in each generation: (**a**) when trained by binary GA; (**b**) when trained by continuous GA

Note that both GAs returned similar results after 30 generations, so neither is better in our case. Also visual inspection of the resulting segmentation in Fig. 4.42c, d confirms this assumption. Both CNNs generate many false positives, so many dark areas are segmented which do not belong to the follicles. The detected follicles just roughly correspond to the actual follicles. Nevertheless, the result looks promising, and we are going to further improve it by the implementation of multiple CNNs as described in Sect. 4.6.5.

Simulated Annealing

Simulated annealing (SA) is yet another of optimisation algorithms based on natural phenomena. Annealing is a term from metallurgy and it describes the graduate cooling of heated material in order to improve its properties such as strength and hardness. The heat causes the atoms to become unstuck from their initial positions (a local minimum of the internal energy) and wander randomly through states of higher energy; the slow cooling gives them more chances of finding configurations with lower internal energy than the initial one (Kirkpatrick et al. 1983).

SA begins the search for optimal solution in arbitrary initial state. The state is actually a vector of the cost function variables. These variables are then randomly modified and, thus, form a new state. The size of modification is proportional to the system temperature, which in fact is an additional control variable. Each generated state is evaluated by the cost function. The obtained result is supposed to equal the energy in SA. According to the acceptance rule, the generated state replaces the current state or it is rejected.

Acceptance rule is usually the Bolzmann probability function $p = \exp(-(E_2 - E_1)/kT)$, where T is the system temperature, $k = 1$, and E_1 and E_2 are the estimated cost function values belonging to the current and the new state, respectively. The new generated state always replaces the current one if its cost is lower. Otherwise the state is accepted according to the calculated probability. An important effect of this rule is that a temporary worse solution can be accepted. This may help escape from a local minimum.

During a training session, the system temperature is gradually lowered according to the cooling schedule. If the temperature is lowered too quickly, the system ends in local minimum. On the other hand, if the temperature decreases too slowly, the SA unnecessarily consumes time. Many strategies as how to decrease the temperature exist (Magnussen et al. 1993). Consider just two of them. The so called arithmetic strategy lowers the temperature by some constant C, so that the new temperature T_n is calculated as difference $T_n = T_{n-1} - C$, where T_{n-1} stands for the previous temperature. Geometric strategy, on the other hand, decrease the temperature faster, using constant α, so that $T_n = \alpha \cdot T_{n-1}$. However, geometric strategy does not guarantee the global minimum convergence.

Different strategies for keeping the temperature at the same level have also been proposed (the inner loop in SA). The temperature can be kept stable for a fixed number of generated states or a fixed number of accepted states. Another, more advanced strategy says the temperature changes only if the average cost of accepted states changes.

The duration of the SA runs (the outer loop) again depends on different strategies. The simplest strategy maintains running until the system temperature exceeds a predefined temperature T_e.

Step 1: Rough Detection of Expressive Follicles. Our segmentation method starts with CNN1 where the given ultrasound image P_i from learning set is transformed into the initial state X_0 and the input U. This CNN is trained to detect follicles $F_i^{(m)}$. To avoid false detected regions, the fitness function depicted in Fig. 4.38 is used in training process. At this step we are interested just in the position of the follicles and not so much in the shape, so the last line in fitness function is changed from fit=(epsilon*N)/(K*L) to fit=(N*N)/(K*L). We train the CNN with the time step $\Delta t = 0.3$ in 30 steps.

Coefficients of the resulting templates **A** and **B**, and bias I obtained using our learning set of four images are as follows:

$$\mathbf{A} = \begin{bmatrix} 4.44 & -2.41 & 3.20 \\ -0.21 & -0.68 & 4.12 \\ 3.56 & -0.37 & 1.65 \end{bmatrix} \mathbf{B} = \begin{bmatrix} -3.15 & 1.32 & -1.75 \\ -4.72 & -0.12 & 0.22 \\ 2.09 & 1.29 & -0.60 \end{bmatrix} I = -4.14. \quad (4.103)$$

denote the resulting image by $S_i^{(1)}$.

It has been established empirically that the CNN1 trained by annotated follicles $F_i^{(m)}$ yields the templates as in (4.73) and is capable of only a rough detection of expressive, i.e. dominant follicles. Therefore, the detection procedure needs additional steps of refinement.

Step 2: Expansion of Detected Expressive Follicles. The detected follicles in the image $S_i^{(1)}$ generated in step 1 just roughly correspond to the actual follicles. They are mainly smaller, but in the correct position, so the ratio $\rho^{(2)}$ is small. To improve this ratio, a new CNN2 is introduced. Of course, CNN2 has to proceed from $S_i^{(1)}$, thus $S_i^{(1)}$ is assigned to the initial state X_0 of CNN2, whilst input U is again P_i. We train the CNN with the time step $\Delta t = 0.3$ in 30 steps.

The output of this CNN is evaluated with the fitness function depicted in Fig. 4.38 during the training phase. In our case, the coefficients of the resulting templates **A** and **B**, and bias I for CNN2 yield:

$$\mathbf{A} = \begin{bmatrix} 4.47 & -0.09 & 3.10 \\ 3.28 & -4.58 & 2.71 \\ 1.48 & 3.91 & 2.84 \end{bmatrix} \mathbf{B} = \begin{bmatrix} 3.20 & 4.53 & 2.83 \\ -5.00 & 5.03 & 3.53 \\ 3.43 & -2.32 & -0.86 \end{bmatrix} I = 2.29. \quad (4.104)$$

The resulting templates expand the regions in $S_i^{(1)}$. Unfortunately, this CNN sometimes joins more regions if the border between them is very dim. Such amalgamations are usually quite detectible, because the resulting area has a typical 8-shape area and could be splitted using some morphological operation in the post-processing phase. One of possible approaches was reported by Cigale and Zazula (2004).

An important contribution of step 2 is a significant improvement of ratio $\rho^{(2)}$ for all the regions in $\mathbf{S}_i^{(1)}$ corresponding to annotated follicles $\mathbf{F}_i^{(m)}$. Again, the output of step 2 is an image which is going to be designated by $\mathbf{S}_i^{(2)}$.

Step 3: Detection of Ovary. The detection of ovary is very important to get reliable results. If the position of the ovary is known, then we can get rid of the phantom follicles lying outside the ovary. As it was mentioned before, the follicles are usually the darkest objects in the ultrasound images of ovary. The position of follicles in ovary varies, but usually some can be found close to the ovarian borders.

To recognise the ovary as reliable as possible, we make use of the fact that the dominant follicles, which are detected in the second step of our algorithm, point out the area where the ovary is to be expected. Therefore, the resulting image from step 2, $\mathbf{S}_i^{(2)}$, is assigned to the initial CNN3 state and the CNN3 input is fed by the original image $\mathbf{P}_i$. We train the CNN with the time step $\Delta t = 0.3$ in 45 steps.

The CNN3 training is done by annotated ovaries $\mathbf{O}_i^{(m)}$. Preliminary investigation showed the obtained results from CNN3 need some improvements, to pinpoint the ovary better. As far as the dominant follicles are concerned, they expand rather well to the borders of the ovary. But all the regions outside the ovary, such as blood vessels, grow up too. The assumption during training was the area of the region covering the ovary is bigger than the areas of other regions, so we eliminate them and thus obtain a coarse position of the ovary. The shape of the ovary is improved by encapsulating it in the convex hull (Barber et al. 1996). These corrections of the CNN3 output are done before the fitness function is computed. Whilst there is just one object to be followed, the Euclidian distance is recognised as an acceptable fitness function. The resulting image is denoted by $\mathbf{S}_i^{(3)}$.

Coefficients of templates $\mathbf{A}$ and $\mathbf{B}$, and bias I obtained by our learning set are as follows:

$$\mathbf{A} = \begin{bmatrix} 4.09 & 3.77 & 0.59 \\ -0.97 & -3.65 & 2.31 \\ 3.94 & 0.98 & 1.46 \end{bmatrix} \quad \mathbf{B} = \begin{bmatrix} -4.42 & 2.49 & 2.23 \\ 2.52 & -1.44 & 2.54 \\ 3.97 & -3.18 & 3.53 \end{bmatrix} \quad I = 4.12. \quad (4.105)$$

Step 4: Post-Processing. In the last step, the results from steps 2 and 3 are joined. The ovary detected in step 3 is taken into the consideration to get rid of the phantom follicles detected in step 2. Each follicle in image $\mathbf{S}_i^{(2)}$ lying outside the ovary region is erased. The simplest approach is to use AND operation on both images. The AND operation can also be implemented using an additional CNN (Roska et al. 1998). The final result is denoted by $\mathbf{S}_i^{(4)}$.

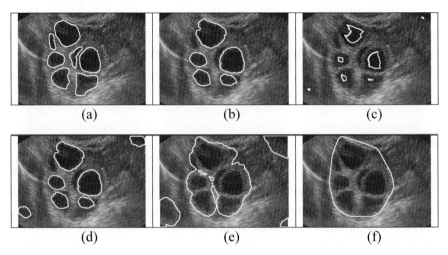

(a)   (b)   (c)

(d)   (e)   (f)

Fig. 4.48. Typical results of multiple-CNN ovarian follicles segmentation: (a) manually annotated positions of follicles, (b) final CNN segmentation image $S_i^{(4)}$, (c) roughly detected expressive follicles $S_i^{(1)}$ using CNN1, (d) expanded expressive follicles $S_i^{(2)}$ using CNN2, (e) position of the ovary as recognised by CNN3 without convex hull, (f) recognised position of the ovary, $S_i^{(3)}$, after implementation of the convex hull

The presented multiple-CNN detection of ovarian follicles is exemplified by Fig. 4.48.

Segmentation Results

The CNN recognition efficiency should be tested independent of learning population. In our case the testing set contained 28 images belonging to different patients. The decision whether a follicle was recognised in its correct position or not is based on ratio $\rho^{(2)}$. We assume the segmented region represent a follicle if $\rho^{(2)} > 0.5$. This criterion guaranties that more than a half of the recognised region overlays a particular annotated follicle. In 28 images 168 follicles were annotated. The algorithm recognised 81 regions of which 63 (78%) belong to the follicles. If the criterion for $\rho^{(2)}$ is changed to $\rho^{(2)} > 0$, the number of recognised follicles raises to 74.

Visual inspection showed that some regions are not big enough, so they are smaller then the annotated follicles which they cover. Some of inexpressive follicles were not recognised. Both problems could be solved by introduction of a new step which would be similar to step 1 but with lower bias in order to recognise the inexpressive follicles. Regions which cover two annotated follicles lying close to each other are also problematic. Some of such regions can be recognised and split in the post-processing step, because they have a characteristic shape of number eight.

In spite of all mentioned drawbacks, the derived multiple-CNN method is quite efficient and reliable when used for the dominant follicles segmentation. A successful implementation of this approach was applied in clinical practice, as reported in Vlaisavljević et al. (2003).

It must be stressed that the obtained templates are not universal and not optimal for all possible ovarian ultrasound images. Results should improve if bigger learning sets would be used, but the CNN training sessions would take longer then.

4.6.6 Conclusion

In this section, we have surveyed the theory of cellular neural networks and demonstrated their potential in the field of digital image processing. Planar CNNs perfectly fit such assignments, not only for their structure which tenders dynamic behaviour of simple cells, each one for an image pixel, but also for their flexibility, training possibilities, local connectivity and processing performance, which cope with real time when applied in the form of VLSI chips.

We experimented with the CNN computer simulators in order to develop and assess approaches to the ultrasound medical image processing and object recognition. The CNN concepts and solutions described in Sect. 4.6.2 pave the derivation of CNN simulators in Sect. 4.6.3. We deal both with continuous-time (CT) and discrete-time (DT) CNN implementations. Whilst a CT CNN simulator code, written in MATLAB language, is revealed and commented in detail, a DT version is left over to the reader in Problem 4.6.7 at the end of this section.

Referring to Hänggi and Moschzty (2000), the stability of CNNs turns out to be equal for both the CT and DT implementation. A difference may appear only in the conditions based on the templates. Whereas symmetrical templates guarantee a stable evolution for the CT CNNs, this condition is not sufficient in the case of DT CNNs. Robustness depends on the set of implemented templates and the output non-linearity as well. Whilst the same templates can be used for CT and DT application, also the robustness issue is independent of the application, except for the output non-linearity where the signum function outperforms the saturation function. However, it is worthwhile to mention that not all problems can be solved by DT CNNs. If the templates are locally irregular, only the CT CNNs guarantee proper results.

Using CT-based simulation, the issues of integration and the integration step size must be considered. In the majority of practical implementations the system dynamics need not be very precise, so an Euler algorithm for numerical integration satisfies thoroughly. It can be combined with the integration step size which equals internal time constant of the CNN cells. This is an optimum decision leading to the correct equilibrium point and, at the same time, reaching it in a finite number of steps with minimum computational effort.

On the other hand, Manganaro et al. (1999) advocates three advantages of discrete-time implementation of CNNs: many systematic approaches have been recommended for an analytical DT CNN template design by solving the corresponding system of inequalities; DT CNN can be very robust when the smallest internal state variations surmount possible changes of the template coefficients; and the computational speed depends only upon the iteration clock cycle.

The most important properties of CNNs viewed in the light of image processing are locality, flexibility and learning capability. The processing knowledge is built in the cells' templates which provide intelligent behaviour. We explained the CNN training in Sect. 4.6.4. We paid particular attention to genetic algorithms owing to their successful imitation of natural selection, as proven elsewhere. A thorough MATLAB code was derived for both the binary and continuous GAs.

Training of CNNs for ultrasound image processing based on GAs, as shown in Sects. 4.6.4 and 4.6.5, has proven to be fast and efficient. At the same time, the examples of ultrasound recordings of women's ovaries enlighten the inadequacy of a single CNN for coping with very complex situations. The ultrasound image processing can be classified very complex because of, firstly, high level of speckle, and other types of, noise as well as low spatial and contrast resolution, and, secondly, similar appearance of the structures and objects to be discerned. Dealing with the ovarian follicles detection, for example, blood vessels, uterine structures or abdominal tissues may look very like follicles, not only concerning their echogeneity but also in their shape.

To overcome this problem we decided to take into account multiple interconnected CNNs. The first step in designing such grid of CNNs must stratify the initial complex situation into simpler layers of tasks that may be allotted to single CNNs. The second step determines the necessary interconnections among these CNNs. Then a learning phase begins. The most straightforwards approach trains every CNN separately. It needs an adequate learning set at a time. As we proceeded in Sect. 4.6.5, separate annotations were prepared for the follicles and for the ovary. Two CNNs were trained separately to be able, one to detect follicles and accidentally all the image objects that resemble the follicles, and the other to determine the region of ovary. The latter actually delineates only those objects residing inside the ovary and recognised by the former CNN, thus being the follicles in fact.

More advanced training techniques may be devised with the grids of CNNs. One, which is a logical upgrade of the simplest separate approach, involves all the CNNs from a grid into the training session at the same time. Fitness functions for individual CNNs still play their role, and so do the learning sets. Additionally, there is also a global fitness function which measures the success of the entire grid of CNNs. It must be sensitive to the partial contributions of individual CNNs, and is very problem-dependent. For example, if a multiple-CNN follicle recognition declares large polygonal and heterogeneous regions as follicles, which should actually be round and homogeneous, the follicle-detection CNN must evidently be retrained. On the other hand, if the final result comprises detected object all over the analysed ultrasound image, the ovarian region has apparently been missed and the corresponding CNN must be retrained. Readers can experiment and try out their own multiple-CNN training proposals by following Problems 4.6.17 and 4.6.18 at the end of this section.

Even a more challenging idea suggests a self-organising grid of CNNs, where the CNNs are selected and linked into the grid according to a global genetic algorithm and a global fitness function. Future research and development in the CNN implementation to image processing will certainly tackle similar ideas.

Problems

Problem 4.6.1

Assume binary input images and the signum non-linear output function. Construct the CNN which will generate a black pixel only if the corresponding input pixel is encompassed by at least six black pixels. Determine the control template **B** and the bias I; put down the conditional expression (see examples in "How to Implement a CNN for Image Processing").

Problem 4.6.2

Elaborate the same assignment as in Problem 4.6.1, with the only difference that an output pixel appears black if the input pixel's diagonal neighbours are black, whilst the other neighbours are white.

Problem 4.6.3

To solve this problem, let us relax the constraints on discrete-time CNNs as imposed in Harrer and Nossek (1992). Imagine the input images with all possible grey-levels normalized within the interval $[-1,+1]$ (white to black). Assume also the saturation output function (see Fig. 4.29). Now, construct a lowpass filter acting on the 1-neighbourhood in such a way that the output image comprises all possible grey-levels between -1 and $+1$ as well (determine **B** and I).

Problem 4.6.4

Combine the solution from Problem 4.6.3 with thresholding (Example 4.6.1, "How to Implement a CNN for Image Processing") in order to create black pixels wherever the lowpass filtered 1-neighbourhood from the input surpasses the grey-level 1.25. Construct the template **B** and the bias *I*; put down the conditional expression. What kind of the output non-linearity do you need?

Problem 4.6.5

To obtain the suggested solution in Example 4.6.3 ("How to Implement a CNN for Image Processing"), one can verify that just two different situations must be taken care of: the output-image pixel $y_n(i,j)$ may change in the $n+1$th iteration only if $y_n(i-1,j)$ is white and $y_n(i,j)$ and $y_n(i+1,j)$ are black, or if $y_n(i-1,j)$ is black and $y_n(i,j)$ is white, whilst for $y_n(i+1,j)$ the colour does not matter. The first situation means deleting upper black pixels in a column with several contiguous black pixels, whilst the second situation corresponds to iterative extending of black pixels down into "white holes". Illustrate these two cases by the following scheme:

Situation 1   Situation 2

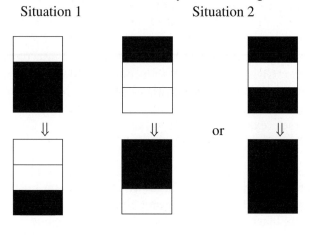

The proposed template $\mathbf{A} = \begin{bmatrix} 0, & 2, & 0 \\ 0, & 1, & 0 \\ 0, & -1, & 0 \end{bmatrix}$ works if the boundary condition on the top and bottom row of $\mathbf{Y}_0$ introduces all white pixels (values -1) and if the signum function is deployed for the non-linearity.

Verify the correctness of **A** by constructing a truth table giving $x_n(i,j)$ and $y_{n+1}(i,j)$ vs. $y_n(i-1,j)$, $y_n(i,j)$ and $y_n(i+1,j)$.

Problem 4.6.6

A discrete-time CNN is given with the following parameters:

$$\mathbf{A} = \begin{bmatrix} 0, & 0, & 0 \\ 0, & 1, & 0 \\ 0, & 0, & 0 \end{bmatrix}, \quad \mathbf{B} = \begin{bmatrix} 0, & -1, & 0 \\ 0, & 1, & 0 \\ 0, & -1, & 0 \end{bmatrix}, \quad \text{and } I = -2.$$

The boundary conditions for all virtual cells (those outside the input image boundaries) are set to −1, and the initial output conditions equal $\mathbf{Y}_0 = \mathbf{0}$.

Write the equations for the evolution of internal states and the CNN's outputs (consider the signum non-linearity). Then draw a simple input image matrix, say 6 × 6, with one horizontal, one vertical, and one diagonal line of at least two pixel lengths. Calculate manually two or three steps of the CNN evolution. What do you notice in the created output image?

Problem 4.6.7

Write your own simulator of discrete-time CNNs. Copy the design presented in Fig. 4.32 and adapt it according to the explanation in "How to Implement a CNN for Image Processing" and (4.86). Bear in mind there is no time-derivative calculation with DTCNNs, so you just have to compute new state values out of the inputs and output feedbacks, and then determine new output values using the corresponding output non-linearity.

Problem 4.6.8

Obtain an arbitrary colour image and get it ready for binarization using continuous-time CNN simulation. First, the RGB image should be converted to greyscale. Note that white pixels transform into value −1, and black pixels into value 1. Experiment with changing the bias I and compare the obtained results. Change the duration of simulation; does the result change?

Problem 4.6.9

Test the following template for optimal vertical edge detection:

$$\mathbf{A} = \begin{bmatrix} 0 & 0 & 0 \\ 0 & 0 & 0 \\ 0 & 0 & 0 \end{bmatrix}, \quad \mathbf{B} = \begin{bmatrix} -0.11 & 0 & 0.11 \\ -0.28 & 0 & 0.28 \\ -0.11 & 0 & 0.11 \end{bmatrix}, \quad I=0.$$

Use the image obtained as a result of Problem 4.6.8 for the input in this assignment. What kind of edges do you detect? Try out also an image with horizontal, vertical and diagonal lines. Change the template to detect horizontal edges (simply transpose matrix **B**).

Problem 4.6.10

Study and implement in the MATLAB code the problem from Example 4.6.5 described in "Combining Several CNNs". Test your MATLAB program with the crt_baby.png image. Note that white pixels transform into value −1 and black pixels into value 1! Put virtual pixels outside the image boundaries to value −1. The image can be downloaded from http://storm.uni-mb.si/CNNExamples.

Problem 4.6.11

The coding function for template parameters into a binary string is depicted in Fig. 4.39. Write a decoding function in MATLAB, which will extract the template parameters from the binary string. Assume the size of templates **A** and **B** is 3 × 3 and each parameter is coded with ten bits. Test the solution by coding and decoding random parameters. The values before coding and after decoding must match.

Problem 4.6.12

Usually the template values are just concatenated into binary strings. Implement the variation of parameter coding where you put the individual bits of consecutive parameters next to each other, so that all first bits are followed by all second bits of all the parameters, etc. Write coding/ decoding function and test it with the binary GA.

Problem 4.6.13

In binary GA depicted in Fig. 4.60, one-point crossover is implemented for mating. Change the mating to (a) two-point crossover and (b) totally random crossover.

Problem 4.6.14

Implement in MATLAB the continuous GA by inserting the changes suggested in Fig. 4.41 appropriately into the binary GA code from Fig. 4.60.

Problem 4.6.15

In the continuous GA code revealed in Solution 4.6.14 the random crossover is implemented for mating. Change it to blending method where factor β is a random variable between 0 and 1. Factor β should be recalculated each time.

Problem 4.6.16

Detect follicles in an ovarian ultrasound image (like in "CNN Training Example") using CNN. Train the CNN with binary and continuous GAs. Employ Euclidian distance as a fitness function. Generate at least 60 generations. Compare the obtained CNN results with annotated images from the learning set! Images for training are stored in the Follicle-LearningSet.zip file and can be downloaded from http://storm.uni-mb.si/CNNExamples. The ultrasound ovarian images are named with prefix L_, whilst for the corresponding annotated images the names begin with prefix A_L_. The follicles in annotated images are marked white (their regions are filled with white). Mind that input images, initial states, annotated images and the number of images in training set must be specified as global variables as defined in Fig. 4.43.

Problem 4.6.17

Implement the procedure described in Sect. 4.6.5 to detect follicles in ovarian ultrasound images. Use templates depicted in that section. Test the detection procedure using the training set images from the FollicleLearningSet.zip file as inputs to the CNNs. The ultrasound images to be used have prefix L_. Compare the obtained result with the result depicted in Fig. 4.48. Then test the procedure with images from the FollicleTestingSet.zip file; their prefix is T_. Are now the results better than the results obtained using the learning set? Why? All files can be downloaded from http://storm.uni-mb.si/CNNExamples.

Problem 4.6.18

Try to obtain better templates for the follicle detection procedure described in Sect. 4.6.5 by using bigger training sets and longer training sessions (at least 100 generations). Enlarge the learning sets by moving a number of images from the testing sets to the learning sets. The learning sets are stored in the FollicleLearningSet.zip file. The ultrasound images begin with prefix L_, whils the images comprising annotated follicles with prefix A_L_. The learning set with annotated positions of ovary can be recognised by the file prefixes A_O_L_. The testing sets are stored in the FollicleTestingSet.zip file. The ultrasound images have prefix T_, whilst the images with follicle annotations prefix A_T_. The position of ovary is annotated in the images with prefix A_O_T_. All files can be downloaded from http://storm.uni-mb.si/CNNExamples. Compare the results reported in Sect. 4.6.5 with the outcomes obtained by you.

Solutions

Solution 4.6.1

$$\mathbf{B} = \begin{bmatrix} 1, & 1, & 1 \\ 1, & 0, & 1 \\ 1, & 1, & 1 \end{bmatrix}, \quad -4 < I < -2,$$

$$I + u(i-1,j) + u(i+1,j) + u(i-1,j-1) + u(i,j-1) + u(i+1,j-1) + u(i-1,j+1) + u(i,j+1) + u(i+1,j+1) > 0$$

The problem can be solved via thresholding in one step, so we do not need the feedback template ($\mathbf{A} = \mathbf{0}$). Bias $-4<I<-2$ forces a negative result in all the cases where there is less than six ones (six black neighbours) around the processed input pixel $p(i,j)$.

Solution 4.6.2

$$\mathbf{B} = \begin{bmatrix} 1, & -1, & 1 \\ -1, & 0, & -1 \\ 1, & -1, & 1 \end{bmatrix}, \quad -8 < I < -6,$$

$$I + u(i-1,j-1) + u(i+1,j-1) + u(i-1,j+1) + u(i+1,j+1) - u(i-1,j) - u(i,j-1) - u(i+1,j) - u(i,j+1) > 0$$

By choosing $-8<I<-6$, the obtained result can be positive only if the weighted sum of the neighbouring pixels equals more than 6 (actually, 8 is the only possible value). This happens only if negative coefficients in $\mathbf{B}$ cover white pixels (values -1) and positive coefficients black pixels (values $+1$) at the same time.

Solution 4.6.3

$$\mathbf{B} = \begin{bmatrix} \dfrac{1}{9}, & \dfrac{1}{9}, & \dfrac{1}{9} \\ \dfrac{1}{9}, & \dfrac{1}{9}, & \dfrac{1}{9} \\ \dfrac{1}{9}, & \dfrac{1}{9}, & \dfrac{1}{9} \end{bmatrix}, \quad I = 0,$$

$$y(i,j) = \frac{1}{9}\{u(i-1,j-1) + u(i,j-1) + u(i+1,j-1) + u(i-1,j) + u(i,j) + u(i+1,j) + u(i-1,j+1) + u(i,j+1) + u(i+1,j+1)\}$$

We have learned the CNNs behave as local image operators if there is no feedback ($\mathbf{A} = \mathbf{0}$); consult "How to Implement a CNN for Image Processing". A well-known fact is that lowpass filtering means averaging. Therefore, the filter's coefficients must all have the same value. The absolute sum of these coefficients defines the filter's gain (Russ 2002). To obtain the resulting filtered values within the span equal to the input image values, they must be divided by this gain.

Solution 4.6.4

$$\mathbf{B} = \begin{bmatrix} \dfrac{1}{9}, & \dfrac{1}{9}, & \dfrac{1}{9} \\[2mm] \dfrac{1}{9}, & \dfrac{1}{9}, & \dfrac{1}{9} \\[2mm] \dfrac{1}{9}, & \dfrac{1}{9}, & \dfrac{1}{9} \end{bmatrix}, \quad I = -1.25,$$

$$-1.25 + \frac{1}{9}\{u(i-1,j-1)+u(i,j-1)+u(i+1,j-1)+u(i-1,j)+u(i,j)+u(i+1,j)+u(i-1,j+1)+u(i,j+1)+u(i+1,j+1)\} > 0$$

To solve the problem a signum function has to be used. This introduces thresholding by the value of given bias I.

Solution 4.6.5

$y_n(i\text{-}1,j)$	$y_n(i,j)$	$y_n(i+1,j)$	$x_n(i,j)$	$y_{n+1}(i,j)$
-1	-1	-1	$-2+I$	-1
-1	-1	+1	$-4+I$	-1
-1	+1	-1	$0+I$	+1
-1	+1	+1	$-2+I$	-1
+1	-1	-1	$2+I$	+1
+1	-1	+1	$0+I$	+1
+1	+1	-1	$4+I$	+1
+1	+1	+1	$2+I$	+1

The truth table takes into account all possible combinations of the processed output pixels in step n. When multiplied by the corresponding coefficients of the template **A**, the internal states in step n are obtained. These trigger the output non-linearity which is in the form of the signum function, so that positive states create black pixels (values +1), white negative states white pixels (values -1). Observing the second and the last column in the earlier table, we see their colour (value) changes only in the fourth, fifth and sixth combination. This corresponds to Situations 1 and 2, as explained in Problem 4.6.5. From the fourth column it is also clear that bias I must lie between 0 and 2, $0<I<2$, if the desired outputs are to be attained.

Solution 4.6.6

$$x_n(i, j) = y_n(i, j) - u(i-1, j) + u(i, j) - u(i+1, j) - 2$$
$$y_{n+1}(i, j) = \text{sgn}[x_n(i, j)]$$

An example of possible test image (top) and the resulting CNN output (bottom):

+1	-1	-1	+1	-1	-1
-1	+1	-1	+1	-1	-1
-1	-1	-1	+1	-1	-1
-1	+1	+1	+1	-1	+1
-1	-1	-1	-1	+1	-1
-1	-1	-1	+1	-1	-1

$$\Downarrow$$

+1	-1	-1	-1	-1	-1
-1	+1	-1	-1	-1	-1
-1	-1	-1	-1	-1	-1
-1	+1	+1	-1	-1	+1
-1	-1	-1	-1	+1	-1
-1	-1	-1	+1	-1	-1

The applied CNN templates cause the vertical lines to disappear.

Solution 4.6.7

```
Function output=linearDTCNN(input,state,...

Atem,Btem,I,steps,ibound,obound);
%
% linearDTCNN - This function simulates the
%               discrete-time CNN operation.
%The meaning of parameters:
% input - the input image
% state - the initial values of internal
states
% Atem  - the feedback template
% Btem  - the control template
% I     - the bias term
% steps - the number of simulation steps
% ibound - boundary values for the input image
% obound - boundary values for the output
image
%The input image and the internal state matrix
must
%be of the same dimensions. The output image
adopts
%these dimensions as well.
% Syntax: output=linearDTCNN(input,state,...
%
Atem,Btem,I,steps,ibound,obound);
%
% If both boundaries are not specified.
if (nargin<7)
   ibound=0;   % Default boundary values
   obound=0;   % equal 0.
elseif (nargin<8)  % If only input boundaries
   obound=0;   % Value for the output
boundaries.
end;
% The constant part of the CNN computation
% depending on the inputs and bias term is
% calculated only once (entire images,
% i.e. all CNN cells are processed here).
binput=imfilter(input,Btem,ibound)+I;

% The output values are clipped according to
the
% signum nonlinear function.
output=sign(state);

% CNN operates in the preselected number of
steps.
% Different termination conditions may be
% introduced by replacing the following FOR-
loop
% with a WHILE-loop.
for n=1:steps,
   % Calculation of new states.
state=imfilter(output,Atem,obound)+binput;

   output=sign(state);   % Computation of new
outputs.
end
```

Solution 4.6.8

The following code loads the colour image called `board.tif`, for example, and converts it into the grey one:

```
image=imread('board.tif');grey_image=rgb2gray(image);
age);
```

Before any further numerical processing, we have to change the type of variable `grey_image` from `uint8` to `double` or `single`:

```
grey_image=double(grey_image);
```

The image pixel values can be transformed into the interval of [−1, 1] using the following relation:

```
input_image=1-(2*grey_image/256);
```

Finally, call the CNN simulator as follows:

```
A=[0 0 0; 0 2 0; 0 0 0];
B=0;
I=-0.5;
output_image=linearCNN(zeros(size(input_image)),...
                       input_image,A,B,I,1,20);
```

The experiment can be repeated with a different bias I (e.g. `I=0.5;`).

Solution 4.6.9

The operator detects primary the vertical edges. It does not perform well on diagonal and horizontal lines. This operator actually corresponds to the Sobel filter. The edges appear with two-pixel widths, which is caused by the form of the **B** template having the middle column all 0. By transposing **B**, the operator becomes sensitive to the horizontal edges, i.e. the changes of the image grey-levels or colour intensity in vertical direction.

Solution 4.6.10

```
function rez=Example5;

sl=imread('crt_baby.png');
sl=single(sl);
sl=1-(sl*2)/255;
% Binarization:
A=[0 0 0;0 2 0;0 0 0];
B=0;
I=-0.4;
t04=linearCNN(zeros(size(sl)),sl,A,B,I,1,10);
% Erosion:
er=t04;
A=0;
B=[1 1 1; 1 1 1; 1 1 1];
```

```
I=-8;
for f=1:10,
  er=linearCNN(er,zeros(size(er)),A,B,I,1,1);
end
%Selected object extraction:
A=[0.5 0.5 0.5; 0.5 4 0.5; 0.5 0.5 0.5];
B=[0 0 0; 0 4 0; 0 0 0];
I=3;
rez=linearCNN(t04,er,A,B,I,1,500);
figure; imshow(rez);
```

Solution 4.6.11

```
function [Atem,Btem,I]=decoding(bstring...
                      ,sizeA,sizeB);
% This function decodes template A, template B
% and bias I from binary string named bstring.
% Each value should be between -5.12 and 5.11
% and is coded by 10 bits.
% The input parameters sizeA and sizeB stand
for
% the number of elements in the templates A
and B.
%
b=10;

bI=bstring(1:b);
bA=bstring(b+1:b+prod(sizeA)*b);
bB=bstring(b+1+prod(sizeA)*b:length(bstring));

bA=reshape(bA,[b prod(sizeA)])';
bB=reshape(bB,[b prod(sizeB)])';

NI=bin2dec(bI);

NtemA=bin2dec(bA);
NtemA=reshape(NtemA,sizeA);

NtemB=bin2dec(bB);
NtemB=reshape(NtemB,sizeB);

Atem=(NtemA-512)/100;
Btem=(NtemB-512)/100;
I=(NI-512)/100;
```

Solution 4.6.12

```
function bstring=coding1(temA,temB,I);
%
% This function codes template A, template B
and
% bias I into binary string bstring. Each value
% should be between -5.12 and 5.11 and is coded
by
% 10 bits. The resulting string contains bit
from
% the subsequent parameters interlaced.
%
b=10;
NtemA=round(100*temA+512);
NtemB=round(100*temB+512);
NI=round(100*I+512);
```

```
bI=dec2bin(NI,b);
bA=dec2bin(NtemA,b);
bB=dec2bin(NtemB,b);

bstring=[bI; bA; bB];
bstring=bstring(:)';


function [temA,temB,I]=decoding(bstring,...
                               sizeA,sizeB);
%
% This function decodes template A, template B
and
% bias I from binary string bstring, where the
% parameters are coded in an interlaced manner.
% Each value should be between -5.12 and 5.11
% and is coded with 10 bits.
% The input parameters sizeA and sizeB stand
for
% the number of elements in the templates A and
B.
%
b=10;

b2string=reshape(bstring,length(bstring)/b,b);
bI=b2string(1,:);
bA=b2string(2:prod(sizeA)+1,:);
bB=b2string(prod(sizeA)+2:length(bstring)/b,:);

NI=bin2dec(bI);

NtemA=bin2dec(bA);
NtemA=reshape(NtemA,sizeA);

NtemB=bin2dec(bB);
NtemB=reshape(NtemB,sizeB);

temA=(NtemA-512)/100;
temB=(NtemB-512)/100;
I=(NI-512)/100;
```

Solution 4.6.13

```
% Two-point crossover:
cross=sort(round(rand(1,2)*(length_c-1)));
% Generation of offspring:
offspring1=[parent1(1:cross(1)) ...
    parent2(cross(1)+1:cross(2))...
    parent1(cross(2)+1:length_c)];
offspring2=[parent2(1:cross(1)) ...
    parent1(cross(1)+1:cross(2))...
    parent2(cross(2)+1:length_c)];

% Totally random crossover:
cross=round(rand(1,length_c));
% Generation of offspring:
offspring1=cross.*parent1+(1-cross).*parent2;
offspring2=cross.*parent2+(1-cross).*parent1;
```

Solution 4.6.14

```
function [chromosome, fit_list]=cont_ga(fcost,
cr);
%
%The function requires external fitness
function.
%Input:
%  fcost - the name of the fitness function
which
%    returns the fitness of two input images
i.e.
%    'efitness' for function
fit=efitness(im1,im2)).
%  cr - optional initial chromosome
%Function returns:
%  chromosome - the fittest chromosome
%  fit_list    - list of the fittest
chromosomes
%     in generation with their fitness.
%
% Written by B. Cigale, August 2005.
%
length_c=19;   % Num. of variables in a
chromosome.
number_c=30;   % Num. of chromosomes in
generation.
number_g=30;   % Number of generated
generations.

% 50% of the fittest chromosomes survive.
natural_selection=0.5;
% Offsping has 10% for mutation.
mutaton_probability=0.10;
% First generation is generated randomly:
generation=10.23*rand(number_c,length_c)-5.12;
% Is optional initial chromosome known?
if (exist('cr'))
  generation(1,:)=cr;
end;
fit_list=[];
for g=1:number_g, % For each generation:
  disp(['Generation ',int2str(g)]);
  % Calculate the fitness. EXTERN function
which
  % returns the fitness for each chromosome.
  fit=fitness(generation,fcost,'cont');
  % Sort the chromosomes by fitness:
  [sfit,ind]=sort(fit,'descend');
  disp(['Fittest: ',num2str(sfit(1))]);
  fit_list=[fit_list sfit(1)];
  % The best chromosome so far.
  chromosome=generation(ind(1),:);
  % Natural selection - only the fittest
remain:
  sfit=sfit(1:number_c*natural_selection);
  % Selection:
% Calculate mating probability for each
chromosome.
```

```
  Pn=sfit/sum(sfit);
  % Create cumulative distribution function
(cdf).
  cdf=cumsum(Pn);
  % Elitism - the best two chromosomes
survive:
  new_generation=[generation(ind(1),:);...
                  generation(ind(2),:)];
  % The rest of the generation is generated in
  % standard way: selection, mating and
mutation.
  for f=3:(number_c/2)+1,
    % Find the first chromosome index whose
    % cumulative probability is greater or
equal
    % to a random number. This is the first
parent.
    i=find(rand<=cdf);
    parent1=generation(ind(i(1)),:);

    % Find parent 2.
    i=find(rand<=cdf);
    parent2=generation(ind(i(1)),:);
    % Make a random string of 0s and 1s to
decide
    % which variable goes where. When 1, the
    % variable goes from parent 1 to offspring 1,
    % otherwise the variable goes from
    % parent 1 to offspring 2.
    swap=round(rand(1,length_c));
    offspring1=swap.*parent1+(1-
swap).*parent2;
    offspring2=(1-
swap).*parent1+swap.*parent2;
    % Mutation - offspring 1:
    if (rand<mutaton_probability)...
        ||(sum(abs(offspring1-
parent1))<0.1)...
        ||(sum(abs(offspring1-parent2))<0.1)
      % In case of mutation the random
      % variable is replaced by a new value.
      p=round(rand*(length_c-1))+1;
      offspring1(p)=10.23*rand-5.12;
    end
    % Mutation - offspring 2:
    if (rand<mutaton_probability) ...
        ||(sum(abs(offspring2-
parent1))<0.1)...
        ||(sum(abs(offspring2-parent2))<0.1)
      p=round(rand*(length_c-1))+1;
      offspring2(p)=10.23*rand-5.12;
    end
    % Add both offspring to new generation:
    new_generation=[new_generation;...
                    offspring1; offspring2];
  end
  % Previous generation is replaced by the new
one:
  generation=new_generation;
end
```

Solution 4.6.15

```
% Mating - blending method:
beta=rand;
offspring1=beta*parent1+(1-beta)*parent2;
offspring2=(1-beta)*parent1+beta*parent2;
```

Solution 4.6.16

You should first load ultrasound images, transform their pixel values to the interval from −1 to 1, and store them as a global variable L_IMAGES. Example for one image:

```
global L_IMAGES;
image=imread('L_001.png');
% Transform values to [-1,+1].
im=1-(2*single(image)/256);
% Inserting the first image.
L_IMAGES(1,:,:)=im;
```

Then load the annotations in a global variable L_MASKS. Example for one image:

```
global L_MASKS;
image=imread('A_L_001.png');
im=logical(image);  % Change the image type.
L_MASKS(1,:,:)=im;  % Inserting the first image
with
                    % annotations.
```

Set global variable L_NO_IMAGES to the number of images in the testing set. An example for four images:

```
global L_NO_IMAGES;
L_NO_IMAGES=4;
```

Finally set the initial states of CNN in global variable L_INITIAL. In our case the initial state equals to the input images, therefore:

```
global L_INITIAL;
L_INITIAL=L_IMAGES;
```

Instruct GA how many generations should it create. The number of generations is defined in variable number_g.

Finally run GA and wait. Training can last for a few hours!

```
% The Eudlidian distance fitness function is
used.
[bstring,fit]=binaryGA('efitness');
```

Solution 4.6.17

The follicle detection results obtained on the testing set images should be worse than those obtained on the learning set images. Such an outcome is expected because the CNN is trained on the learning set images, hence it recognises them most reliably.

```
function [result,ovary]=...

follicle_segmentation(image);
%
image=1-(2*double(image)/256);
% Step 1:
A=[ 4.44 -2.41   3.20;
   -0.21 -0.68   4.12;
    3.56 -0.37   1.65];
B=[-3.15  1.32 -1.75;
   -4.72 -0.12   0.22;
    2.09  1.29 -0.60];
I=-4.14;
rough_folicles=linearCNN(image,image,A,B,I,0.3
,30);
% Step 2:
A=[ 4.47 -0.09   3.10;
    3.28 -4.58   2.71;
    1.48  3.91   2.84];
B=[ 3.20  4.53   2.83;
   -5.00  5.03   3.53;
    3.43 -2.32 -0.86];
I=2.29;
expanded folicles=linearCNN(...

image,rough_folicles,A,B,I,0.3,30);
% Step 3:
A=[ 4.09  3.77   0.59;
   -0.97 -3.65   2.31;
    3.94  0.98   1.46];
B=[-4.42  2.49   2.23;
    2.52 -1.44   2.54;
    3.97 -3.18   3.53];
I=4.12;
ovary1=linearCNN(...

image,expanded_folicles,A,B,I,0.3,45);
ovary=zeros(size(ovary1));
% Test whether ovary was detected or not:
[lovary,no]=bwlabel(ovary1>0);
if (no>0) % If the ovary detect:
   % find the biggest area.
   a=regionprops(lovary,'Area');
   [maxa,inda]=max([a.Area]);
```

```
ovary_core=lovary==inda;
% Put a convex hull around this area:
o=regionprops(uint8(ovary_core),...
              'ConvexImage','BoundingBox');
cor=o.BoundingBox;
ovary(ceil(cor(2)):floor(cor(2))+cor(4),...
      ceil(cor(1)):floor(cor(1))+cor(3))=...
         o.ConvexImage;
% Step 4:
result=min(ovary,expanded_folicles>0);
else
  result=expanded_folicles>0;
end
```

Solution 4.6.18

Bigger learning sets and longer training sessions should produce better templates.

When training the first CNN, load images as described in Solution 4.6.16. Then copy fitness function `czfitness` (Fig. 4.38) to function called `czfitness1`. You have to change the last line of the function to `fit=(N*N)/(K*L)`.

After you set the number of generations, run the training.

```
[bstring_cnn1, fit_cnn1]=binaryGA('czfitness1');
```

For CNN2, you must decode the result of CNN1 first, because the initial states for CNN2 have to be created. The following function would be useful:

```
function create_initial_states(A,B,I);
% Global variables:
global L_IMAGES;
global L_NO_IMAGES;
global L_INITIAL;
% Process all given images:
for g=1:L_NO_IMAGES,
  im=squeeze(L_IMAGES(g,:,:));
  in=squeeze(L_INITIAL(g,:,:));
  output=linearCNN(im,in,A,B,I,0.3,30);
  L_INITIAL(g,:,:)=output;
end
```

After the initial states are created, train CNN2.

```
[bstring_cnn2, fit_cnn2]=binaryGA('czfitness2');
```

Finally, train CNN3. Decode the template parameters first, and then call function `create_initial_states`, because the initial state for CNN3 is taken from the output of CNN2. CNN3 is trained to detect ovary, therefore new images must be loaded in global variable L_MASKS (images with prefix A_O_L_)! The fitness function should be changed before training, because only the biggest object recognised by CNN3 must be evaluated in order to converge to the assumed region of ovary. Before the evaluation this object (presumably the ovary) must also be encapsulated by a convex hull. The following code should be added for such fitness function:

```
output=linearCNN(im,in, A, B, I,0.3, 40);
[lovary,n]=bwlabel(output>0);
a=regionprops(lovary,'Area');
[maxa,inda]=max([a.Area]);
ovary_core=lovary==inda;
o=regionprops(uint8(ovary_core),...
              'ConvexImage','BoundingBox');
ovary=zeros(size(output));
cor=o.BoundingBox;
ovary(ceil(cor(2)):floor(cor(2))+cor(4),...
      ceil(cor(1)):floor(cor(1))+cor(3))=...
        o.ConvexImage;
output=ovary;
```

Use Euclidian distance for the CNN3 training.

```
[bstring_cnn3, fit_cnn3]=binaryGA('efitness');
```

4.7 Computer-Aided Diagnosis for Virtual Colonoscopy

Hiroyuki Yoshida

Abstract

This chapter describes a practical paradigm for the detection of colonic polyps, a precursor of colon cancer, in virtual colonoscopy (VC), called computer-aided detection (CAD) of polyps. VC is an emerging technique for screening of colon cancers. It uses computed tomography (CT) scanning to obtain a series of cross-sectional images of the colon. CAD automatically detects the locations of suspicious polyps and masses in VC and presents them to radiologists, typically as a second opinion. Despite its relatively short history, CAD is becoming one of the mainstream techniques that could bring VC to prime time for screening of colorectal cancer. Rapid technical developments have advanced CAD substantially during the last several years, and a fundamental scheme for the detection of polyps has been established, in which 3-dimensional image processing, analysis, and display techniques play a pivotal role. The latest CAD systems based on this scheme yield a clinically acceptable high sensitivity and a low false-positive rate, and observer studies indicates the benefits of these systems in improving radiologists' diagnostic performance. This chapter describes the fundamental CAD scheme for the detection of polyps in VC, details the key techniques, diagnostic performance of CAD and that of radiologists aided by CAD, and the pitfalls in CAD.

4.7.1 Introduction

During the past decade, numerous attempts have been made to develop computerized methods that process, analyze, and display multidimensional medical images in radiology. A typical example is the 3-dimensional (3-D) visualization of semiautomatically segmented organs (e.g. segmentation of the liver, endoluminal visualization of the colon and bronchus), or image processing of a part of an organ for generation of an image that is more easily interpreted by human readers (e.g. peripheral equalization of the breast in mammograms, digital subtraction bowel cleansing in virtual colonoscopy). These computerized methods often automate only one of the image-processing tasks and depend on user interaction for the remaining tasks.

Computer-aided diagnosis (CAD) goes beyond these semiautomated image-processing applications, and enters the area of *medical image understanding or interpretation*. In its most general form, CAD can be defined as a diagnosis made by a radiologist who uses the output of a computerized scheme for automated image analysis as a diagnostic aid. Conventionally, CAD acts as a "second reader", pointing out abnormalities to the radiologist that might otherwise have been missed. This definition emphasizes the intent of CAD to support rather than substitute for the human reader, and the final diagnosis is made by the radiologist.

The concept of CAD is applicable to all modalities and disease types. Historically, CAD has been most popular in the diagnosis of breast cancers, such as the detection of microcalcifications and the classification of masses in mammograms (Astley and Gilbert 2004). It is not unique, however, to these types of diseases and modalities. CAD is more important and beneficial for examinations that became feasible only recently because of the advancement of digital imaging technologies such as the detection of lung nodules in computed-tomographic (CT) scans of the lungs and the detection of polyps in virtual colonoscopy (VC), in which a large number of images need to be interpreted rapidly for finding a lesion with low incidence.

VC, also known as *CT colonography*, is an emerging alternative technique for screening of colon cancers (Levin et al. 2003; Morrin and LaMont 2003). VC uses CT scanning to obtain a series of cross-sectional images of the colon for detection of polyps and masses. The CT images are reformatted into a simulated 3-D "endoluminal view" of the entire colon that is comparable to that seen with optical colonoscopy (Fig. 4.49).

Radiologists can "fly through" the virtual colon, from the rectum to the cecum and back, searching for polyps and masses. By virtual colonoscopy, therefore, one can non-invasively examine the interior of the colon without physically invading it; thus, VC is a safer procedure than is optical colonoscopy.

CAD for VC typically refers to a computerized scheme for automated detection of polyps in VC scans (Yoshida and Dachman 2004, 2005). It reveals the locations of suspicious polyps to radiologists. This offers a second opinion that has the potential of improving radiologists' detection performance, and of reducing variability of the diagnostic accuracy among radiologists, without significantly increasing the reading time.

Despite its relatively short history, CAD is becoming a major area of investigation and development in VC. Rapid technical progress has established the fundamental CAD scheme for the detection of polyps during the last several years. Academic institutions are demonstrating prototype CAD systems, and commercial systems that implement the full CAD scheme or a part of it are becoming available. The latest prototype CAD systems yield clinically acceptable high sensitivity and low false-positive rates. Some technical and clinical problems linger, however, that impede CAD's becoming a clinical reality.

This section describes the fundamental CAD scheme for the detection of polyps in VC, details of the key techniques, the diagnostic performance of CAD and that of radiologists aided by CAD, and the pitfalls in CAD.

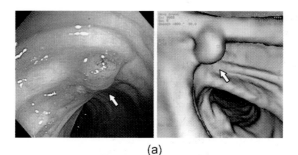

(a)

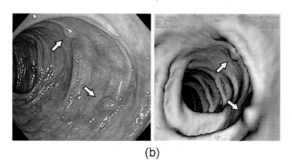

(b)

Fig. 4.49. Examples of endoluminal views of polyps in VC. (**a**) The left image shows an optical colonoscopy view of a 15-mm polyp (arrow) in the ascending colon. The right image shows an endoluminal view of the same polyp (arrow) in VC. (**b**) The same views for two 10-mm polyps in the ascending colon. The shapes of the polyps are similar in the optical colonoscopy and VC views

A similar CAD approach has been developed for the detection of masses; readers interested in CAD for masses can turn to reference (Näppi et al. 2002b).

4.7.2 CAD Scheme for the Detection of Polyps

Technical advances in CAD during the last several years have led to the establishment of a fundamental CAD scheme for the detection of polyps in VC. To date, most of the CAD schemes have been based on a 3-D volume generated from the transverse CT images of the colon by interpolation of the CT images along the transverse direction. These schemes comprise the following four fundamental steps (1) extraction of the colonic wall from the VC images, (2) detection of polyp candidates in the extracted colon, (3) characterization of false positives, and (4) discrimination between false positives and polyps. Key techniques for each of these steps are described later.

Extraction of the Colonic Wall

Colonic polyps appear as objects that adhere to the colonic wall and protrude into the colonic lumen (Fig. 4.49). Therefore, the first step of CAD is the extraction of the colonic wall, including folds. Several fully automated methods for extraction of the colon have been developed (Chen et al. 2000; Hong et al. 1997; Iordanescu et al. 2005; Li and Santago 2005; Masutani et al. 2001; Näppi et al. 2002a, 2004; Wyatt et al. 2000). Most of these methods use the CT values characteristic of the colonic wall and the contrast between the colonic wall and the air in the colonic lumen as a means of extracting the colon from a 3-D volume.

Some of these methods are fully automated (Masutani et al. 2001; Näppi et al. 2002a, 2004; Wyatt et al. 2000) and thus are suitable for CAD, whereas others are semi-automated (Chen et al. 2000; Iordanescu et al. 2005; Summers et al. 2000) and are more suitable for interactive visualization of the colon. In addition, two types of approaches have been proposed: a surface-generation method for extraction of the inner surface of the colonic wall (Chen et al. 2000; Hong et al. 1997; Iordanescu et al. 2005; Li and Santago 2005; Wyatt et al. 2000) and a volume-generation method for extraction of a thick region that encompasses the entire colonic wall (Frimmel et al. 2005; Masutani et al. 2001; Näppi et al. 2002a). The former is quick, and the algorithms for generation of a surface are well established; however, the method has the risk of losing a part of a polyp. With the latter approach, one can extract entire polyps, including their internal structure; however, the algorithm tends to be complicated and thus tends to be computationally expensive.

Some of the segmentation methods assume that the colons are well cleansed and distended (Iordanescu et al. 2005); however, in real VC examinations, many colons contain collapsed regions. Thus, the main technical challenge in colon segmentation is to extract collapsed regions whilst minimally extracting extracolonic structures. An approach that effectively address this problem is the reconstruction of a 3-D isotropic volume from CT images in VC examinations, followed by knowledge-guided colon segmentation (KGS) (Näppi et al. 2002a) (Fig. 4.50).

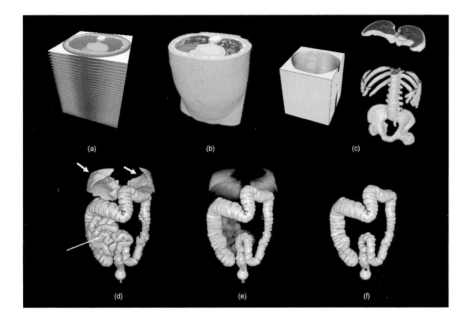

Fig. 4.50. Illustration of the KGS segmentation process. (**a**) Axial CT images from a VC examination. (**b**) 3-D isotropic volume of the abdomen generated from the CT image in (a). (**c**) Extracolonic structures removed in the anatomy-based extraction process. The process first eliminates the region outside the body (left), the lung bases (upper right), and the osseous structures (lower right). (**d**) Result of the anatomy-based extraction. This process extracts the entire colon; however, a portion of the lung bases (*thick arrows*) and small bowel (*thin arrow*) remain unremoved. (**e**) Segmentation of the colon by use of colon-based analysis. This process identifies the air-filled colonic lumen by use of volume growing. The identified colonic lumen is highlighted, whereas the extracolonic structures are dimmed. (**f**) Final segmented colon obtained by intersection of (d) and (e). Note that the extracolonic structures, which are included in (d), are not present in the final segmentation

Reconstruction of 3-D isotropic Volume

To allow a 3-D analysis of the colon and polyps, we generate a 3-D volume that is isotropic in all directions, i.e. the x, y, and z directions, from the axial CT images in a VC examination. For this purpose, the axial CT images are interpolated along the axial direction by linear interpolation of the CT values of the pixels at the same locations in the neighbouring two slices for elimination of the differences between the spatial and axial resolutions in a single data set (Fig. 4.50a, b).

The linear interpolation process is expected to serve to eliminate the differences in the CT parameter settings – in particular, the reconstruction intervals – among different data sets. This is important because, although the CT images will be acquired under a standard CT parameter setting (see section on "Performance of CAD in the detection of polyps"), a small difference in parameter settings used in different examinations may possibly affect the appearance of polyps, thus affecting the detection performance. Similarly, the spatial resolution is equalized by use of linear interpolation, because the spatial resolution (in-plane pixel dimension) varies from 0.5 to 0.75 mm per voxel due to the varying field of view across patients. A typical VC examination consists of 350–500 axial CT images in a 512×512 matrix. After the interpolation, the x, y, z dimensions will be $512 \times 512 \times (700{-}1{,}000)$ voxels, with 0.5 mm/voxel to cover the entire colon.

Reconstruction of the 3-D isotropic volume also addresses the problem of image noise. The tube current in the CT scanning ranges from 20 mA (low X-ray dose) to 300 mA (high dose). Typically, the lower the dose, the higher the image noise level; that is, a lower tube current yields a lower signal-to-noise ratio. For reducing the CT image noise, the isotropic volume is filtered by a noise reduction filter such as a Gaussian filter (Lohmann 1998) and the edge-preserving non-linear Gaussian filter chains (Aurich et al. 1999). Typically, the kernel size of the filter can be adjusted so that it reduces the noise effectively, and so that the remainder of the CAD processes, when they are applied to the isotropic volume, can yield a high performance in the detection of polyps.

Knowledge-Guided Colon Segmentation

The KGS has three major steps: anatomy-based extraction (ABE), colon-based analysis (CBA), and an intersection step. The ABE step extracts nearly the entire visible colonic wall, but the segmented region could contain extracolonic structures (Fig. 4.50c, d). In particular, the segmented region could contain small bowel or stomach that adheres to the colonic wall. Therefore, in the second step, the CBA step uses ABE segmentation to extract the colonic lumen whilst excluding the extracolonic structures (Fig. 4.50e). The final KGS segmentation is obtained by intersecting of the ABE and CBA segmentations (Fig. 4.50f).

The ABE step segments the colonic wall by first removing (1) the air surrounding the body region, (2) the osseous structures, and (3) the lung bases from the isotropic volume (4.50c). These structures are extracted by thresholding of the isotropic volume with the following ranges: outer air: $(0, (L_A + L_F)/2)$; bony structures: $(2L_M/3 + 100, \infty)$; lung bases: $((L_A + L_L)/2, L_L + 150)$, followed by 26-point connected component analysis. Here, L_A, L_L, L_F, and L_M are the CT values corresponding to air, lung, fat, and muscle, respectively, as determined adaptively by detection of the characteristic peak values CT values of these structures (Masutani et al. 2001). Morphological dilation and erosion with a spherical kernel are applied to the extracted regions for extraction of structures that have a 1–5 mm thickness. After removal of the surrounding anatomic structures, the colonic wall is segmented from the remaining region by thresholding of the CT values with a range of $[L_M - 800, L_M - 50]$ and gradient magnitude values with a range of $[L_M - 950, \infty]$. Let $P = \{P_i\}$ be the set of the resulting connected components, and let $|P_i|$ denote the number of voxels in a component P_i. The largest component, $P_L \in P$, represents the principal region of the colon. Additional components are included in the colon segmented by ABE, C^+, as follows (Fig.4.50d):

$$C^+ = P_L \cup \{P_j; |P_j| \geq 0.01 \,|P_L|, j \neq L, P_j \in P\}.$$

The CBA step segments the colonic lumen by use of a volume-growing method based on C^+ (Fig. 4.50e). A first seed voxel is set at the rectum, because the location of the rectum can be estimated quite reliably from the ABE segmentation. To determine the location of the seed voxel, we first extract a volume of interest V_R that represents the bottom 20% region of the isotropic volume covering C^+. Let $R_{C^+} = V_R \cap C^+$. Then R_{C^+} divides $V_R \setminus R_{C^+}$ into two disconnected components: the colonic lumen and the region outside the colon. The rectum is identified as the second-largest connected component within $V_R \setminus R_{C^+}$, and the seed is set at the centre of the rectum.

Once a seed voxel has been established at the rectum, the volume-growing process adds the surrounding 26-connected voxels with a low CT value to the expanding region, unless they belong to the ABE segmentation or the previously identified extracolonic structures. The process terminates when the number of voxels added to the expanding region decreases by more than 75% from the previous iteration.

Let C^- denote the segmentation produced by the CBA. The final segmentation of the colon is obtained by intersection of the two segmentations C^+(ABE) and C^-(CBA): $C = C^+ \cap C^-$ (Fig. 4.50f). Six conditions are used for testing whether C represents the complete colon. First, the number of voxels removed from C^+ is limited according to $C_\Delta = (|C^+| - |C|)/|C^+| > 0.50$, where $|C|$ denotes the number of voxels within the segmented region. Second, the amount of segmented voxels within the region of the ascending and the descending colon should not differ by more than 20%. The third condition tests the presence of the segmented colon within the expected location of the rectum. The fourth condition checks whether the colon represented by C passes within 10–25% from the boundary of the isotropic volume that covers C^+. The fifth condition tests whether there are very large air-filled components that have not been segmented. Finally, the sixth condition terminates the segmentation process if $C_\Delta > 0.95$.

The volume-growing step may need to be continued if the complete colon is not segmented. Let r_i denote the region that has been segmented by the most recent volume-growing step. The updated CBA segmentation is the region $r_i' \cup C^-$, where r_i' represents r_i expanded to cover C^+ in the desired thickness within the local neighbourhood of r_i. If the C thus obtained is considered to be still incomplete, the volume-growing process is continued by choice of a seed point from within the largest non-segmented air-filled region.

On average, the regions extracted by KGS method covered approximately 98% of the visible surface region of the colonic wall (Näppi et al. 2002a).

4.7.3 Detection of Polyp Candidates

On the colonic wall that has thus been extracted, polyps tend to appear as bulbous, cap-like structures adhering to the wall, whereas folds appear as elongated, ridge-like structures, and the colonic wall itself appears as a large, nearly flat, cup-like structure (Fig. 4.51). Therefore, shape analyses that differentiate among these distinct types of shapes and scales are effective in the detection of polyps. To this end, various methods have been developed, including use of a volumetric shape index and curvedness (Yoshida and Näppi 2001), surface curvature with a rule-based filter (Summers et al. 2000), sphere fitting (Kiss et al. 2002), and an overlapping surface normal method (Paik et al. 2004), each of which has been shown to be effective in detecting polyps.

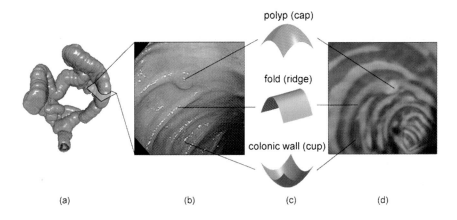

polyp (cap)

fold (ridge)

colonic wall (cup)

(a)   (b)   (c)   (d)

Fig. 4.51. Schematic illustration of the geometric modelling of the structures in the colonic lumen. (**a**) Colon extracted from the VC data. (**b**) Colonoscopy image of a 6-mm polyp in the sigmoid colon. (**c**) Geometric models of the structures in the colonic lumen. (**d**) Pseudo-colouring of the colonic lumen based on the shape index values corresponding to the geometric models (see Fig. 4.52). The polyp (green) is clearly differentiated from the folds (pink) and the colonic wall (brown)

3-D Geometric Features: Volumetric Shape Index and Curvedness

The effective approaches to characterizing of polyps is a 3-D geometric feature called the *volumetric shape index* and the *volumetric curvedness* (Dorai and Jain 1997; Koenderink 1990). The volumetric shape index characterizes the topologic shape of the volume in the vicinity of a voxel.

Let $h(p)$ denote a voxel value (the CT value) at a voxel $p = (x, y, z) \in \mathbf{R}^3$. A 3-D iso-surface P at a level a is given by $P = \{p = (x, y, z) \in \mathbf{R}^3; h(p) = a\}$. To obtain an infinitely differentiable 3-D function, we convolve the volume with a Gaussian filter. The implicit-function theorem indicates that, at each voxel p, there exists a small neighbourhood of p in which z can be expressed by a smooth function ϕ of x and y. By denoting (x, y) by (u, v) in this neighbourhood, we can represent the iso-surface P in a parametric 2-D form as $P = \{(u, v) \in \mathbf{R}^2; h(u, v, \phi(u, v)) = a\}$.

The *principal curvatures* are obtained as the eigenvalues of the following matrix (Kobayashi and Nomizu 1963, 1969), called the *Weingarten endomorphism*:

$$W = \begin{pmatrix} E & F \\ F & G \end{pmatrix}^{-1} \begin{pmatrix} L & M \\ M & N \end{pmatrix} = \frac{1}{EG - F^2} \begin{pmatrix} GL - FM & GM - FN \\ EM - FL & EN - FM \end{pmatrix}. \qquad (4.106)$$

Here, E, F, and G are the *first fundamental forms* (Kobayashi and Nomizu 1963, 1969) defined at a voxel p as

$$E \equiv P_u \cdot P_u, \quad F \equiv P_u \cdot P_v, \quad G \equiv P_v \cdot P_v, \qquad (4.107)$$

where P_u and P_v are the partial derivatives of P in terms of u and v at p, respectively. In the following, a suffix of u, v, x, y, or z indicates the partial derivative in terms of these variables, and all of the quantities are defined on a single voxel p unless mentioned otherwise. L, M, and N are the *second fundamental forms* (Kobayashi and Nomizu 1963, 1969), defined as:

$$L \equiv P_{uu} \cdot Q, \quad M \equiv P_{uv} \cdot Q, \quad N \equiv P_{vv} \cdot Q, \qquad (4.108)$$

where Q is a vector normal to the iso-surface P.

To calculate the principal curvatures, we use the implicit differentiation and the chain rule to obtain $P_u = \partial P / \partial u = (1, 0, \partial \phi / \partial u) = (1, 0, -h_x / h_z)$. Substituting similar expressions for P_v, P_{uu}, P_{uv}, and P_{vv} in (4.106)–(4.108), we can compute the determinant of the matrix W, called the *Gaussian curvature*, K, and half of the trace of W, called the *mean curvature*, H, as follows:

$$K \equiv \det(W) = \frac{LN - M^2}{EG - F^2} = \frac{1}{|h|^2} \sum_{(i,j,k)=\mathrm{Perm}(x,y,z)} \left\{ h_i^2 (h_{jj} h_{kk} - h_{jk}^2) + 2 h_j h_k (h_{ik} h_{ij} - h_{ii} h_{jk}) \right\},$$

$$H \equiv \frac{1}{2} \mathrm{trace}(W) = \frac{EN - 2FM + GL}{2(EG - F^2)} = \frac{1}{|h|^{3/2}} \sum_{(i,j,k)=\mathrm{Perm}(x,y,z)} \left\{ -h_i^2 (h_{jj} + h_{kk}) + 2 h_j h_k h_{jk} \right\}. \qquad (4.109)$$

Here, $\mathrm{Perm}(x, y, z)$ denotes a permutation of (x, y, z), i.e. $\mathrm{Perm}(x, y, z) \equiv \{(x, y, z), (y, z, x), (z, x, y)\}$. Because the principal curvatures κ_1 and κ_2 are the eigenvalues of the Weingarten endomorphism of (4.106), the Gaussian curvature and the mean curvature can also be defined by (Kobayashi and Nomizu 1963, 1969)

$$K = \kappa_1 \kappa_2, \quad H = \frac{\kappa_1 + \kappa_2}{2}. \tag{4.110}$$

From (4.110), it is easy to derive that the principal curvatures at the voxel p can be expressed as:

$$\kappa_i(p) = H(p) \pm \sqrt{H^2(p) - K(p)}, \quad i = 1, 2.$$

Let us denote $\kappa_{min} = \min\{\kappa_1, \kappa_2\}$ and $\kappa_{max} = \max\{\kappa_1, \kappa_2\}$. Then the *volumetric shape index* $S(p)$ at voxel p is defined by (Koenderink 1990)

$$S(p) = \frac{1}{2} - \frac{1}{\pi} \arctan \frac{\kappa_{max}(p) + \kappa_{min}(p)}{\kappa_{max}(p) - \kappa_{min}(p)}. \tag{4.111}$$

The local shape index measures the shape of a local iso-surface patch at a voxel. Every distinct shape corresponds to a unique value of the shape index with a value range of [0,1]. Because the transition from one shape to another occurs continuously, the shape index can describe subtle shape variations effectively (Dorai and Jain 1997).

As shown in Fig. 4.52, the five well-known shape classes have the following shape index values: cup (0.0), rut (0.25), saddle (0.5), ridge (0.75), and cap (1.0). However, the transition from one shape class to another occurs continuously, and thus the shape index can describe subtle shape variations effectively. For example, a shape index value of 0.875 represents the "dome" class, which is a transient shape from ridge to cap. One can define the shape index at a voxel because the partial derivatives of the CT value at the voxel, which are needed for calculation of κ_{max} and κ_{min}, can be computed directly from the volumetric data without the need to extract the iso-surface explicitly (Lohmann 1998; Monga and Benayoun 1995; Thirion and Gourdon 1995). Therefore, the volumetric shape index can be defined at every point in a volume, and it captures the intuitive notion of the local shape of an iso-surface at a voxel.

The volumetric curvedness at voxel p is defined as the local curvedness of the iso-surface that passes the voxel, defined as:

$$C(p) = \frac{2}{\pi} \ln \sqrt{(\kappa_{min}(p)^2 + \kappa_{max}(p)^2)/2}.$$

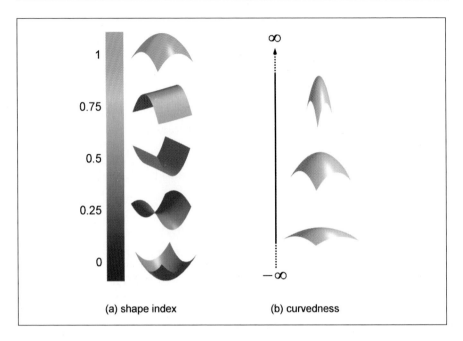

(a) shape index   (b) curvedness

Fig. 4.52. Schematic illustration of the volumetric shape index and curvedness. (**a**) The shape index maps the topological shape of a surface patch into a numerical value. The figure shows five sample shapes with their shape index values: cup, rut, saddle, ridge, and cap. Colour coding of these different types of shapes can differentiate among the structures on the colonic wall (see Fig. 4.51). (**b**) The curvedness represents the size and scale of the shape. For example, a gentle cap-like shape and a sharp cap-like shape have the same shape index value, but have a different curvedness values

The local curvedness is a measure of the size or scale of the shape and represents how gently curved the iso-surface is. The dimension of the curvedness is that of the reciprocal of the length, and its range is $]-\infty, \infty[$. Curvedness is a "dual" feature to the shape index in that the shape index measures "which" shape the local neighbourhood of a voxel has, whereas the curvedness measures "how much" shape the neighbourhood includes. The curvedness also provides scale information: a large negative value implies a very gentle change, whereas a large positive value implies a very sharp knife-like edge. Generally, points on a large sphere-like object have small curvedness values.

Because of the characteristic shape of polyps, a combination approach in which the shape index and curvedness are used can differentiate effectively among polyps, folds, and the colonic wall (Fig. 4.51).

Segmentation of Polyp Candidates

For segmentation of polyp candidates, first, voxels that have shape index and curvedness values between a predefined minimum and maximum are extracted as *seed regions*. We set the minimum and maximum threshold values for the shape index to 0.9 and 1.0, respectively, so that we can select the regions that are in the cap class. The minimum and maximum threshold values for curvedness are set so that the curvedness is within the range of targeted polyps. Starting with seed regions, *hysteresis thresholding* (Lohmann 1998) based on the shape index and curvedness is applied to the extracted colon for obtaining polyp candidates. The hysteresis thresholding extracts a set of voxels spatially connected to the seed regions having shape index and curvedness values within the predefined minimum and maximum values. This process is intended to extract a large connected component that corresponds to the major portion of a polyp.

Because hysteresis thresholding does not necessarily extract the complete visually perceived region of a polyp candidate, we subject each detected region to *iterative conditional morphological dilation* (Näppi and Yoshida 2003). In this method, a region segmented by hysteresis thresholding is used as a seed for a morphological volume-growing process. At each step of the iteration, the method will add a layer of voxels to the surface of the seed region, except for the voxels that have CT values lower than that characteristic of the colonic lumen. As a result, the extracted region will cover the entire polyp region, but not the colonic wall. The rate of increase of the volume of the expanded region will be monitored, and the dilation will be terminated when the growth rate becomes less than 1% (Fig. 4.53). The dilation will also be limited by the radius (7.5 mm) of the largest polyp we wish to detect, and the resulting region will be identified as the region for the polyp candidates. Examples of detected polyp candidates after the extraction step are shown in Fig. 4.54.

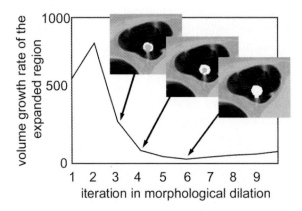

Fig. 4.53. Illustration of the extraction of the conditional morphological dilation. The region of initial detection (the white region in the left-most image) is expanded, by conditional morphological dilation, until the growth rate of the expanded region reaches a minimum, at which the entire region corresponding to the polyp has been segmented (the white region in the right-most image). Reprinted, with permission, from Yoshida and Dachman (2005)

4.7.4 Characterization of False Positives

Typically, the polyp candidates thus detected include a large number of false positives. Studies have shown that stool and prominent folds are major sources of false positives in CAD (Yoshida et al. 2002a, b). Various methods characterizing false positives have been developed for reduction of their number. These methods include volumetric texture analysis (Näppi and Yoshida 2002), CT attenuation (Summers et al. 2001), random orthogonal shape section (Gokturk et al. 2001), and optical flow (Acar et al. 2002).

3-D Texture Analysis for Discrimination of Stool and Folds

Stool differentiation is often based on the difference in the internal density variations between polyps and stool. These density variations are caused by the tendency of stool to contain air bubbles that can be recognized in CT images as an inhomogeneous textural pattern, or a mottled pattern. In contrast, polyps tend to have a homogeneous textural pattern, or solid

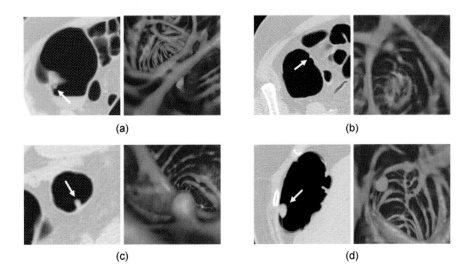

(a)   (b)

(c)   (d)

Fig. 4.54. Example of polyps detected by CAD. For each pair, the left image shows an axial CT image containing a polyp, indicated by an arrow, and the right image shows its 3-D endoluminal view by perspective volume rendering. The colour coding is based on the shape index values (see Fig. 4.52), in which polyps, folds, and the colonic wall are shown in green, pink, and brown, respectively. (a) A 6-mm sessile polyp in the cecum and (b) a 5.3-mm polyp in the cecum. These polyps were missed by a radiologist at first reading. (c) A 7-mm polyp in the transverse colon, and (d) an 11-mm sessile polyp in the hepatic flexure. Polyps are clearly differentiated from folds and the colonic wall. Reprinted, with permission, from Yoshida and Dachman (2005)

pattern, without intratumoral air. Thus, the use of volumetric texture analysis that characterizes the homogeneity of the CT density within a polyp, such as the variance of CT values, is effective for differentiating between stool and polyps (Näppi and Yoshida 2002).

Generally, folds appear to be much more elongated objects than polyps, and thus differentiation of folds from polyps is often based on the difference in appearance between polyps and folds. A technique called directional gradient concentration (DGC) (Näppi and Yoshida 2002) measures the degree of concentration of the gradient orientations of the CT values for differentiation between polyps and folds.

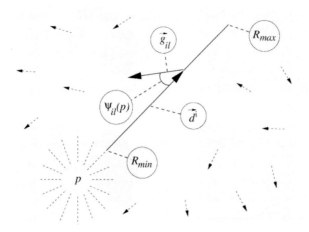

Fig. 4.55. Illustration of the GC and DGC calculation methods. The calculation of GC and DGC values at operating point p is based on considering the directions (angle $\psi_{il}(p)$) of the gradient vectors $\vec{g}_{il}$ over a distance range $[R_{min}, R_{max}]$ in selected directions $\vec{d}^i$ as described in the text. Reprinted, with permission, from Yoshida and Näppi (2001)

Briefly, the GC feature at a point p characterizes the overall direction of the gradient vectors around p. It is defined by:

$$GC(p) = \frac{1}{D}\sum_{i=1}^{D} e_i^{max}(p), \quad e_i^{max}(p) = \max_{R_{min} \leq n < R_{max}} \left\{ \frac{1}{n+1}\sum_{l=0}^{n} \cos\psi_{il}(p) \right\}, \quad (4.112)$$

where D is the number of the symmetrically 3-D-oriented direction vectors $\vec{d}^i$ originating from p. The angle $\psi_{il}(p)$ is calculated between $\vec{d}^i$ and $\vec{g}_{il}$, where $\vec{g}_{il}$ is the gradient vector located at distance l from p in direction $\vec{d}^i$ (Fig. 4.55). The value of the GC feature is maximal at the centre of a Gaussian sphere. To maximize its response to the polyps adhering to the colonic wall which typically appear hemispherical, the DGC is defined as follows:

$$DGC(p) = \frac{1}{2D}\sum_{i=1}^{D/2} \begin{cases} | e_i^{max}(p) - e_{i+D/2}^{max}(p) |; & e_i^{max}(p), e_{i+N/2}^{max}(p) > 0, \\ e_i^{max}(p) + e_{i+D/2}^{max}(p); & \text{otherwise.} \end{cases}$$

Here, e_i^{max} and $e_{i+D/2}^{max}$ are computed in opposite directions, $\vec{d}^i$ and $\vec{d}^{i+D/2}$.

Because of the hemispherical nature of polyps, the gradient vectors at voxels on the boundary between the polyp and luminal air converge to a

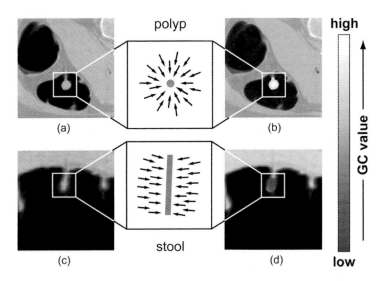

Fig. 4.56. Differentiation between folds and polyps by use of gradient concentration. Gradient concentration characterizes the average direction of the 3D gradient vectors surrounding a voxel. The vectors converge (green colour) towards the centre of a polyp (top), but tend to disperse (red colour) along a fold (bottom). Reprinted, with permission, from Yoshida and Dachman (2005)

point deep in the centre of the polyp. Moreover, small polyps have an apparently increasing soft-tissue density towards their centre, due to the partial volume effect. Therefore, DGC(p) at each voxel p in a polyp tend to point towards the central region of the polyp. On the other hand, folds are elongated, ridge-like objects. The gradient vectors at voxels on the boundary between a fold and luminal air converge to a line. Moreover, folds have no central region towards which the density increases. Therefore, DGC in a fold tends to concentrate nowhere. Because of these differences, DGC yields a higher response for polyps than for folds, and thus the DGC can be an effective means for differentiating between folds and polyps (Fig. 4.56).

4.7.5 Discrimination from False Positives and Polyps

The final output of CAD, that is, a set of polyps detected by CAD, is obtained by application of a statistical classifier, which is based on the image features extracted in the previous steps, to the differentiation of true polyps from false positives. Investigators use parametric classifiers such as quadratic discriminant analysis (Yoshida and Näppi 2001), non-parametric classifiers such as neural networks (Jerebko et al. 2003b; Kiss et al. 2002; Näppi et al. 2004), a committee of neural networks (Jerebko et al. 2003a), and a support vector machine (Gokturk et al. 2001). In principle, any combination of features and classifiers that provides a high classification performance should be sufficient for the differentiation task. The classifiers can be based on either supervised or unsupervised learning; however, thus far, only the former learning method has been used in CAD for VC.

One of the effective approaches is to employ a Bayesian neural network (BNN) to reduce FP polyp candidates based on 3-D shape and texture features. The Bayesian neural network is a feed-forwards network where the weights are assigned by Bayesian estimation (Kupinski et al. 2001). The BANN has the same structure as does a multi-layer perceptron, but it is unique in that the objective function includes a regularization term, equivalent to a Bayesian prior probability, to reduce the likelihood that the BNN will be over-trained (Bishop 1995; Haykin 1999; Kupinski et al. 2001). In the limit of infinite training data, a BNN can yield an ideal observer (i.e. the Bayes optimal) decision function for that data population. Moreover, on empirical observations, given a finite sample of training data, a BANN can estimate an ideal observer decision function reasonably well (Kupinski et al. 2001). Furthermore, in practical situations where the data probability density functions are unknown or difficult to determine, a BNN has been found empirically to have better performance than other classifiers (Edwards et al. 2001). BNN is thus capable of generating a highly generalizable decision boundary based on a small number of training cases.

The input to the BNN consists of feature statistics (FS) values of the image features described in the previous sections. Examples of the FS of a feature are: mean, minimum, maximum, variance, skew, kurtosis, entropy, contrast, and the average of ten maximum values of the feature. Let n denote the number of FSs used in discriminant analysis, and let $\mathbf{F}_i = (F_i^1, F_i^2, ..., F_i^n)$ denote an n-dimensional feature vector of a polyp candidate s_i, in which the component F_i^j represents the jth FS value of the polyp candidate. Given a training set with known classes, the BNN generates a quadratic *decision boundary* that optimally partitions the feature space spanned by the n features into two classes, i.e. a true-positive class denoted by C_{TP} and a false-positive class denoted by C_{FP}. To this end, a *discriminant function* $g(\mathbf{F}_i) : \mathbf{R}^n \rightarrow \mathbf{R}$ is generated, which projects the n-dimensional feature space to a scalar *decision variable space*.

Generally, the larger the value of $g(\mathbf{F}_i)$, the more likely it is that s_i is a polyp. In other words, $g(\mathbf{F}_i)$ is proportional to the ranked ordering of the likelihood that s_i is a polyp. Therefore, we classify the polyp candidates into classes C_{TP} and C_{FP} by partitioning the feature space through thresholding of the decision variable as follows:

$$C_{TP} = \{F_i; g(F_i) \geq t\}, \ C_{FP} = \{\mathbf{F}_i; g(\mathbf{F}_i) < t\},$$

where t is a threshold value. Those candidates that are classified in the polyp class C_{TP} are reported as the final detected polyps by the CAD system.

The CAD output is displayed, in a 3-D workstation, as a list of detected polyps and/or integrated in 2-D multiplanar reconstruction and 3-D endoluminal views of the colon by use of, for example, the colouring scheme that delineates the detected polyps and the normal structures in the colonic lumen (Näppi et al. 2005) (Fig. 4.57).

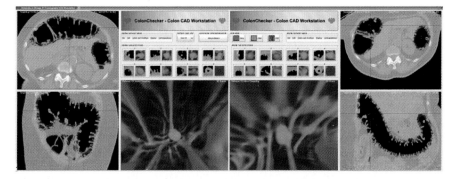

Fig. 4.57. Prototype colon CAD workstation (Näppi et al. 2005). The left and right images show the 2D multiplanar reconstruction (MPR) views of the supine and prone scans of a patient, respectively, with the computer-extracted colonic wall superimposed. The lower middle two images show the 3-D endoluminal views of the colon. The CAD output is integrated into the 2D MPR and 3-D endoluminal views by use of the colouring scheme that delineates the detected polyps and the normal structures in the colonic lumen. Polyps detected by CAD are shown as a list of icons in the middle rows in the figure. By clicking on one of these icons, one can jump to the corresponding polyp on a 3-D endoluminal view and/or an MPR view. The polyp (green) is displayed in both supine and prone views if it is found in the corresponding regions in these two views

4.7.6 Performance of CAD in the Detection of Polyps

Several academic institutions have conducted clinical trials to demonstrate the performance of their CAD schemes (Kiss et al. 2002; Näppi et al. 2004; Näppi and Yoshida 2003; Paik et al. 2004; Summers et al. 2001; Yoshida et al. 2002a, b; Yoshida and Näppi 2001). In these studies, optical colonoscopy was used as the gold standard, that is, the locations of the polyps detected by CAD were compared with the "true" locations of polyps that were determined visually in VC data sets based on colonoscopy and pathology reports. In most of theses studies, the performance of CAD was evaluated on VC cases that were acquired with a protocol that is currently widely used for VC.

An example of such an evaluation database is the following (Näppi and Yoshida 2003): The VC database consists of cases obtained from 72 clinical VC examinations. Each patient underwent standard pre-colono-scopy cleansing with a polyethylene glycol based solution or with a phosphate based diarrheal agent combined with magnesium citrate, and the colon was insufflated with room air.

The VC scanning was performed in the prone and supine positions with a helical CT scanner (GE 9800 CTi or LightSpeed QX/i; GE Medical Systems, Milwaukee, WI). Thus, there were 144 VC data sets. The collimation was 2.5–5.0 mm, the pitch was 1.5–1.7, and the reconstruction intervals were 1.5–2.5 mm. The matrix size of the axial images was 512×512, with a spatial resolution of 0.5–0.7 mm/pixel. A reduced current of 60 or 100 mA with 120 kVp was used for minimizing the radiation exposure. Each VC data set covered the entire region of the abdomen, from diaphragm to rectum, and consisted of 150–300 CT images with a matrix size of 512×512. After linear interpolation along the axial direction, the z-dimension of the resulting isotropic volumes contained between 500 and 700 voxels, and the physical resolution of these volumes was 0.5–0.75 mm/voxel.

Optical colonoscopy was performed on the same day as the VC. Experienced radiologists established the locations of the polyps in the VC data sets by use of the colonoscopy reports, pathology reports, and multiplanar reformatted views of the VC data. The results were used as the gold standard.

Fourteen of the cases (28 data sets) contained a total of 21 colonoscopy-confirmed polyps at least 5 mm in diameter. This size is the lower limit of the size range for polyps that are considered to be clinically significant (Johnson and Dachman 2000). One polyp was not visible in a supine data set of a patient because of a collapsed region of the colon.

Seventeen of the polyps were 5–10 mm, three polyps were 11–12 mm, and one polyp was 25 mm in diameter.

In a by-polyp analysis, in which all polyps were treated independently, the CAD system yielded 95% sensitivity, with an average of 1.5 false positives per patient (0.7 false positives per data set). In a by-patient analysis, the sensitivity was 100% (i.e. the scheme detected at least one polyp in an abnormal patient), with 1.3 false positives per patient (Näppi and Yoshida 2003).

In other studies, a group at Stanford University reported a 100% sensitivity with 7.0 false positives per data set (only the supine data set of each patient was used) based on eight patients that included a total of seven polyps >10 mm in four patients (Paik et al. 2004). The sensitivity was less than 50% at the same false positive rate for 11 polyps 5–9 mm that were found in three of the above eight patients. A group at the National Institutes of Health reported a 90% sensitivity with 15.7 false positives per data set, based on 40 patients (80 data sets) that included a total of 39 polyps ≥3 mm in 20 patients (Jerebko et al. 2003b).

These studies indicate that CAD is likely to succeed in detecting polyps with high sensitivity and a low false-positive rate. Generally, studies indicate that the performance of CAD schemes ranges between 70 and 100% by-patient sensitivity for polyps ≥6 mm (Yoshida and Dachman 2004, 2005). A meta-analysis of the reported performance of VC showed that, for human readers, the pooled by-patient sensitivity for polyps ≥10 mm and for those 6–9 mm was 85% and 70%, respectively, (Mulhall et al. 2005). Comparing this performance with that of CAD, it appears that the sensitivity of CAD is approaching that of an average human reader.

4.7.7 Improvement of Radiologists' Detection Performance by use of CAD

The ultimate goal of CAD is to improve the performance of radiologists in the detection of polyps and masses. Thus, establishing the sensitivity and specificity of CAD is only the first step in the evaluation of the benefit of CAD; CAD must be shown to improve the performance of radiologists.

A recent observer performance study, which evaluated the effect of CAD on radiologists in an environment that closely resembled a clinical interpretation environment of VC, showed that CAD could substantially improve radiologists' detection performance (Okamura et al. 2004). Four observers with different levels of reading skill (two experienced radiologists, a gastroenterologist, and a radiology resident) participated in the study, in which an observer read 20 VC data sets (including 11 polyps

5–12 mm in size), first without and then with CAD. The observer rated the confidence level regarding the presence of at least one polyp ≥5 mm in the colon. The detection performance, measured by the area under the receiver-operating characteristic curve (A_z) (Metz 2000), increased for all of the observers when they used CAD. The average A_z values without and with CAD were 0.70 and 0.85, respectively, and the difference was statistically significant ($p = 0.025$). The increase in the A_z value was the largest for the gastroenterologist (0.21) among the four observers (Fig. 4.58). Another observer study was conducted by Mani et al. (2004) based on 41 VC cases in which the average by-polyp and by-patient detection performance for three observers increased from 63 to 74% and from 73 to 90%, respectively, for 12 polyps ≥ 10 mm in ten subjects, although the differences were not statistically significant.

These small-scale studies show the potential of CAD in increasing radiologists' detection performance, especially for those with limited experience, as indicated by the second study. A larger-scale study needs to be conducted for showing more convincingly the benefits of CAD in improving the detection performance, in reducing the variability of the detection accuracy among readers, and in bringing the detection accuracy of inexperienced readers up to that of experienced readers.

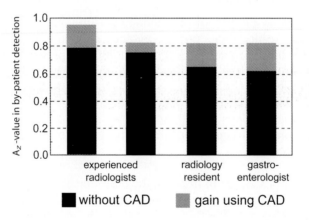

Fig. 4.58. Example of the gain in human reader performance with use of CAD (Okamura et al. 2004). Regardless of the different levels of reading skill, the detection performance, measured by the area under the ROC curve (vertical axis), increased for all of the readers when they used CAD. Among the four observers, the increase in performance was the largest for the gastroenterologist

4.7.8 CAD Pitfalls

CAD False Negatives

Knowing the pattern of CAD false negatives (actual lesions missed by CAD) is important for improved sensitivity when the output of a CAD scheme is used as a detection aid. The types of false negatives included in CAD results are similar to those encountered by radiologists (Yoshida et al. 2002a, b). Most of the CAD techniques depend on a shape analysis that assumes that polyps appear to have a cap-like shape, that is, they appear as polypoid lesions. Therefore, polyps that do not protrude sufficiently into the lumen (e.g. diminutive polyps and flat lesions (Fig. 4.59a)), whose shapes deviate significantly from polypoid (e.g. infiltrating carcinoma), those that lose a portion due to the partial volume effect, those that are located in a collapsed region of the colon, or those that are submerged in fluid, may be missed by CAD. Improvement of the CAD scheme for reliable detection of these types of polyps remains for future investigation.

CAD False Positives

False positives, that is, erroneous detections of actually normal anatomic structures, may lead to unnecessary further workups such as polypectomy

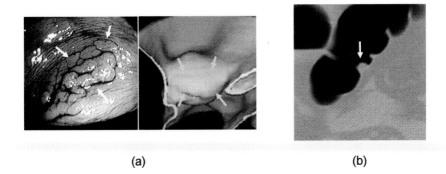

(a)   (b)

Figure 4.59. Examples of CAD pitfalls. (a) False negative detection (flat lesion). The left image shows an optical colonoscopy view of a large flat lesion (arrows), which is another type of precursor of colon cancer; however, unlike polyps, it has a low vertical elevation from the colonic wall and a large width. The right image shows an endoluminal view of the same lesion (arrows) in VC. Because of its low height, it was not well depicted in VC and thus was not detected by CAD. (b) False-positive detection (solid stool). This polyp-mimicking piece of stool has a cap-like appearance and a solid internal texture pattern, and thus it was detected incorrectly as a polyp.

by colonoscopy; therefore, knowledge about the pattern of CAD false positives is important for them to be dismissed. Studies showed that most of the false positives detected by CAD tend to exhibit polyp-like shapes, and that the major causes of CAD false positives are the following (Yoshida et al. 2002a, b): Approximately half (45%) of the false positives are caused by folds or flexural pseudotumors. They consist of sharp folds at the sigmoid colon, folds prominent on the colonic wall, two converging folds, ends of folds in the tortuous colon, and folds in the not-well-distended colon. One fifth (20%) are caused by solid stool (Fig. 4.59b), which is often a major source of error for radiologists as well. Approximately 15% are caused by residual materials inside the small bowel and stomach, and 10% are caused by the ileocecal valve. Among other causes of false positives are rectal tubes, elevation of the anorectal junction by the rectal tube, and motion artifacts, each amounting to less than 3%.

Studies show that radiologists can dismiss the majority of these false positives relatively easily based on their characteristic locations and appearance (Dachman et al. 2002). However, there are types of false positives, such as solid stool that mimics the shape of polyps and adheres to the colonic wall, which are difficult to differentiate from polyps even for an experienced radiologist. More research is required for establishing how radiologists can remove these false positives to make reliable, correct final diagnoses.

4.7.9 Conclusion

CAD techniques for VC have advanced substantially during the last several years. As a result, a fundamental CAD scheme for the detection of polyps has been established, and commercial CAD products for VC are now appearing. Thus far, CAD shows the potential for detecting polyps and cancers with high sensitivity and with a clinically acceptable low false-positive rate. However, CAD for VC needs to be improved further for more accurate and reliable detection of polyps and cancers. There are a number of technical challenges that CAD must meet, and the resulting CAD systems should be evaluated based on large-scale, multicentre, prospective clinical trials. If the assistance in interpretation offered by CAD is shown to improve diagnostic performance substantially, CAD is expected to make VC a cost-effective clinical procedure, especially in the screening setting.

Exercises

1. Derive the expressions for the Gaussian and mean curvatures, K and H, in (4.109) and (4.110).

2. There is one shape class that cannot be expressed by the volumetric shape index as defined by (4.111). Which shape class is it?

3. Draw the shape classes between the adjacent two shape classes shown in Fig. 4.52.

4. Gradient concentration, $GC(p)$ in (4.112), can be defined in 2D space. Calculate the value of the gradient concentration for a circle, an ellipse, and a strip, and show that the GC value is highest for a circle.

5. In section "Discrimination of false positives from polyps", $\mathbf{F}_i = (F_i^1, F_i^2, ..., F_i^n)$ denotes a feature vector of a polyp candidate s_i in an n-dimensional feature space. Explain why the discriminant function, $g(\mathbf{F})$, can be interpreted geometrically, in the feature space, as proportional to the signed distance from $\mathbf{F}_i$ to the decision boundary given by $g(\mathbf{F}_i) = 0$. Also, explain why $g(\mathbf{F}_i)$ can be interpreted as proportional to the likelihood that s_i is a polyp.

6. Describe an alternative CAD scheme for VC that addresses the following problems in the existing CAD schemes (1) reliable segmentation of the collapsed portions of the colon where the air-filled colonic lumen becomes very narrow; (2) reliable detection of a flat lesion, another type of colon cancer precursor, which is an object with less than 3 mm of vertical elevation from the colonic wall and whose width is twice as large as its height; (3) reliable differentiation of residual solid stool in the colon, which has a homogeneous internal textural pattern, from polyps.

Acknowledgements

Our test maps were produced by the *Planck* technical working group 2.1 (Diffuse emission separation methods). Some of the processing has been carried out through the HEALPix pixelization scheme (http://www. eso.org/science/healpix), by Krysztof M. Górski et al. Our work has been partially supported by the Italian Space Agency (ASI) under contracts I/R/217/01 and 1/R/073/01.

References

Acar, B., Beaulieu, C.F., Gokturk, S.B., Tomasi, C., Paik, D.S., Jeffrey, R.B., Jr., et al. (2002). Edge displacement field-based classification for improved detection of polyps in CT colonography. *IEEE Trans. Med. Imaging, 21*(12), 1461–1467

Ahmed, A., Andrieu, C., Doucet, A., & Rayner, P.J.W. (2000). On-line non-stationary ICA using mixture models. In *Proc. ICASSP 2000.* IEEE Signal Processing Society, *5,* 3148–3151

Amari, S.I., & Cichocki, A. (1998). Adaptive blind signal processing – Neural network approaches. *Proc. IEEE, 86,* 2026–2048

Andrieu, C., & Godsill, S.J. (2000). A particle filter for model based audio source separation. In P. Pajunen & J. Karhunen (Eds.), *2nd International workshop on ICA and blind signal separation.* Helsinki, FL: Helsinki University of Technology, pp. 381–386

Astley, S.M., & Gilbert, F.J. (2004). Computer-aided detection in mammography. *Clin. Radiol., 59*(5), 390–399

Attias, H. (1999). Independent factor analysis. *Neural Comput., 11,* 803–851

Aurich, V., Winkler, G., Hahn, K., Martin, A., & Rodenacker, K. (1999). Noise reduction in images: Some recent edge-preserving methods. *J. Pattern Recog. Image Anal., 9,* 749–766

Avilo, G., Broda, K., & Gabbay, D. (2001). Symbolic knowledge extraction from trained neural networks. *Artif. Intell., 125*(1), 153–205

Baccigalupi, C., Bedini, L., Burigana, C., De Zotti, G., Farusi, A., Maino, D., Maris, M., Perrotta, F., Salerno, E., Toffolatti, L., & Tonazzini, A. (2000). 43. *Mon. Not. R. Astron. Soc., 318,* 769–780

Baccigalupi, C., Perrotta, F., De Zotti, G., Smoot, G.F., Burigana, C., Maino, D., Bedini, L., & Salerno, E. (2004). Extracting cosmic microwave background polarization from satellite astrophysical maps. *Mon. Not. R. Astron. Soc., 354,* 55–70

Bach, F.R., & Jordan, M.I. (2002). Tree-dependent component analysis. In *Proceedings of the eighteenth conference on uncertainty in artificial intelligence.* San Francisco: Morgan Kauffmann

Bäck, T. (1994). Selective pressure in evolutionary algorithms: A characterization of selection mechanisms. In *Proceedings of the 1st IEEE conference on evolutionary computation.* Piscataway, NJ: IEEE Press, pp. 57–62

Bader, D.A., Jaja, J., Harwood, D., & Davis, L.S. (1996). Parallel algorithms for image enhancement and segmentation by region growing with experimental study. *Proc. IEEE IPPS-96,* 414

Barber, C.B., Dobkin, D.P., & Huhdanpaa, H.T. (1996). The quickhull algorithm for convex hulls. *ACM Trans. Math. Soft., 22*(4), 469–483

Barreiro, R.B., Hobson, M.P., Banday, A.J., Lasenby, A.N., Stolyarov, V., Vielva, P., & Górski, K.M. (2004). Foreground separation using a flexible maximum-entropy algorithm: An application to COBE data. *Mon. Not. R. Astron. Soc., 351,* 515–540

Barros, A.K. (2000). The independence assumption: Dependent component analysis. In M. Girolami (Ed.), *Advances in independent component analysis*, Berlin Heidelberg New York: Springer, pp. 63–71

Barros, A.K., & Cichocki, A. (2001). Extraction of specific signals with temporal structure. *Neural Comput.*, *13*, 1995–2003

Bedini, L., Bottini, S., Baccigalupi, C., Ballatore, P., Herranz, D., Kuruoglu, E.E., Salerno, E., & Tonazzini, A. (2003). *A semi-blind second-order approach for statistical source separation in astrophysical maps. ISTI-CNR, Technical Report*, 2003-TR-35, Pisa, Italy: ISTI-CNR

Bedini, L., Herranz, D., Salerno, E., Baccigalupi, C., Kuruoglu, E.E., & Tonazzini, A. (2005). Separation of correlated astrophysical sources using multiple-lag data covariance matrices. *EURASIP J. Appl. Signal Process.*, *2005*(15), 2400–2412

Bell, A.J., & Sejnowski, T.J. (1995). An information-maximization approach to blind separation and blind deconvolution. *Neural Comput.*, *7*, 1129–1159

Belouchrani, A., Abed-Meraim, K., Cardoso, J.-F., & Moulines, E. (1997). A blind source separation technique based on second order statistics. *IEEE Trans. Signal Process.*, *45*, 434–444

Bennett, C., Hill, R.S., Hinshaw, G., Nolte, M.R., Odegard, N., Page, L., Spergel, D.N., Weiland, J.L., Wright, E.L., Halpern, M., Jarosik, N., Kogut, A., Limon, M., Meyer, S.S., Tucker, G.S., & Wollack, E. (2003). First-year Wilkinson microwave anisotropy probe (WMAP) observations: Foreground emission. *Astrophys. J. Suppl. Ser.*, *148*, 97–117

Bishop, C.M. (1995). *Neural network for pattern recognition*. Oxford: Oxford University Press

Blake, C.L., & Merz, C.J. (1998). UCI Repository of machine learning databases http://www.ics.uci.edu/~mlearn/MLRepository.html. Irvine, CA: University of California, Department of Information and Computer Science

Bouchet, F.R., Prunet, S., & Sethi, S.K. (1999). Multifrequency Wiener filtering of cosmic microwave background data with polarization. *Mon. Not. R. Astron. Soc.*, *302*, 663–676

Bow, S.-T. (2002). *Pattern recognition and image processing*. New York, USA: Dekker

Breiman, L., Friedman, J., Olshen, R., & Stone, C. (1984). *Classification and regression trees*. Belmont, CA: Wadsworth

Brodley, C., & Utgoff, P. (1995). Multivariate decision trees. *Mach. Learn.*, *19*(11), 45–77

Bronzino, J.D. (1995). *The biomedical engineering handbook*. Boca Raton, USA: CRC

Brugge, M.H., Stevens, J.H., Nijhuis, J.A.G., & Spaanenburg, L. (1998). License plate recognition using DTCNNs. *Fifth IEEE International workshop on cellular neural networks and their applications proceedings*, pp. 212–217

Cardoso, J.-F. (1998). Blind signal separation: Statistical principles. *Proc. IEEE*, *86*, 2009–2025

Cardoso, J.-F. (1999). High-order contrasts for independent component analysis. *Neural Comput.*, *11*, 157–192

Cardoso, J.-F., Snoussi, H., Delabrouille, J., & Patanchon, G. (2002). Blind separation of noisy Gaussian stationary sources. Application to cosmic microwave background imaging. In *Proceedings of the EUSIPCO 2002*. European Association for Signal, Speech and Image Processing, *1*, 561–564

Casella, G., & Robert, C.P. (1999). *Monte Carlo statistical methods*. Berlin Heidelberg New York: Springer

Castleman, K.R. (1996). *Digital image processing*. Upper Saddle River, USA: Prentice-Hall

Cayón, L., Sanz, J.L., Barreiro, R.B., Martínez-González, E., Vielva, P., Toffolatti, L., Silk, J., Diego, J.M., & Argüeso F. (2000). Isotropic wavelets: A powerful tool to extract point sources from cosmic microwave background map. *Mon. Not. R. Astron. Soc.*, *315*, 757–761

Chandler, B., Rekeczky, C., Nishio, Y., & Ushida, A. (1999). Adaptive simulated annealing in CNN template learning. *IEICE Trans. Fundam.*, *E82*(2), 398–402

Chen, D., Liang, Z., Wax, M.R., Li, L., Li, B., & Kaufman, A.E. (2000). A novel approach to extract colon lumen from CT images for virtual colonoscopy. *IEEE Trans. Med. Imaging*, *19*(12), 1220–1226

Chipman, H., George, E., & McCullock R. (1998a). Bayesian CART model search. *J. Am. Stat.*, *93*, 935–960

Chipman, H., George, E., & McCulloch, R. (1998b). Making sense of a forest of trees. *Proceedings of the symposium on the interface*, Fairfax station, VA: Interface Foundation

Chua, L.O., Gulak, G., Pierzchala, E., & Rodríguez-Vázquez, A. (Ed.) (1998). *Cellular neural networks and analog VLSI*. Boston, USA: Kluwer

Chua, L.O., & Roska, T. (2001). *Cellular neural networks and visual computing: Foundation and applications*. Cambridge: Cambridge University Press

Chua, L.O., & Yang, L. (1988). Cellular neural networks: Theory. *IEEE Trans. Circuits Syst.*, *35*(10), 1257–1272

Cichocki, A., & Amari, S. (2002). *Adaptive blind signal and image processing*. New York: Wiley

Cigale, B., Divjak, M., & Zazula, D. (2002). Application of simulated annealing to biosignal classification and segmentation. *15th IEEE Symposium on Computer-Based Medical Systems*, 165–170

Cigale, B., & Zazula, D. (2004). Segmentation of ovarian ultrasound images using cellular neural networks. *Int. J. Pattern Recognit. Artif. Intell.*, *18*(4), 563–581

Comon, P. (1994). Independent component analysis: A new concept? *Signal Process.*, *36*, 287–314

Costagli, M., Kuruoglu, E.E., & Ahmed, A. (2004). Astrophysical image separation using particle filters. *Lect. Notes Comput. Sci.*, *3195*, 930–937

Crounse, K.R., & Chua, L.O. (1995). Methods for image processing and pattern formation in cellular neural networks: A tutorial. *IEEE Trans. Circuits Syst.*, *42*(10), 583–601

Dachman, A.H., Näppi, J., Frimmel, H., & Yoshida, H. (2002). Sources of false positives in computerized detection of polyps in CT colonography. *Radiology*, *225*(P), 303

Delabrouille, J., Cardoso, J.-F., & Patanchon, G. (2002). Multidetector multicomponent spectral matching and applications for cosmic microwave background data analysis. *Mon. Not. R. Astron. Soc.*, *346*, 1089–1102

De Lathauwer, L. (1997). *Signal processing based on multilinear algebra*. Ph. D. thesis, Heverlee, Belgium: Katholieke Universiteit Leuven

Dempster, E.J., Laird, N.M., & Rubin, D.B. (1977). Maximum likelihood from incomplete data via EM algorithm. *Ann. R. Stat. Soc.*, *39*, 1–38

Denison, D., Holmes, C., Malick, B., & Smith, A. (2002). Bayesian methods for nonlinear classification and regression. Chichester, UK: Wiley

De Zotti, G., Toffolatti, L., Argüeso, F., Davies, R.D., Mazzotta, P., Partridge, R.B., Smoot, G.F., & Vittorio, N. (1999). The Planck surveyor mission: Astrophysical prospects. In *3K Cosmology. Proceedings of the EC-TMR Conference*. Woodbury, NY: American Institute of Physics, p. 204

Dietterich, T. (2000). *Ensemble methods in machine learning*. Proceedings of the multiple classifier systems. *Lecture Notes in Computer Science*. Berlin Heidelberg New York: Springer, pp. 1–15

Domingos, P. (1998). Knowledge discovery via multiple models. *Intell. Data Anal.*, *2*, 187–202

Domingos, P. (2000). Bayesian averaging of classifiers and the overfitting problem. *Proc. Mach. Learn.* Stanford: Morgan Kaufmann, pp. 223–230

Dorai, C., & Jain, A.K. (1997). Cosmos – a representation scheme for 3D free-form objects. *IEEE Trans. Pattern Anal. Mach. Intell.*, *19*, 1115–1130

Doucet, A., De Freitas, J.F.G., & Gordon, N.J. (2001). *Sequential Monte Carlo methods in practice*. Berlin Heidelberg New York: Springer

Duda, R.O., & Hart, P.E. (2001). *Pattern classification*. 2nd edn. New York: Wiley Interscience

Edwards, D.C., Papaioannou, J., Jiang, Y., Kupinski, M.A., & Nishikawa, R.M. (2001). *Eliminating false-positive microcalcification clustgers in a mammography cad scheme using a bayesian neural network*. Paper presented at the Proc SPIE

Everson, R.M., & Roberts, S.J. (2000). Particle Filters for Non-stationary ICA. In M. Girolami (Ed.), *Advances in independent components analysis*. Berlin Heidelberg New York: Springer, pp. 23–41

Farlow, S. (1984). *Self-organizing methods in modeling: GMDH-Type Algorithms*. New York: Dekker

Fieldsend, J.E., Bailey, T.C., Everson, R.M., Krzanowski, W.J., Partridge, D, & Schetinin, V. (2003). Bayesian inductively learned modules for safety critical systems. *Proceedings of the symposium on the interface*. Computing Science and Statistics, Fairfax Station, USA: Interface Foundation

Frimmel, H., Nappi, J., & Yoshida, H. (2005). Centerline-based colon segmentation for CT colonography. *Med. Phys.*, *32*(8), 2665–2672

Galant, S. (1993). *Neural network learning and expert systems*. Cambridge, MA: MIT

Gao, J., Zhou, M., & Wang, H. (2001). A threshold and region growing method for filament disappearance area detection in solar images. *Proceedings of the information science and systems*. Laurel, MD: Johns Hopkins University

Gokturk, S.B., Tomasi, C., Acar, B., Beaulieu, C.F., Paik, D.S., Jeffrey, R.B., Jr., et al. (2001). A statistical 3D pattern processing method for computer-aided detection of polyps in CT colonography. *IEEE Trans. Med. Imaging*, *20*(12), 1251–1260

Green, P. (1995). Reversible jump Markov chain Monte Carlo computation and Bayesian model determination. *Biometrika*, *82*, 711–732

Handschin, J.E., & Mayne, D.Q. (1969). Monte Carlo techniques to estimate the conditional expectation in multi-stage non-linear filtering. *Int. J. Control*, *9*, 547–559

Hänggi, M., & Moschzty, G.S. (2000). *Cellular neural network: Analysis, design and optimisation*. Boston, USA: Kluwer

Harrer, H., & Nossek, J.A. (1992). Discrete-time cellular neural networks. *Int. J. Circ. Theor. App. 20*, 453–467

Haupt, R.L., & Haupt, S.E. (2004). *Practical genetic algorithms*. New Jersey, USA: Wiley Interscience

Haykin, S. (1999). *Neural networks: A comprehensive foundation*. New Jersey: Prentice Hall

Haykin, S., (Ed.) (1994). *Blind deconvolution*. Englewood Cliffs, USA: Prentice-Hall

Haykin, S., (Ed.) (2000). *Unsupervised adaptive filtering, Vol. II: Blind deconvolution*. New York: Wiley

Holland, J.H. (1992). Genetic algorithms. *Sci. Am. 267*, 66–72

Hobson, M.P., Jones, A.W., Lasenby, A.N., & Bouchet, F.R. (1998). Foreground separation methods for satellite observations of the cosmic microwave background. *Mon. Not. R. Astron. Soc.*, *300*, 1–29

Hong, L., Liang, Z., Viswambharan, A., Kaufman, A., & Wax, M. (1997). Reconstruction and visualization of 3D models of colonic surface. *IEEE Trans. Nuclear Sci.*, *44*, 1297–1302

Hopfield, J.J. (1982). Neural networks and physical systems with emergent computational abilities. *Proc. Natl. Acad. Sci. USA*, *79*, 2554–2558

Hyvärinen, A. (1998). Independent component analysis in the presence of gaussian noise by maximizing joint likelihood. *Neurocomputing*, *22*, 49–67

Hyvärinen, A. (1999a). Fast and robust fixed-point algorithms for independent component analysis. *IEEE Trans. Neural Networks*, *10*, 626–634

Hyvärinen, A., (1999b). Gaussian moments for noisy independent component analysis. *IEEE Signal Process. Lett.*, *6*, 145–147

Hyvärinen, A., Karhunen, J., & Oja, E. (2001). *Independent component analysis*. New York: Wiley

Hyvärinen, A., & Oja, E. (1997). A fast fixed-point algorithm for independent component analysis. *Neural Comput.*, *9*, 1483–1492

Hyvärinen, A., & Oja, E. (2000). Independent component analysis: Algorithms and applications, *Neural Network*, *13*, 411–430

Ingber, L. (1989). Very fast simulated re-annealing. *J. Math. Comput. Model.*, 12, 967–973

Iordanescu, G., Pickhardt, P.J., Choi, J.R., & Summers, R.M. (2005). Automated seed placement for colon segmentation in computed tomography colonography. *Acad. Radiol.*, *12*(2), 182–190

Jerebko, A.K., Malley, J.D., Franaszek, M., & Summers, R.M. (2003a). Multiple neural network classification scheme for detection of colonic polyps in CT colonography data sets. *Acad. Radiol.*, *10*(2), 154–160

Jerebko, A.K., Summers, R.M., Malley, J.D., Franaszek, M., & Johnson, C.D. (2003b). Computer-assisted detection of colonic polyps with CT colonography using neural networks and binary classification trees. *Med. Phys.*, *30*(1), 52–60

Johnson, C.D., & Dachman, A.H. (2000). CT colonography: The next colon screening examination? *Radiology*, *216*(2), 331–341

Kirkpatrick, S., Gelatt, C.D., & Vecchi, M.V. (1983). Optimization by simulated annealing. *Science*, *220*(4598), 671–680

Kiss, G., Van Cleynenbreugel, J., Thomeer, M., Suetens, P., & Marchal, G. (2002). Computer-aided diagnosis in virtual colonography via combination of surface normal and sphere fitting methods. *Eur. Radiol.*, *12*(1), 77–81

Knuth, K. (1998). Bayesian source separation and localization. *Proceedings of the SPIE: Bayesian inference for inverse problems*, pp. 147–158

Kobayashi, S., & Nomizu, K. (1963). *Foundations of differential geometry I*, Vol. 1. New York: Interscience

Kobayashi, S., & Nomizu, K. (1969). *Foundations of differential geometry II*, Vol. 2. New York: Interscience

Koenderink, J.J. (1990). *Solid shape*. Cambridge, MA: MIT

Kozek, T., Roska, T., & Chua, L.O. (1993). Genetic algorithm for CNN template learning. *IEEE Trans. Circuits Syst. I*, *40*(6), 392–402

Kuncheva, L. (2004). *Combining pattern classifiers: Methods and algorithms*. New York: Wiley

Kupinski, M.A., Edwards, D.C., Giger, M.L., & Metz, C.E. (2001). Ideal observer approximation using bayesian classification neural networks. *IEEE Trans. Med. Imaging*, *20*(9), 886–899

Kuruoglu, E.E., Bedini, L., Paratore, M.T., Salerno, E., & Tonazzini, A. (2003). Source separation in astrophysical maps using independent factor analysis. *Neural Network*, *16*, 479–491

Kuruoglu, E., Tonazzini, A., & Bianchi, L. (2004). Source separation in astrophysical images modelled by Markov random fields. In *Proceedings of the ICIP 2004*. IEEE Signal Processing Society , pp. 2701–2704

Laarhoven, P.J.M., & Aarts, E.H.L. (1987). *Simulated annealing: Theory and applications*. Dordrecht: Reidel

Lee, T., Lewicki, M., & Sejnowski, T. (1999). Independent component analysis using an extended infomax algorithm for mixed sub-Gaussian and super-Gaussian sources. *Neural Comput.*, *11*, 409–433

Lee, S.E., & Press, S.J. (1998). Robustness of Bayesian factor analysis estimates. *Commun. Stat. A. Theor.*, *27*, 1871–1893

Levin, B., Brooks, D., Smith, R.A., & Stone, A. (2003). Emerging technologies in screening for colorectal cancer: CT colonography, immunochemical fecal occult blood tests, and stool screening using molecular markers. *CA Cancer J. Clin.*, *53*(1), 44–55

Li, H., & Santago, P. (2005). Automatic colon segmentation with dual scan CT colonography. *J. Digit. Imaging*, *18*(1):42–54

Lohmann, G. (1998). *Volumetric image analysis.* New York: Wiley

Loncar, A., Kunz, R., & Tetzlaff, R. (2000). SCNN 2000 - Part I: Basic structure and features of the simulation system for cellular neural networks. *Proceedings of the 6th IEEE International workshop on cellular neural networks and their applications (CNNA)*, Catania, pp. 123–128

Madala, H., & Ivakhnenko, A. (1994). *Inductive learning algorithms for complex systems modeling.* Boca Raton: CRC

Manganaro, G., Arena, P., & Fortuna, L. (1999). *Cellular neural network: Chaos, complexity and VLSI processing.* Berlin Heidelberg New York: Springer

Magnussen, H., Nossek, J.A., & Chua, L.O. (1993). The learning problem for discrete-time cellular neural networks as a combinatorial optimization problem. *Simulated annealing technical reports UCB//ERL-93-88.* Berkeley: University of California

Maino, D., Farusi, A., Baccigalupi, C., Perrotta, F., Banday, A.J., Bedini, L., Burigana, C., De Zotti, G., Gorski, K.M., & Salerno, E. (2002). All-sky astrophysical component separation with Fast Independent Component Analysis (FastICA). *Mon. Not. R. Astron. Soc.*, *334*, 53–68

Mani, A., Napel, S., Paik, D.S., Jeffrey, R.B., Jr., Yee, J., Olcott, E.W., et al. (2004). Computed tomography colonography: Feasibility of computer-aided polyp detection in a "first reader" paradigm. *J. Comput. Assist. Tomogr.*, *28*(3), 318–326

Masutani, Y., Yoshida, H., MacEneaney, P.M., & Dachman, A.H. (2001). Automated segmentation of colonic walls for computerized detection of polyps in CT colonography. *J. Comput. Assist. Tomogr.*, *25*(4), 629–638

Metz, C.E. (2000). Fundamental roc analysis. In J. Beutel, H.L. Kundel, & R.L.V. Metter (Eds.), *Handbook of medical imaging* Vol. 1, Bellingham, WA: SPIE, pp. 751–770

Mohammad-Djafari, A. (2001). A Bayesian approach to source separation. *Proc. AIP Conf.*, *567*, 221–244

Monga, O., & Benayoun, S. (1995). Using partial derivatives of 3D images to extract typical surface features. *Comput. Vis. Image Und.*, *61*, 171–189

Morrin, M.M., & LaMont, J.T. (2003). Screening virtual colonoscopy – ready for prime time? *N. Engl. J. Med.*, *349*(23), 2261–2264

Moulines, E., Cardoso, J.-F., & Gassiat, E. (1997). Maximum likelihood for blind separation and deconvolution of noisy signals using mixture models. *Proceedings of the ICASSP 1997.* IEEE Signal Processing Society, *5*, 3617–3620

Mulhall, B.P., Veerappan, G.R., & Jackson, J.L. (2005). Meta-analysis: Computed tomographic colonography. *Ann. Intern. Med.*, *142*(8), 635–650

Müller, J.A., & Lemke, F. (2003). Self-organizing data mining: Extracting knowledge from data. Canada: Trafford

Näppi, J., Dachman, A.H., MacEneaney, P., & Yoshida, H. (2002a). Automated knowledge-guided segmentation of colonic walls for computerized detection of polyps in CT colonography. *J. Comput. Assist. Tomogr.*, *26*(4), 493–504

Näppi, J., Frimmel, H., Dachman, A.H., & Yoshida, H. (2002b). Computer aided detection of masses in CT colonography: Techniques and evaluation. *Radiology*, *225*(P), 406

Näppi, J., Frimmel, H., Dachman, A.H., & Yoshida, H. (2004). *A new high-performance cad scheme for the detection of polyps in CT colonography.* Paper presented at the Medical Imaging 2004: Image Processing

Näppi, J., Frimmel, H., & Yoshida, H. (2005). Virtual endoscopic visualization of the colon by shape-scale signatures. *IEEE Trans. Inf. Technol. Biomed.*, *9*(1), 120–131

Näppi, J., & Yoshida, H. (2002). Automated detection of polyps with CT colonography: Evaluation of volumetric features for reduction of false-positive findings. *Acad. Radiol.*, *9*(4), 386–397

Näppi, J., & Yoshida, H. (2003). Feature-guided analysis for reduction of false positives in cad of polyps for computed tomographic colonography. *Med. Phys.*, *30*(7), 1592–1601

Okamura, A., Dachman, A.H., Parsad, N., Näppi, J., & Yoshida, H. (2004). *Evaluation of the effect of cad on observers' performance in detection of polyps in CT colonography.* Paper presented at the CARS. Chicago, IL: Computer Assisted Radiology and Surgery

Paik, D.S., Beaulieu, C.F., Rubin, G.D., Acar, B., Jeffrey, R.B., Jr., Yee, J., et al. (2004). Surface normal overlap: A computer-aided detection algorithm with application to colonic polyps and lung nodules in helical ct. *IEEE Trans. Med. Imaging*, *23*(6), 661–675

Patanchon, G., Snoussi, H., Cardoso, J.-F., & Delabrouille, J. (2003). Component separation for Cosmic Microwave Background data: a blind approach based on spectral diversity. *Proceedings of the PSIP 2003.* Grenoble, France (extended version in astro-ph/0302078), pp. 17–20

Perry, S.W., Wong H.-S., & Guan L. (2002). *Adaptive image processing: A computational intelligence perspective.* Boca Raton, USA: CRC

Pope, K.J., & Bogner, R.E. (1996). Blind signal separation. I. Linear, instantaneous combinations. *Digit. Sign. Process.*, *6*, 5–16

Potočnik, B., & Zazula, D. (2002). Automated analysis of a sequence of ovarian ultrasound images, Part I – segmentation of single 2D images. *Image Vis. Comput.*, *20*(3), 217–225

Qahwaji, R., & Green, R. (2001). Detection of closed regions in digital images. *J. Comput. Appl.*, *8*(4), 202–207

Quinlan, J. (1993). *C4.5: Programs for machine learning.* Los Altos, CA: Morgan Kaufmann

Ripley, B. (1994). Neural networks and related methods for classification. *J. Roy. Stat. Soc. B*, *56*(3), 409–456

Roska, T., Kék, L., Nemes, L., Zarándy, Á., Brendel, M., & Szolgay, P. (1998). *CNN software library (templates and algorithms) version 7.2*. Research report of the analogic (dual) and neural computing systems laboratory, Budapest, Hungary: MTA SZTAKI

Russ, J.C. (2002). *The image processing handbook*. Boca Raton, USA: CRC

Salerno, E., Baccigalupi, C., Bedini, L., Burigana, C., Farusi, A., Maino, D., Maris, M., Perrotta, F., & Tonazzini, A. (2000). *Independent component analysis approach to detect the cosmic microwave background radiation from satellite measurements*. Pisa, Italy: ISTI-CNR, technical report B4-04-04-00

Salzberg, S., Delcher, A., Fasman, K., & Henderson, J. (1998). A decision tree system for finding genes in DNA. *Comput. Biol.*, *5*, 667–680

Schetinin, V. (2003). A learning algorithm for evolving cascade neural networks. *Neural Process Lett.*, *17*, 21–31

Schetinin, V., Fieldsend, J.E., Partridge, D., Krzanowski, W.J., Everson, R.M., Bailey, T.C., & Hernandez, A. (2004). The Bayesian decision tree technique with a sweeping strategy. *Proceedings of the IEEE conference on advances in intelligent systems - theory and applications, (AISTA 2004)*, Luxembourg: IEEE Computer Society

Schetinin, V., & Schult, J. (2005). A neural network technique for learning concepts from electroencephalograms. *Theor. Biosci.*, *124*, 41–53

Sethi, I., & Yoo, J. (1997). Structure-driven induction of decision tree classifiers through neural learning. *Pattern Recogn.*, *30*(11), 1893–1904

Shulman, D., & Herve, J.Y. (1989). Regularization of discontinuous flow fields. *Proceedings of the workshop on visual motion*. IEEE Computer Society Press, pp. 81–85

Smoot, G.F., Bennett, C.L., Kogut, A., Wright, E.L., Aymon, J., Boggess, N.W., Cheng, E.S., Deamici, G., Gulkis, S., Hauser, M.G., Hinshaw, G., Jackson, P.D., Janssen, M., Kaita, E., Kelsall, T., Keegstra, P., Lineweaver, C., Loewenstein, K., Lubin, P., Mather, J., Meyer, S.S., Moseley, S.H., Murdock, T., Rokke, L., Silverberg, R.F., Tenorio, L., Weiss, R., & Wilkinson, D.T. (1992). Structure in the COBE differential microwave radiometer 1st-year maps. *Astrophys. J.*, *396*(1), L1

Snoussi, H., Patanchon, G., Macias-Pérez, J., Mohammad-Djafari, A., & Delabrouille, J. (2001). Bayesian blind component separation for cosmic microwave background observation. *Proceedings of the AIP workshop on Bayesian and maximum-entropy methods*. American Institute of Physics, pp. 125–140

Stolyarov, V., Hobson, M.P., Ashdown, M.A.J., & Lasenby, A.N. (2002). All-sky component separation for the Planck mission. *Mon. Not. R. Astron. Soc.*, *336*, 97–111

Stolyarov, V., Hobson, M.P., Lasenby, A.N., & Barreiro, R.B. (2005). All-sky component separation in the presence of anisotropic noise and dust temperature variations. *Mon. Not. R. Astron. Soc.*, *357*, 145–155

Stone, J.V., Porrill, J., Porter, N.R., & Wilkinson, I.W. (2002). Spatiotemporal independent component analysis of event-related fMRI data using skewed probability density functions. *Neuroimage, 15*, 407–421

Summers, R.M., Beaulieu, C.F., Pusanik, L.M., Malley, J.D., Jeffrey, R.B., Jr., Glazer, D.I., et al. (2000). Automated polyp detector for CT colonography: Feasibility study. *Radiology, 216*(1), 284–290

Summers, R.M., Johnson, C.D., Pusanik, L.M., Malley, J.D., Youssef, A.M., & Reed, J. E. (2001). Automated polyp detection at CT colonography: Feasibility assessment in a human population. *Radiology, 219*(1), 51–59

Tegmark, M., Eisenstein, D.J., Hu, W., & de Oliveira-Costa A. (2000). Foregrounds and forecasts for the cosmic microwave background. *Astrophys. J., 530*, 133–165

Tenorio, L., Jaffe, A.H., Hanany, S., & Lineweaver, C.H. (1999). Applications of wavelets to the analysis of cosmic microwave background maps. *Mon. Not. R. Astron. Soc., 310*, 823–834

Thirion, J.-P., & Gourdon, A. (1995). Computing the differential characteristics of isointensity surfaces. *Comput. Vis. Image Und., 61*, 190–202

Tonazzini, A., Bedini, L., Kuruoglu, E.E., & Salerno, E. (2003). *Blind separation of auto-correlated images from noisy data using MRF models*. Nara, Japan: Proceedings of the fourth international symposium on independent component analysis and blind source separation, pp. 675–680

Tonazzini, A., Bedini, L., & Salerno, E. (2006). A Markov model for blind image separation by a mean-field EM algorithm. *IEEE Transactions on image processing, 15*, 473–482

Tonazzini, A., & Gerace, I. (2005). Bayesian MRF-based blind source separation of convolutive mixtures of images. *Proceedings of the EUSIPCO 2005*, 4–8 September 2005, Antalya, Turkey: *EUSIPCO*

Tong, L., Liu, R.W., Soon, V.C., & Huang, Y.-F. (1991). Indeterminacy and identifiability of blind identification. *IEEE Transactions on circuits and systems, 38*, 499–509

Torkkola, K. (1996). Blind separation of delayed sources based on information maximization. Atlanta, USA: *IEEE International Conference on Acoustics, Speech & Signal Processing*, 7–10

Vielva, P., Martínez-González, E., Cayón, L., Diego, J.M., Sanz, J.L., & Toffolatti L. (2001). Predicted Planck extragalactic point-source catalogue. *Mon. Not. R. Astron. Soc., 326*, 181–191

Vlaisavljević, V., Reljič, M., Lovrec, V.G., Zazula, D., & Sergent, N. (2003). Measurement of perifollicular blood flow of the dominant preovulatory follicle using 3D power Doppler. *Ultrasound Obst. Gyn., 22*(5), 520–526

Web page of AnaLogic Computers Ltd. www.analogic-computers.com

Wyatt, C.L., Ge, Y., & Vining, D.J. (2000). Automatic segmentation of the colon for virtual colonoscopy. *Comput. Med. Imaging Graph, 24*(1), 1–9

Yang, T. (2002). *Handbook of CNN image processing: All you need to know about cellular neural networks*. Tucson, USA: Yang's Scientific Research Institute LLC

Yeredor, A. (2002). Non-orthogonal joint diagonalization in the least-squares sense with application in blind source separation. *IEEE Transactions on Signal Processing, 50,* 1545–1553

Yoo, T.S. (Ed.) (2004). *Insight into images: Principles and practice for segmentation, registration, and image analysis.* Wellesey, USA: A K Peters

Yoshida, H., & Dachman, A.H. (2004). Computer-aided diagnosis for CT colonography. *Semin. Ultrasound CT, 25*(5), 419–431

Yoshida, H., & Dachman, A.H. (2005). Cad techniques, challenges, and controversies in computed tomographic colonography. *Abdom. Imaging, 30*(1), 26–41

Yoshida, H., Masutani, Y., MacEneaney, P., Rubin, D.T., & Dachman, A.H. (2002a). Computerized detection of colonic polyps at CT colonography on the basis of volumetric features: Pilot study. *Radiology, 222*(2), 327–336

Yoshida, H., & Näppi, J. (2001). Three-dimensional computer-aided diagnosis scheme for detection of colonic polyps. *IEEE Trans. Med. Imaging, 20*(12), 1261–1274

Yoshida, H., Näppi, J., MacEneaney, P., Rubin, D.T., & Dachman, A.H. (2002b). Computer-aided diagnosis scheme for detection of polyps at CT colonography. *Radiographics, 22*(4), 963–979

Zharkova, V.V., Ipson, S.S., Qahwaji, R., Zharkov, S., & Benkhalil, A. (2003a). An automated detection of magnetic line inversion and its correlation with filaments elongation in solar images. Barcelona, Spain: Proceedings of the SMMSP-2003, pp. 115–121

Zharkova, V.V., Ipson, S.S., Zharkov, S.I., Benkhalil, A., Aboudarham, J., & Bentley, R.D. (2003b). A full disk image standardization of the synoptic solar observations at the Meudon Observatory. *Solar Phys., 214*(1), 89

Zharkova, V.V., & Schetinin, V. (2003). *The recognition of filaments in solar images with an artificial neural network.* Bruges, Belgium: Proceedings of the ESANN-2003

Zharkova, V.V., & Schetinin, V. (2003, 2005). The recognition of filaments in solar images with an artificial neural network. *Solar Phys., 228*(1–2), 363–377

5 Feature Recognition and Classification Using Spectral Methods

K. Revathy

5.1 Feature Recognition and Classification with Wavelet Transform

5.1.1 Introduction

Astronomical image processing refers to a large variety of numerical methods to extract scientific information from the observation of celestial objects. Most of the astronomical images are crowded with point sources like stars, quasi point-like objects like faint diffuse galaxies and clusters, nebulae, etc. Digital imaging and image processing form the core of modern astronomy. Much of what we know about the structure and behavior of planets, the Sun, stars, interstellar clouds, galaxies, and the universe as a whole is learned from the processing and analysis of astronomical images. Usually it is not the image, but the objects or structures in images that are of interest. Astronomical images are large in size and are susceptible to noise from different sources. The image intensity which is often of the order of 10^4 and the dynamic range of most of the detectors is at least 16 bits. The computational techniques used in astronomical image processing are applied in a sequence that starts with the acquisition of the image formed in the focal plane of the telescope. The subsequent steps involve the removal of noise and artifacts followed by the extraction of scientific information using processing techniques like enhancement, analysis, and classification. The quality and nature of processing depends mainly on the applications which may vary from a simple display of a galaxy image to most complex analysis leading to some insight into spatial structures. Image processing in astrophysics involves recording and subsequent processing of raw images as well as enhancement, classification, and structural analysis of these images. This necessitates the requirement and development of specialized image processing techniques exclusively for astronomical image processing.

K. Revathy: *Feature Recognition and Classification Using Spectral Methods,* Studies in Computational Intelligence (SCI) **46**, 339–374 (2007)
www.springerlink.com

Wavelets have been found to be extremely useful in analyzing astronomical images. Over the last decade wavelets have had a growing impact on signal processing theory and practice, both because of their unifying role and their success in applications. The subject of wavelets is expanding at such a tremendous rate that it is difficult to cover here all aspects of its theory or applications. Wavelets have emerged in the last fifteen years as a synthesis of ideas from fields as different as electrical engineering, statistics, physics, computer science, mathematics, and astrophysics. Currently, the most popular image transform used in many applications is the wavelet transform. We shall introduce the wavelet transform and deal with image processing algorithms in the transform domain and finally discuss the wavelet based techniques such as image visualization, filtering, image fusion, compression, and zooming applied on a set of astronomical images. Wavelets are functions that satisfy certain mathematical requirements and are used in representing data or other functions. Wavelet transformation is established as a powerful tool for the analysis and synthesis of signals (Chui 1992; Grape 1995; Coifman et al. 1992; Chan 1995; Daubechies 1992; Kaiser 1994; Meyer and Ryan 1993; Mallat 1989). Localization of signal characteristics in spatial (or temporal) and frequency domains can be accomplished very efficiently with wavelets. The real strength of wavelet transform representation is that functions (signals or images) that look like the wavelet function at any scale may be well represented by only a few of the wavelet basis functions. The wavelet transform therefore provides an efficient representation for functions which have similar characteristic to the functions of the wavelet basis. Wavelets are becoming a popular image processing and data compression tool because they are base functions of finite length and can represent sharp edges by a small number of coefficients. Because of their multiresolution nature, wavelet coding schemes are suitable for applications where scalability and tolerable degradation are important (Antonini et al. 1992; Mandal et al. 1994). The basic structure of a wavelet transform is composed of recursive filtering and decimation, both of which are relatively easy to implement. The kernel separability property in wavelet transform theory is an interesting feature that renders multidimensional wavelet transform as a powerful tool for multidimensional signal processing.

5.1.2 Introduction to Wavelets

In any image processing system, we first perform measurements to obtain information about a physical quantity and analyze this information to build some knowledge about the system. Fourier transform (FT) is an important tool for the analysis of signals but it is localized in time/space. This makes it unsuitable for analyzing nonstationary signals. For nonstationary signals such as images, a windowed Fourier Transform (STFT – short time Fourier transform) can be used. Gabor (1946) introduced a local Fourier analysis, taking into account a sliding Gaussian window. Such approaches provide tools for investigating time as well as frequency. In wavelet analysis, a fully scalable modulated window is shifted along the signal and for every position, the spectrum is evaluated. The result will be a collection of time frequency representation of the signal, all with different resolutions. In wavelet transform terminology, instead of time–frequency, we use time–scale representation. Large time scales correspond to low frequencies, and short time scales correspond to high frequencies. Very common tool in signal analysis is the spectrogram defined as the square modulus of the STFT. It provides a distribution of the energy of the signal in the time–frequency plane. A similar distribution can be defined in the wavelet case. One can define the wavelet spectrogram or scalogram as the squared modulus of the wavelet transform. In contrast to spectrogram, the energy of the signal is distributed with different resolutions. The formulation of the scalogram makes it a more convenient tool than the Fourier transform because relationships between the different time scales become easier to visualize.

Features of Wavelets

Wavelets are localized in both the space/time and scale/frequency domains. Hence they can easily detect local features in a signal. They are capable of providing time and frequency information simultaneously, hence giving a time–frequency representation of signals. Wavelets are based on a multiresolution analysis (MRA). Their decomposition allows one to analyze a signal at different resolution levels (scales). Wavelets are smooth, which can be characterized by their number of vanishing moments. The higher the number of vanishing moments, the more smooth signals can be approximated in a wavelet basis. Fast and stable algorithms (e.g., Pyramid algorithms) exist to calculate wavelet transform and its inverse transform. Different wavelet basis functions offer very good decorrelation of data and hence sparse representation. A wavelet transform has a computational complexity of $O(n)$, since they are compact. In a wavelet transform, higher frequencies are better resolved in time, and lower frequencies are better resolved in frequency. This means that, a certain high-frequency component can be located better in time than a low-frequency component. On the contrary, a low-frequency component can be located better in frequency than a high-frequency component. A wavelet transform has an infinite set of possible basis functions. Thus wavelet analysis provides immediate access to information that can be obscured by methods such as Fourier analysis. There has been active research in applying wavelets to various aspects of image processing like enhancement, segmentation, analysis, classification, retrieval, compression, etc. The wavelet transform decomposes a signal into frequency bands and this particular way of dividing matches the statistical properties of most images very well. Extensive literature exists on the wavelet transforms (Chui 1992; Daubechies 1992; Meyer 1991) and their application (Nayar et al. 2002; Gallagher et al. 1998; Blanco et al. 1999; Bocchialini et al. 1995; Christopoulou et al. 2003). The wavelet transform provides a powerful and versatile framework for astronomical image processing. The 2D continuous wavelet transform (CWT) and the discrete wavelet transform (DWT) are useful for the analysis and feature detection in images.

Continuous Wavelet Transform

The CWT is given by (Starck and Murtagh 2001):

$$\mathrm{CWT}_x^\psi = \Psi_x^\psi\left(\tau,s\right) = \frac{1}{\sqrt{|s|}} \int x(t)\psi^*\left(\frac{t-\tau}{s}\right) dt \qquad (5.1)$$

where * indicates complex conjugate. The CWT maps a one-dimensional function of t to a function of two variables, τ the translation parameter and s, the scale parameter. $\Psi(t)$ is the transformation function and it is called the mother wavelet (Daubechies 1992). The term wavelet means a small wave. The smallness refers to finite length. The term mother implies that the functions with different region of support that are used in the transformation process are derived from one main function, or the mother wavelet. The scaling parameter s either dilates (s > 1) or compresses (s < 1) a signal. The CWT can be thought of as the inner product of the test signal x(t) with the basis functions $\Psi_{t,s}(t)$:

$$\mathrm{CWT}\, T_x^\psi\left(\tau,s\right) = \Psi_x^\psi\left(\tau,s\right) = \int x(t).\Psi_{\tau,s}^* dt \qquad (5.2)$$

Where

$$\Psi_{\tau,s}(t) = \frac{1}{\sqrt{s}}\, \Psi\left(\frac{t-\tau}{s}\right) \qquad (5.3)$$

This definition of the CWT shows that the wavelet analysis is a measure of similarity between the basis functions (wavelets) and the signal itself. The calculated CWT coefficients refer to the closeness of the signal to the wavelet at the current scale.

Discrete Wavelet Transform

The wavelet series expansion for any function f(x) e $L^2(R)$ can be written as

$$f(x) = \sum\nolimits_{j=-\infty}^{\infty} \sum\nolimits_{k=-\infty}^{\infty} C_{j,k} \, \Psi_{j,k}(x) \qquad (5.4)$$

where the transform coefficients are given by the inner products:

$$C_{j,k} = <f(x) \, \Psi_{j,k}(x)> = 2^{1/2} \int\limits_{-\infty}^{\infty} f(x)\Psi(2^j x - k)dx \qquad (5.5)$$

The wavelet series is simply a sampled version of the CWT over a dyadic grid defined by $s = 2^j$ and $t = 2^j k$, where j and k are integers. The information it provides is highly redundant as far as the reconstruction of the signal is concerned. This redundancy, on the other hand, requires a significant amount of computation time and resources. The DWT, provides sufficient information both for analysis and synthesis of the original signal, with a significant reduction in the computation time. The DWT analyzes the signal at different frequency bands with different resolutions by decomposing the signal into a coarse approximation and detail information. It employs two sets of functions, called scaling functions and wavelet functions, which are associated with low-pass and high-pass filters, respectively. The decomposition of the signal into different frequency bands is simply obtained by successive high pass and low pass filtering of the time domain signal. The frequencies that are most prominent in the original signal will appear as high amplitudes in that region of the DWT signal with these particular frequencies. The frequency bands that are not very prominent in the original signal will have very low amplitudes, and that part of the DWT signal can be discarded without any major loss of information allowing data reduction. This is how DWT provides a very effective data reduction scheme. The DWT is conveniently formulated as a recursion. Each recursion comprises a filtering, matrix multiplication and a vector permutation (Daubechies 1988; Mallat 1999).

5.1.3   Multiresolution Analysis

Many of the developments preceding wavelet analysis came in a field generally called MRA. A DWT approach can be obtained from multiresolution analysis (Mallat 1989). MRA analyzes the signal at different frequencies with different resolutions. It is designed to give good time resolution and poor frequency resolution at high frequencies and good frequency resolution and poor time resolution at low frequencies. Signals that are encountered in practical applications, often contain high-frequency components for short duration and low-frequency components for long durations and hence suitable for MRA.

The CWT can be computed by changing the scale of the analysis window, shifting the window in time, multiplying by the signal, and integrating over all times. In the discrete case, filters of different cut-off frequencies are used to analyze the signal at different scales. The signal can then be passed through a series of high-pass filters to analyze the high frequencies, and through a series of low-pass filters to analyze the low frequencies. A compact way of describing MRA is via Mallat's definition (Mallat 1989; Starck and Murtagh 1998; Blackledge and Turner 2000). Figure 5.1 shows an example of MRA using wavelets.

5.1.4   Wavelets in Astronomical Image Processing

Wavelet transform is useful for both deterministic and statistical kinds of information measurement. The deterministic kind indicates what is in the image and the locations of its contents. Whereas, the statistical kind gives the amount of information. Application of wavelet transform in astronomical

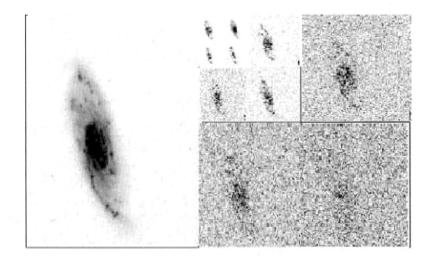

Fig. 5.1. Multiresolution analysis of galaxy M51 using wavelets

image Processing includes visualization, image fusion, noise filtering, multiresolution support (MRS), object detection, compression, and zooming. The time frequency localization property gives rise to new algorithm for edge detection and noise reduction in images. The ability of WT to sort information into unrelated levels of structures at different scales has been used for coding and progressive transmission. The CWT was first applied for the analysis of galaxy clusters (Slezak et al. 1990, 1993). Describing structures as a function of their characteristic scale is a precise property of the wavelet transform. This property has been used to define a multiscale vision model to detect features in images (Bijaoui and Ruea 1995; Bijaoui et al. 1996). The CWT has been used in the detailed analysis of individual galaxies (Frick et al. 2001) and in the analysis of X-ray structure of various objects like clusters of galaxies. Again cosmic background explorer (COBE) data on the cosmic microwave background radiation was analyzed using wavelet methods (Sanz et al. 1999). The extreme ultraviolet telescope (EIT) onboard the solar heliospheric observatory (SOHO) satellite which observe the Sun in four wavelengths could give a large number of images by which astronomers can study different events appearing at different altitude locations and scales. Several studies in this direction already exist, e.g., discriminating cosmic ray hits and the bright points in the solar corona (Antoine et al. 2002), detecting the active region etc. Well suited algorithms for automatically analyzing and studying these images are required.

The Á Trous Algorithm

In addition to the discrete transform resulting from the MRA, there exists a large range of discrete transform: Morlet transform, the pyramidal transform and á trous algorithm. The á trous algorithm is a popular form of wavelet transform (Starck et al. 1998). This is a "stationary" or redundant transform and is popular in astrophysical image processing. In astronomical images, an isotropic or symmetric analysis always produces better results (Starck and Murtagh 2001). For most applications like detection, filtering, deconvolution, etc. Á trous algorithm is widely used while for compression DWT is often preferred and Morlet transform is preferred for time frequency analysis (Nayar et al. 2002). The results obtained in areas like visualization, object detection, filtering, compression, and zooming are considerably better than traditional approaches and are briefly discussed in in "Visualization," "Image Fusion," "Image Filtering," "Multiresolution Support," "Multi Scale Vision Model," "Wavelet Transform and Image Compression," and "Image Zooming".

Fig. 5.2. Decomposed image (three levels) and residue image (*bottom right*) of NGC 2403 using á trous algorithm

Visualization

Visualization using wavelet transform helps to have better knowledge of the structure of the image. MRA using DWT has been implemented on a galaxy image (Raju 2002) and one can visualize that under-sampling reduces the size to one-fourth of previous level. Whereas, transformationl using á trous algorithm leaves the size of the image unaltered. The image of NGC 2403 is decomposed using á trous algorithm into three levels; the three transformed images and the residue image are depicted in Fig. 5.2.

Image Fusion

The wavelet transform splits the information on the scale axis so that data fusion may be done scale by scale. It is common to observe the same astronomical objects at different wavelengths and at different epochs. The comparison of these observations requires adapted image fusion tools. Probably one of the simplest image fusion methods is to average the set of images pixel by pixel. Weighted addition of pixel values is another simple alternative. Often such pixel based methods lead to a feature contrast reduction (Gross et al. 2000; Berg 1999; Rockinger and Benz 1996). The wavelet function has a null mean and the background variations are automatically removed (Li et al. 1995; Bijaoui and Giudicelli 1991). Larger absolute values of wavelet transform coefficients correspond to sharp changes in brightness.

The fusion rule at edges, lines and region boundaries can be a selection of the larger absolute value of the two coefficients at the same point. A more complicated fusion rule can be considered, based on the fact that the human visual system detects features at points where the local energy reaches its maximum. The local energy is the sum of the squared responses of a quadrature pair of odd-symmetric and even-symmetric filters. The most commonly used fusion methods are the pixel-based fusion rule, window-based fusion rule, and region-based fusion rule. To perform pixel level fusion successfully, all input images must be exactly spatially registered, i.e., the pixel positions of all input frames must correspond to the same location in real world. The fusion process should preserve all relevant information of the input imagery in the composite image, while suppressing irrelevant image parts and noise in the fusion result. The fusion scheme should not introduce artifacts or inconsistencies which would distract the human observer or following processing stages. To illustrate the image fusion, two images of NGC 3938, obtained with two different filters (J and R) are taken. The fused images corresponding to four different methods of fusion are given in Fig. 5.3. The four methods are (a) two images are fused by taking the average of pixel intensities. (b) The two images are decomposed using DWT with DAUB4 filter to four levels. To get a single set of wavelet coefficients, the smooth 16×16 subimages corresponding to two wavelet coefficient sets are averaged and for the rest of the sub-bands, highest wavelet coefficient (absolute) for a given position are chosen. The new set of coefficients are then inverse transformed. (c) Same as (b), but the coefficient set is formed by taking the average of the two coefficient sets. (d) Á trouse algorithm is used for decomposition in to four levels. For the first three levels, new coefficient set is obtained by taking the highest absolute coefficient for a given position from the two sets of coefficients and for the residue set, the average of the two residue images is taken.

Image Filtering

Filtering is one of the most important tools in image processing. Filtering is often used for image enhancement, noise-removal, image restoration and image analysis. In the pixel domain, a large variety of linear and statistical filters are available. Filtering is also performed in transform domains such as FFT and Wavelet. In order to enhance fine structures, low-pass and high-pass filters are often applied. For detecting and enhancing structures in galaxies, some simple techniques can be used like blurring the original image by convolution with a kernel (median, modal, Gaussian or rectangular)

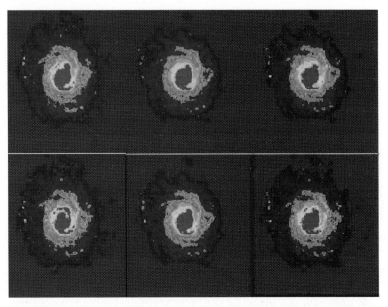

Fig. 5.3. Image Fusion using Wavelets (NGC 3938). Original images (*top left* and *middle*) and others are fused images obtained by methods (**a–d**) described in text

and further subtraction of this new image from the original one to find features otherwise hidden in the overall signal of the galaxy (Abans and Abans 1998). Wavelets are used for filtering galaxy images, with a purpose to bring out more details (Raju 2002).

Unsharp masking can enhance the contrast of smaller features in an image relative to the larger features. Unsharp-masking combines an image with a blurred version of itself to increase detail. Blurring is usually done by smoothing the given image with an n × n smoothing box (n depends on size of features of interest). A similar technique can be implemented in the wavelet domain also. For example, in the á trous algorithm, the residue image is a blurred version of the original image. The amount of smoothness depends on the number of levels of decomposition. The reconstruction of a transformed image without the residue part is equivalent to unsharpmasking. The results of applying this to NGC 3938, first with four levels of decomposition and later with six levels of decomposition, are shown in Fig. 5.4. High-pass and low-pass filters can be easily implemented in the wavelet domain, simply by reconstructing the image without the lowest or highest sub-band. Low-pass filtering removes fine details from the image, whereas high-pass filtering emphasizes the fine details. Using á trous algorithm, four level decomposition high pass and low pass filtered NGC 3938 image is given in Fig. 5.5.

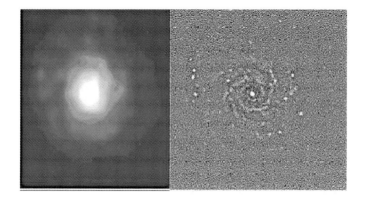

Fig. 5.4. Unsharp Masking Using Wavelet Transform (Image : N3938) for Four Level Decomposition (*left*) and for Six Level Decomposition (*right*)

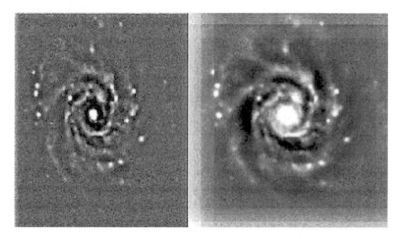

Fig. 5.5. Low-pass (*left*) and High-pass (*right*) filtered image NGC 3938, using a Wavelet transform

Astronomical images, like many other classes of images, contain noise. In general, astronomical images are contaminated by Gaussian noise, Poisson noise or a combination of the two. Even when the noise does not follow a known distribution, we can assume it to be locally Gaussian. Noise reduction can be easily implemented in the wavelet domain since the wavelet coefficients of a noisy image are also noisy (Berg 1999; Starck et al. 1995a; 1998; Starck and Murtagh 1998, 2001). Applying a hard thresholding to each wavelet scale and taking the inverse transform, the noise-free image could be obtained as shown in Fig. 5.6.

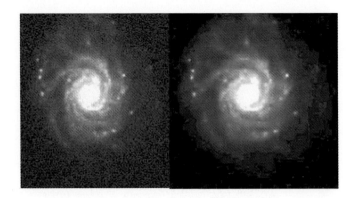

Fig. 5.6. NGC 3938 added with known noise (*left*) and after denoising with wavelet thresholding (*right*)

Multiresolution Support

The MRS of an image describes in a logical or boolean way if an image I contains information at a given scale j and at a given position (x, y) (Starck and Murtagh 1998; Starck et al. 1995b, 1998; Murtagh et al. 1995). If M(I)(j, x, y) = 1, then I contains information at scale j and at the position (x, y). M depends on several parameters like (a) input image (b) algorithm used for the multiresolution decomposition (c) noise and other additional constraints. In the wavelet domain, MRS can be defined as follows: M (j, x, y) = 1, if the wavelet coefficient w j (x, y) is significant; M(j, x, y) = 0, otherwise. The algorithm to create MRS is given by (a) computing the wavelet transform of the image, (b) estimating the noise standard deviation and hence deduce the statistically significant level at each scale, (c) the binarization of each scale, and (d) modification using a priori knowledge. Step (d) is optional. A typical use of a priori knowledge is the suppression of isolated pixels in the MRS in the case where the image is obtained with a point spread function (PSF) of more than one pixel. In order to visualize MRS, an image S can be defined as

$$S\,(x, y) = \sum_{j=1}^{p} 2^{j} M(j, x, y)$$

Visualization of MRS of NGC 3938 is given in Fig. 5.7. MRS is used in image filtering, image restoration and in image visualization.

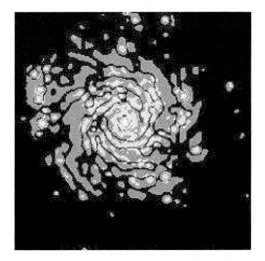

Fig. 5.7. NGC 3938 Multiresolution support

Multiscale Vision Model

A multiscale vision model was developed to detect and characterize all the sources of different sizes in an astronomical image (Bijaoui 1997). In an image, an object occupies a physically connected region and pixel of this region can be linked to the others. The connectivity in the direct space can be extended to wavelet transform space (WTS) also. All structures form a 3D connected set which is hierarchically organized, the structures at a given scale are linked to smaller structures of the previous scale. This set gives the description of an object in the WTS. After applying the wavelet transform to the image, a thresholding in the WTS is performed in order to identify the statistically significant pixels. These pixels are regrouped by a scale-by-scale segmentation procedure in order to define the object structures. Then, an inter-scale connectivity graph is established. The object identification procedure extracts each connected subgraph that corresponds to 3D connected sets of pixels in the WTS. By referring to the object definition, the subgraph can be associated with the objects. From each set of pixels an image of the object can be reconstructed using this reconstruction algorithm.

Wavelet Transform and Image Compression

An image can contain complex features at several different scales and hence representing it at an adequate resolution often requires a very large matrix. Representation, storage and transmission of images are thus costly in terms of computation, memory and time. A feature commonly found in images is the high correlation between the values of neighboring pixels, often resulting in high redundancy of information content in the images. The principal task of an efficient coding scheme for images is, therefore, elimination of this redundancy leading to significant data compression. Over the past several years, the wavelet transform has gained widespread acceptance in image compression. In many applications wavelet-based compression schemes, especially for high compression ratios, outperform other coding schemes, both in terms of signal-to-noise ratio and visual quality. At high compression levels, wavelets allow for a graceful degradation of the whole image quality, while preserving the important details of the image. The possibility to combine image processing and compression is a very appealing factor. The multiresolution aspect of wavelet transform makes it most suitable for transmitting compressed images since a progressive reconstruction is possible at the receiving station.

Availability of a large number of orthogonal and biorthogonal families of wavelets provides a good base to choose from. Region of interest (ROI) coding is easy to implement in wavelet domain. A number of advanced coding techniques such as Lifting (Calderbank et al. 1998), EZW (Shapiro 1993) and SPIHT (Said and Perlman 1996) were proposed over the past few years with an objective to improve compression performance. The compression standard JPEG 2000 is based on wavelets (Adams and Ward 2001). Performance of wavelet compression depends on various parameters like the type of wavelets used, the quality of picture, the compression requirements etc. A large number of research papers has come out on the application of wavelet transform as a tool for image compression. Based on Daubechies wavelets Revathy et al. (2000) have investigated the performance of wavelet based image compression of astronomical images and analyzed the different parameters describing the characteristics of different test images and their relation to the compression based on different types of Daub wavelets. The performance of wavelet compression was studied based on objective parameters mean normalized error (MNE) and root-mean-square error (RMS) evaluated from the reconstructed images. It was found that the activity level of images measured with spectral flatness measure (SFM) which is defined as the ration of the arithmetic and geometric mean of the Fourier coefficients, plays an important role in deciding the filter to be used and optimum compression level (Revathy et al. 1998, 2000). Figure 5.8 shows original image (M51) and image reconstructed using only 5% of the transformed wavelet coefficients.

Image Zooming

Image over-sampling/subsampling is an important image processing activity, often required in many applications. Over-sampling (zooming) facilitates image interpretation and better visualization of images. Subsampling is useful in applications such as fitting of large images into small display area and creating icons. There exist a number of techniques for image zooming (over/subsampling). Zooming is basically an interpolation operation, for which spline interpolation methods are most widely accepted.

The simplest over-sampling is the nearest-neighbor interpolation (NNI), which duplicates the original pixel values. In practice, the most frequently used over-sampling is the linear interpolation. This interpolator is based on a local hypothesis of luminance signal continuity and calculates, by averaging a value at a subpixel position. Such lower order methods are simplest and most rapid to implement but they produce artifacts. NNI is extremely cheap computationally, but typically introduces very noticeable block-like artifacts. Bilinear interpolation, which calculates each new pixel value from its four closest neighbors, is somewhat better but tends to blur small image details. Higher order interpolation methods – splines, in particular – produce much better outcome but require more computation. Wavelet based image zooming was studied in detail with different filters like Daub, Coiflet and biorthogonal by Revathy et al. (2000) and it was shown that wavelet methods gives a smoother image compared to classical methods. The bi-orthogonal filter has been found to be the best choice for image

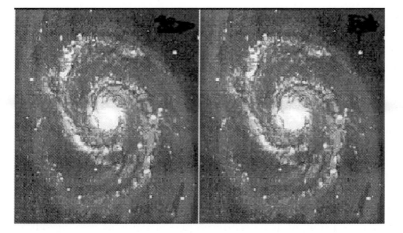

Fig. 5.8. Image of galaxy M51 (*left*) and image reconstructed using 5% of the transformed wavelet coefficients (*right*)

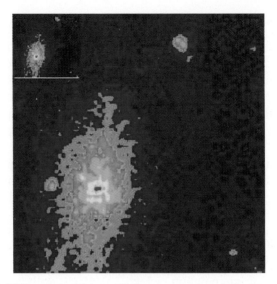

Fig. 5.9. NGC 2541: original image (*left top corner*) and the image zoomed with BIOR wavelet transform

zooming. The zooming of a portion of galaxy NGC 2541 (Fig. 5.9) reveals that wavelet zooming can be a good choice for zooming astronomical images for the purpose of visual examination.

5.2 Feature Recognition and Classification with Fractal Analysis

5.2.1 Introduction

The term "fractal" was coined by Mandelbrot (1977, 1983) to describe the irregular structure of many natural objects and phenomena. Fractal geometry is a new language used to describe, model and analyze complex forms found in nature. Over the past few years fractal geometry has been used as a language in theoretical, numerical, and experimental investigations. It provides a set of abstract forms that can be used to represent a wide range of irregular objects. Fractal objects contain structures that are nested within one another. Each smaller structure is a miniature form of the entire structure. The use of fractals as a descriptive tool is diffusing into various scientific fields, from astronomy to biology. Fractal concepts can be used not only to describe the irregular structures but also to study the dynamic properties of these structures. In simple terms, a fractal is an object made of parts which are self-similar. The notion of self-similarity is the basic property of fractal objects. Self-similarity means that the pattern of the whole shape is similar to the pattern of an arbitrary small part of the shape. In principle, a theoretical or mathematically generated fractal is self-similar over an infinite range of scales, while natural fractal images have a limited range of self-similarity. Fractal objects share certain characteristics which distinguish them from traditional objects defined by Euclidean geometry. They have the property of self-similarity or statistical self-similarity. The main idea of fractal geometry with the concept of self-similarity, is that an object appears to look similar at different scales. This concept has only relatively recently started to be developed mathematically and applied to various branches of science and engineering. This can be applied to systems of varying physical size, depending on the complexity and diversity of the fractal model considered. The universe itself can be included as a fractal, the self-similar parts of which have yet to be fully classified.

Simple rules can often describe the structures of many natural objects in a way that conventional techniques cannot. A fractal description may be a deterministic fractal such as the Von Koch snow flake. The Von Koch's curve introduced by the Swedish mathematician, Helge Von Koch, in early 1900s can be generated by a step-by-step procedure that takes a simple initial figure and turns it into an increasingly more ragged form. The first six stages in constructing a Koch snowflake are given in Fig. 5.10 (Stevens 1995). Deterministic fractals show repeated structures at all scales and each subsequent magnification reveals even more fine structures. The Von Koch curve illustrates the features of the whole class of deterministic fractals.

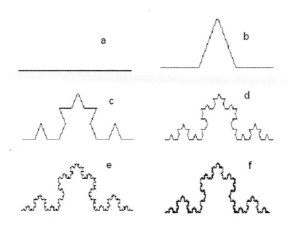

Fig. 5.10. Representation of Von Koch curve in varying scales

5.2.2 Fractal Dimension

Natural fractals do not in general possess deterministic self similarity over all scales. The scale dependence of fractal objects provides the basis for the definition of dimensionality. Fractal geometry characterizes the way in which a quantitative data set grows in mass, ith linear size. The fractal dimension (D) is a measure of nonlinear growth, which reflects the degree of irregularity over multiple scales. It is very often a noninteger and is the basic measure of fractals. The scale dependence of fractal objects provides the basis for the definition of dimensionality. The fractal dimension provides an objective means and a powerful tool for analyzing and extracting various features of images. It reflects the degree of irregularity or roughness of any surface. The fractal dimension can be defined in connection with real world data and can be measured. The curves, surfaces and volumes are complex objects for which ordinary measurements become meaningless. Different techniques have been proposed to measure the degree of complexity, by evaluating how fast the length, surface, or volume increases with respect to smaller and smaller scales. The fundamental idea is to assume that the two quantities like length or surface and the scale, which do not vary arbitrarily but rather are related by a law of the form of $y \; \alpha \; x^d$, which allows to compute one quantity from the other. Based on the self-similarity of geometric forms, one can find the power-law describing the number of pieces a versus $1/s$, where s is the scale factor which characterizes the parts as copies of the whole. The exponent d of this law is the fractal dimension.

For a D-dimensional object, the number of identical parts, N divided by a scale ratio, r can be calculated from $N = 1/r^D$. However natural objects show only statistical self-similarity. A mathematical fractal has an infinite amount of details. This means that magnifying it adds additional details, thereby increasing the overall size. In nonfractals, the size remains the same in spite of the applied magnification. The graph of log (fractal size) against log (magnification factor) gives a straight line. If the object is nonfractal, then this line is horizontal since the size does not change. If the object is fractal, the line is no longer horizontal since the size increases with magnification. The geometric method of calculating the fractal dimension is by computing the slope of the above plotted line. The self-similarity dimension D can be obtained as (Luo 1998):

$$D = \log (N) / \log (1/r) \qquad (5.6)$$

There are many advantages of using the fractal dimension over other image features. The fractal dimension is insensitive to different image transformations such as changes in local intensity, zooming, and local scaling of gray level values. Fractal dimension is also insensitive to multiplicative noise (Kasparis et al. 2001). Surfaces with different scales of roughness may have same the fractal dimension. This can be due to the combined differences in coarseness and directionality. To discriminate such surfaces, a set of fractal dimensions are proposed by Chaudhuri and Sarkar (1995). They are the fractal dimension computed on overlapping windows, fractal dimensions of high and low gray-valued images, fractal dimensions of horizontally and vertically smoothed images etc. It is appropriate and necessary to study the nature of fractal dimension for such modified images. The variation of the fractal dimension with (a) change in resolution, (b) change in brightness, (c) change in contrast, (d) edge detected image, (e) high and low gray-valued images, and (f) filtering have been studied by Lekshmi (2003).

Computation of Fractal Dimension

The different methods for calculating fractal dimension are walking-divider method, Box counting method, Prism counting method, Epsilon–Blanket method, Perimeter–Area Relationship, Fractional Brownian Motion, Power spectrum method, and Hybrid method (Blackledge 1989; Falconer 1989; Feder 1998; Turner et al. 1998). Box Counting: is the most popular algorithm for computing the fractal dimension of one-dimensional and two-dimensional data. The box counting method was originally developed by Voss (1985). In this method, the fractal surface is covered with a grid of n-dimensional boxes or hyper-cubes with side length δ and counting the number of boxes that contains a part of the fractal N (δ). For signals, the grid consists of squares and for images, the grid consists of cubes. The fractal surface is covered with boxes of recursively different sizes. An input signal with N elements or an image of size N × N is used as input where N is a power of 2. The slope β is obtained from the logarithmic plot of the number of boxes used to cover the fractal against the box size and the fractal dimension DB is given by Dβ = - β. This dimension is also known as the box or Minkowski dimension. For a smooth one-dimensional curve, it is expected that N(δ) ≈ (L/δ) where L is the length of the curve, N(δ) is the number of nonempty boxes and δ is the step size. The generalization to the box counting measure is given as:

$$N(\delta) \; \alpha \; \frac{1}{\delta^{D\beta}}$$

(5.7)

$$D_\beta = \frac{\lim}{\delta \to 0} - \frac{\log N(\delta)}{\log(\delta)} \tag{5.8}$$

Epsilon–Blanket: In this method the fractal dimension of curves/surfaces are computed using the area/volume measured at different scales (Peleg et al. 1984). The concept is based on the analysis of multiscale construction. For example, for a surface, the upper and lower boundary surfaces are to be generated. The blanket is defined by a band of thickness 2ε and the multiscale analysis is done for different values of ε. The estimation of the fractal dimension involves taking the logarithm of the difference of the upper and lower bounding surfaces divided by the scale factor ε and fitting a line to it in a log–log plot as a function of scale. In the case of curves, the set of points whose distance lies with in ε is considered. This gives a strip of width 2ε that surrounds the curve. The length of the curve $L(\varepsilon)$ is calculated from the strip area $A(\varepsilon)$ by

$$L\left(\varepsilon\right) = \frac{A(\varepsilon)}{2\varepsilon} \tag{5.9}$$

The fractal dimension, D can be computed using the relation

$$L\left(\varepsilon\right) \alpha \, \varepsilon^{(1-D)} \tag{5.10}$$

In the case of surfaces, the set of points in three-dimensional space which is not more than ε from the surface, gives a blanket of volume $V(\varepsilon)$ whose width is again 2ε. The surface area is then given by

$$A\left(\varepsilon\right) = \frac{V(\varepsilon)}{2\varepsilon} \tag{5.11}$$

and D can be computed as

$$A\left(\varepsilon\right) = \varepsilon^{2-D} \tag{5.12}$$

5.2.3 Fractal Signature

It has been found that the fractal dimension alone is not sufficient to quantify the structures in images. Other fractal structures have been proposed. For any measured property M which is a function of scale ε, the fractal dimension can be computed by Richardson's equation,

$$M(\varepsilon) = c\, \varepsilon^{D_\tau - D} \qquad\qquad (5.13)$$

where c is a constant and D_τ is the topological dimension. Using the fact that $M(1) = c$, we can calculate D for $\delta = 2, 3$, etc. A plot of D vs. δ gives the fractal signature.

5.2.4 Concept of Multifractals

To the degree that the fractal dimension is a statistical measure of the whole image, it represents a measure of its global complexity and hence it quantifies the fractal properties of the image as a whole. But there can be local differences in complexity and hence the fractal dimension can vary as a function of location within an image (Turner et al. 1998). A structure can be considered as a mixture of different fractals, each one with a different box-counting dimension. In such a case, the conglomerate will have a dimension that is the dimension of the largest component. In such situations, it is necessary to define a spectrum of numbers which gives information about the distribution of fractal dimensions in a structure. This is the underlying concept in multifractals (Petgen et al. 1991; Vehel and Mignot 1994; O'Neil 1997).

5.2.5 Fractals in Astronomical Image Processing

Fractal is a mathematical set with a high degree of complexity which can model many classes of images. Its applications are getting more and more numerous in many domains. Since Mandelbrot (1977, 1983) proposed the concept of fractal, the subject of fractal geometry has drawn a great deal of attention from scientists from various areas of research. As fractal geometry is able to describe the world around us, it would be an ideal tool for image processing. The applications of fractals can be divided into two groups. One is to analyze data sets with completely irregular structures and the other is to generate data by recursively duplicating certain patterns. In image processing, fractals have been used for image compression, texture analysis (Pentland 1984), image encoding, pattern recognition (Chaudhuri and Sarkar 1995; Baldoni et al. 1998; Tang et al. 2002), modeling of objects, representation and classification of images (Luo 1998; Falconer 1998; Sato et al. 1996; Li et al. 1997; Cochran et al. 1996; Lee and Lee 1998). A novel approach to feature extraction based on fractal theory was studied by Tang et al. (2002). Fractals and multifractals are applied to model hierarchical inhomogeneous structures in several areas of astrophysics (Stern 1997). Fractal techniques are applied in the investigation of the distribution of galaxies and voids (Gaite and Manrubia 2002), structure of galaxies Milovanov (1994), giant molecular clouds and nebulae (Tarakanov 2004; Datta 2003) and classification of galaxies (Lekshmi 2003; Revathy and Lekshmi 2003). In solar images, fractal analysis has been applied to study the characteristics of spatial distribution of magnetic fields on the solar surface (Tao et al. 1995; Lawrence et al. 1993), sunspots, active regions (Revathy et al. 2005) and boundaries of UV emission structures (Gallagher et al. 1998).

Fractals can be used to model the underlying process in a variety of applications. Astronomy, in particular cosmology, was one of the fields where fractals have been found and applied to study various phenomena. The largest subsystem studied by means of fractals is the distribution of galaxies. The map of the distribution of the galaxies obeys a power-law which is the characteristic behavior expected in hierarchical fractals (Perdang 1990; Heck and Perdang 1991; Elmegreen and Elmegreen 2001). Fractals have proved to be successful in the detection of edge points. The boundaries between homogeneous regions do not fit well into a fractal model. The boundaries give rise to nonfractal intensity surface and this provides an efficient method of detecting the edge points from an image. Fractals are used in other image processing tasks like classification of textures (Turner et al. 1998), determination of shape from texture, image compression and analysis (Lekshmi et al. 2003). Fractal coding is one of the promising image coding techniques. It is based on the work of Jacquin (1993), fractals and iterated function systems (IFS) to describe them. Here, blocks of an image are considered as affine transformations of the other blocks taken from the image itself.

Image Analysis

Image analysis has an important role in many applications ranging from medical imaging to astronomy. The purpose of image analysis is to extract symbolic information from an image. Many of the image processing techniques which are valuable as visualization tools are useful at preprocessing stages in the image analysis task. It is different from other image processing applications like restoration, enhancement and coding where the output is another image. The image is processed in such a way that it removes all unwanted information retaining only that part of the image which contributes significantly to the analysis task. The analysis task creates a mapping from a digital image to the description of image content. The image description can either be a number which would represent the number of objects in the image, or it can be a degree of anomaly which would define the shape variation of an object or it can be labeling of pixels to classify different regions of the image. There are three levels in image analysis. In the low level processing, functions that may be viewed as automatic reactions are dealt with, whereas intermediate level processing deals with the task of extracting regions in an image that results from a low level process. Image analysis involves the study of feature extraction, segmentation, and classification (Jain 1997).

Segmentation: Euclidean Techniques

Segmentation is the process of identifying regions of pixels in an image. The main goal is to divide an image into parts that have a strong correlation with objects or areas in the real world. There are three classes of segmentation procedures; global knowledge, edge based and region based. Global knowledge methods rely on knowing something about the image such as the expected pixel intensity. Edge-based methods find the borders between regions. Region based methods use different properties of the image to define regions such as gray level, color, texture, shape, etc. Over the past thirty years or so, classification and segmentation has been studied using human and vision perspectives. The three decades have witnessed a slow but steady evolution of techniques ranging from co-occurrence matrices to Markov random fields. The main aim here is to recognize homogeneous regions within an image as distinct and belonging to different objects. The segmentation process can be based on finding the maximum homogeneity in gray levels within the regions identified. Issues related to segmentation involve choosing good segmentation algorithms, measuring their performance, and understanding their impact on the scene analysis system.

A segmented image consists of two regions namely homogeneous region and transition region. The most difficult task in a vision system is to identify the subimages that represent objects. The partitioning of images into subimages is what is done in segmentation. In other words, segmentation is grouping of pixels into regions (Ri) such that adjacent regions possess different characteristics.

$$\bigcup_{i=1}^{k} R_i = \text{Entire Image} \qquad (5.14)$$

$$R_i \cap R_i = \Phi \quad i \neq j \qquad (5.15)$$

There are different techniques for finding object regions in gray-level images. They are histogram thresholding, edge detection, tree/graph based approach, region growing, clustering. In histogram thresholding, the picture is thresholded at its most clearly separated peak. The process iterates for each segmented part of the image until no separate peaks are found in any of the histograms. The criteria to separate peaks was based on the ratio of peak maximum to peak minimum to be greater than or equal to two.

In edge based segmentation, pixel neighborhood elements are used for image segmentation. For each pixel, its neighbors are first identified in a window of fixed size. A vector of these neighbors as individual gray values or vector of average gray levels in windows of size 1×1, 3×3 and 5×5 is determined. Then a weight matrix is defined which when multiplied with these vectors will yield a discriminant value that allows the classification of pixel in one of the several classes (Cheriet et al. 1998). In tree/graph based approaches segmentation is derived from the consensus of a set of different segmentation outputs on one input image. Instead of statistics characterizing the spatial structure of the local neighborhood of a pixel, for every pair of adjacent pixels their collected statistics are used for determining local homogeneity. Region growing algorithms take one or more pixels, called seeds, and grow the regions around them based upon a certain homogeneity criteria. If the adjoining pixels are similar to the seed, they are merged with them within a single region. The process continues until all the pixels in the image are assigned to one or more regions. Cluster analysis allows the partitioning of data into meaningful subgroups and it can be applied for image segmentation or classification purposes. Clustering either requires the user to provide the seeds for the regions to be segmented or it uses nonparametric methods for finding the salient regions without the need for seed points (Jain and Dubes 1988). In a number of segmentation techniques, such as fuzzy c-means clustering, the number of clusters present in the image have to be specified in advance. An automatic segmentation and classification method for outdoor images using neural networks was proposed and images were segmented using color and texture information of the object by Campbell et al. (1997). The images are segmented using self-organizing feature maps (SOFM) based on texture and color information of the objects.

Segmentation: Fractal Techniques

Different techniques are available for performing image analysis, of which fractal methods are promising in certain applications. Fractal techniques have been extensively used to characterize data in the fields of medicine, atmospheric physics, geophysics, astrophysics, etc. Kasparis et al. (2001) describe a new approach to the segmentation of textured gray-scale images using fractal features. The technique used is the variation method followed by filtering of the image to generate the feature set. Dimensionality reduction is applied to reduce the feature set and clustering is performed to obtain the segmented regions. Chaudhuri and Sarkar (1995) have defined a technique to compute fractal dimension of images known as differential box counting method. This technique has proved to be successful in classifying textures by means of unsupervised k-means clustering. Klonowski (2000) describes how fractal models may be used for image segmentation, texture classification and estimation of roughness from image data. Barbara and Chen (2000) designed a fractal clustering algorithm that places points incrementally in the cluster, for which the change in fractal dimension after adding the point, is the least. Vuduc (1997) introduced a local fractal operator with his algorithm, which is a modified form of Sarkar method to find the fractal dimension of a single pixel. Fractal techniques are widely used in solar physics to identify fractal clusters in sunspot penumbrae and solar active regions, to study the ratio of facular area to sunspot area (Chapman et al. 2001), to identify faculae on continuum images (Watson and Preminger 2000; Revathy et al. 2004) and to obtain features for modeling of the total solar irradiance (Oritz et al. 2002).

Computation of Pixelwise Fractal Dimension

Image segmentation using fractal is done in two steps. First by computing the pixelwise fractal dimension and then followed classification. There exist different methods of computing pixelwise fractal dimension (Kasparis et al. 2001; Vuduc 1997). They are simple thresholding, Sarkar method, Beavor method, Variation method, and the modified variation method. These methods are described briefly in the following section. Simple thresholding is a direct implementation of the box counting method applied locally to each pixel. For a pixel (i, j), a small window, W of size m × m surrounding it is considered. The fractal dimension corresponding to the pixel is the slope obtained by fitting a line of log N(r) versus log r where N(r) is the number of nonempty boxes and r is the size of the box. In the Sarkar method the fractal dimension corresponding to the pixel is the slope obtained by fitting a line of log N (r) versus log(r). N(r) is the summation of n(r) for various grid sizes, r that surrounds the pixel and n(r) is the difference between the bin numbers corresponding to the maximum and minimum intensity values present in grid size, r. The Beavor method is essentially the same as that of the Sarkar method except that the intensities of the grid are rescaled locally before proceeding with the box counting method. The fractal dimension corresponding to the pixel is the slope obtained by fitting a line of log N (r) versus log r. N(r) is the summation of n(r) for various grid sizes, r that surrounds the pixel. n(r) is the difference between the bin numbers corresponding to the maximum and minimum intensity values present in grid size, r.

In the variation method, the fractal dimension corresponding to the pixel is the slope obtained by fitting a line of log (R/ε) versus log(R/ε)3Eε, for ε = 1, 2, 3.. εmax and R is the size of the window. Eε is the average of over the window R, and Vε is the difference between the maximum and minimum intensity values computed over a window of size T that surrounds the pixel, where T = 2ε + 1 for ε = 1, 2, 3.. εmax. Another method is a modified form of the variation method (Lekshmi et al. 2003). Here, the fractal dimension corresponding to the pixel is the slope obtained by fitting a line of log (R/ε) versus log(R/ε)3Eε, for ε = 1, 2, 3... εmax and R is the size of the window. Eε is the average of Vε over the window R and Vε is the average of the intensity values computed over a window of size T (<R) that surrounds the pixel, where T is same as above, for ε = 1, 2, 3.. εmax. This method proved to be better in segmenting certain images. Once pixelwise fractal dimensions are found out, the image intensities are replaced by these values. Then one of the segment a segmentation methods namely (a) iterative thresholding, (b) region growing, and (c) histogram based thresholding can be applied. A portion of the chromospheric image from BBSO was studied using all the methods discussed above. The results are given in Fig. 5.11. In Fig. 5.11 the original image and images obtained by applying three segmentation techniques on the FD map corresponding to three algorithms are depicted. It can be seen that the modified variation method with histogram based thresholding gives the best result in segmenting the image.

The automatic identification of solar features, such as faculae and plages, is becoming increasingly important as the resolution and the size of solar data sets increases. The introduction of space-borne solar telescopes, in addition to the ground based observations has increased the solar image data set many fold. The identification of solar features is required for the quantitative studies of the solar activity, which includes locations, lifetimes, contrasts, and other characteristics of sunspots, faculae etc. An active region of the Sun was investigated by using modified variation method followed by fuzzy-based segmentation. In this approach, depending on the grade of the membership function, the image can be segmented to different

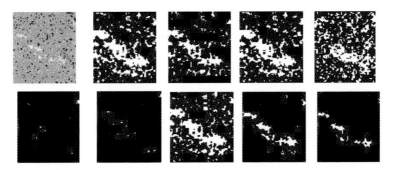

Fig. 5.11. *Top left* original image, *top second, third, fourth*: images obtained by applying techniques (a), (b), and (c) on Sarkar algorithm, *top rightmost* and *bottom first* and *second*: images obtained by applying techniques (a), (b), and (c) on variation method, *bottom third, fourth* and *fifth*: images obtained by applying techniques (a), (b), and (c) on modified variation method algorithm

regions. Solar images taken at different wavelengths by the SOHO satellite, subjected to the fuzzy-based segmentation approach (Revathy et al. 2004), is given in Fig. 5.12. From these results the segmented areas and their time evolution can be studied.

Fractals in Image Compression

Image compression is required for preview functionality in large image data bases such as the Hubble space telescope (HST) archive with interactive sky atlases, linking image and catalogue information and for image transmission. The basic principle in fractal theory is that an image can be reconstructed by using self-similarities in the image itself. Digital images possess the characteristics of being both data extensive and relatively redundant objects which make them an obvious target for compression. The redundancy of the digital images is assumed to be exploitable through self-transformability on a block-wise basis, i.e., the image can be built of transformed copies of parts of itself. Basically, a fractal code consists of three ingredients: a partitioning of the image region into portions called ranges, an equal number of other image regions called domains, and for each domain-range pair two transformations, a geometric transformation which maps domain into range, and an affine transformation, that adjusts the intensity value in the domain to those in the range. Nonlinear transformations could also be used as long as they are contractive. The term contractive means that any two points on the transformed image are closer together than they are in the original image. In fractal compression the objective is to find approximations to the generation of a self-similar object, when given a digital image of the object. Examples of few galaxy images and the reconstructed images are given in Fig. 5.13. The compression ratio is about 90%. Fractals can be important in progressive transmission of images, especially in satellite imagery. Decompression speed, resolution independence and the compression ratio distinguish the fractal compression from other compression methods.

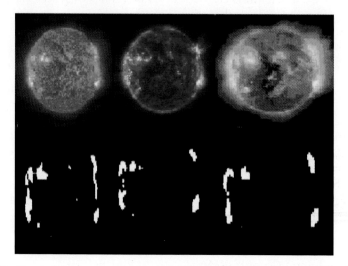

Fig. 5.12. Fuzzy-based segmentation of SOHO EIT images taken at 304 Å, 171 Å and 284 Å

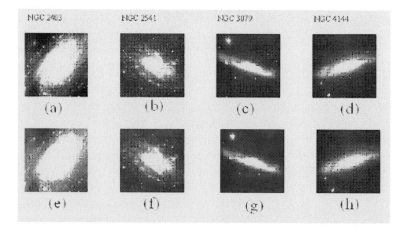

Fig. 5.13. Image compression original (*top*) and reconstructed (*bottom*) images

Fractal Techniques in Classification

Classification is necessary for a complete understanding of objects and in the design of a pattern recognizer. By classification, we mean putting together objects that possess common features. Usually, discriminant features are used for the classification purpose. Classifier design consists of establishing a mathematical basis for classification procedure. The classification of galaxies has been the subject of studies for a number of years. A morphological classification scheme of galaxies was introduced by Hubble in early 1900s. The morphological type describes the global appearance of a galaxy. Also, it provides information about the evolution of galaxies, its structure and stellar content. In the conventional approach, galaxies were classified by visual inspection (Burda and Feitzinger 1992). The drawbacks of this scheme are that they are slow, ambiguous and could not be applied to distant galaxies.

The multivariate nature of galaxies makes classification of galaxies difficult and less accurate. A highly automated classification system is required which can separate galaxy properties better than the conventional systems. Different classification procedures have been introduced in the past few decades (Odewahn 1995; Naim et al. 1995; Molinari and Smareglia 1998; Zaritsky 1995). The techniques mainly employed are either statistical or neural network based. In the statistical approach, the decision boundaries are determined by the probability distribution of the patterns belonging to each class. A more recent method is based on neural networks. Here the most commonly used type of network is the feed-forward network which includes multilayer perceptron and radial basis function. Several modifications are available for this algorithm, of which the most recent is the one developed by Philip et al. (2002), known as difference-boosting neural network for star–galaxy classification. If the features are distributed in a hierarchical manner, a syntactic approach can be made. Different types of parameters are defined for the classification purpose. They are galaxy axial ratio, surface brightness, radial velocity, flocculence, etc. (Burda and Feitzinger 1992). Classifications like morphological, structural, or spectral are done based on these parameters. A morphological classification of galaxies was investigated by Naim et al. (1995) using artificial neural networks with the parameters like galaxy axial ratio, surface brightness, etc. Spectral classification has been found to correlate well with morphological classification. In spectral classification, the galaxies are classified by comparing the relative strengths of the components with those of galaxies of known morphological type (Philip et al. 2002). Andreon (1997) has classified galaxies structurally using their structural components. Classification of galaxies can also be done based on luminosity functions (Han 1995). Molinari and

Smareglia (1998) have used Kohonen Self Organizing Map to classify galaxies based on luminosity functions.

The two main steps that precede classification of an image are feature extraction and feature selection. Feature extraction involves deciding which properties of the object (size, shape, etc) best distinguish among the various classes of objects and this should be computed. The feature extraction phase identifies the inherent characteristics of an image. By using these characteristics, the subsequent task of classification is done. Features are based on either the external properties of an object (boundary) or on its internal properties.

Feature selection refers to algorithms that select the best subset of the input feature set. Algorithms which create new features based on either transformations or combinations of the original feature set are called feature extraction algorithms. However the terms feature extraction and feature selection are used interchangeably in the literature. Often feature extraction precedes feature selection. There are many fractal features that can be generated from an image (Chaudhuri and Sarkar 1995). The most commonly used the fractal feature is the fractal dimension. In some applications, fractal dimension alone is capable of discriminating one object from one another. Revathy and Lekshmi (2003) have used fuzzy concepts and neural network architecture for classifying galaxies using fractal dimension alone. A fuzzy membership function which follows the model of a parabola was constructed and the grades of membership function for both classes (elliptical and spiral) were obtained. Applying a threshold to the grades of membership function a given galaxy could be classified as elliptical or spiral. The classification rate was 74%. But in certain applications, the fractal dimension alone may not be sufficient in identifying the desired object. Here, a set of fractal features obtained on simple image transformations can be used as the feature set. The performance could be improved by taking another fractal feature named fractal signature. The Fractal signature was valuated using epsilon blanket method. It assigns a unique signature to each sample. Samples belonging to each class have similar signatures which makes them distinguish from one another easily. The classification of galaxies have been a subject of study for a number of years. Fractal signature of nearby galaxies has been studied by Lekshmi et al. (2003) with the aim of classification. The fractal signature over a range of scales proved to be an efficient feature set with good discriminating power. Some of the results are given in Fig. 5.14 which shows the variation of a fractal signature with surface area. It could be observed that the fractal signature of spirals and ellipticals vary between the groups, though they are similar within the group. The classification rate using a fractal signature and a neural net has been found to be 95% (Lekshmi et al. 2003).

5.2.6 Conclusion

The notion of wavelets and fractals and their use in various image processing algorithms are briefly described. The fractal based methods together
with wavelet methods can lead to better and efficient algorithms. The
modified fractal dimension methods based on wavelet analysis can be the
future direction for the image analysis, coding as well as classification.
We expect hybrid schemes incorporating wavelets and fractals to achieve
greater success. Theoretical work inspired by wavelets has led to new
techniques which are more promising. Explorations on these frontiers are
likely to bring us more successful applications in astrophysical image
processing.

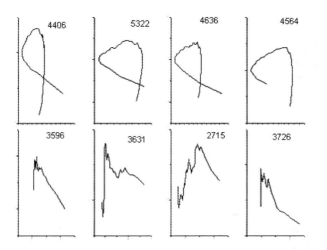

Fig. 5.14. Fractal signature S(ε) with surface area A(ε) for elliptical(top) and
spiral (*bottom*) galaxies

References

Abans, F., & Abans, O. (1998). Looking for fine structures in galaxies. *Astron. Astrophys. Supp. Ser. 128,* 289–297

Adams, M.D., & Ward, R. (2001). *Wavelet transforms in the JPEG-2000 standard.* mdadams@ieee.org

Andreon, S., & Davoust, E. (1997). Structural classification of galaxies in clusters.
Astron. Astrophys., 319, 747–756

Antoine, J.P., Demanet, L., Hochedez, J-F., Jaques, L., Terrier, R., & Verwichte,
E. (2002). Application of 2D wavelet transform to astrophysical images. *Phys.
Mag., 24,* 93–116

Antonini, A., Barland, M., Mathier, P., & Daebechies, I. (1992). Image coding using wavelet transform. *IEEE Trans. Image Process., 1,* 205–220

Baldoni, M., Baroglio, C., Cavagnino, D., & Saitta, L. (1998). Towards automatic fractal feature extraction for image recognition. In H. Liu, & H. Motada (Eds.), *Feature extraction, construction and selection,* Dordritcht: Kluwer

Barbara, D., & Chen, P. (2000). *Using the fractal dimension to cluster data sets.* Proceedings of the sixth international conference on knowledge discovery and data mining (KDD-2000), 260–264

van den Berg, J.C. (1999). *Wavelets in physics.* Cambridge: Cambridge University Press

Bijaoui, A., & Giudicelli, M. (1991). Optimal image addition using wavelet transform. *Exp. Astron., 1,* 347–363

Bijaoui, A., & Ruea, F. (1995). A multi scale vision model adapted to astronomical images. *Signal Process., 46,* 345–362

Bijaoui, A., Slezak, E., Rue, F., & Lega, E. (1996). Wavelet and the study of distant universe. *Proc. IEEE, 84,* 670–679

Bijaoui, A. (1997). *Vision modeling in astronomy: Recent results.* CCMA Conference: ESF Network

Blackledge, J.M., & Turner, M.J. (2000). *Image processing II, mathematical methods, algorithms and applications.* Chichester, UK: Ellis Horwood

Blackledge, J.M. (1989). *Quantitative coherent imaging: Theory methods and some applications.* London: Academic

Blanco, S., Bocchialine, K., Costa, A., Domenech, G., Rovira, M., & Vial, J.C. (1999). Multi-resolution wavelet analysis of SUMER-SOHO observations in a solar prominence. *Solar Phys., 186,* 281–290

Bocchialini, K., & Baudin, F. (1995). Wavelet analysis of chromospheric solar oscillations. *Astron. Astrophys., 299,* 893–896

Burda, P., & Feitzinger, J.V (1992). Galaxy classification using pattern recognition methods. *Astron. Astrophys., 261,* 697–705

Calderbank, R.C., Daubechies, I., Sweldens, W., & Yeo, B.L. (1998). Wavelet transform that map integers to integers. *Appl. Comput. Harmon. Anal., 5,* 332–369

Campbell, N.W., Thomas, B.T., & Troscianko, T. (1997). Automatic segmentation and classification of outdoor images using neural networks. *Int. J. Neur. Syst., 8,* 137–144

Chan, Y.T. (1995). *Wavelet basics.* Dordrecht: Kluwer

Chapman, G.A., Cookson, A.M., Dobias, J.J., & Walton, S.R. (2001). Improved determination of the area ratio of faculae to sunspots. *Astrophys. J., 555,* 462–465

Chaudhuri, B.B., Sarkar N. (1995). Texture segmentation using fractal dimension. *IEEE Trans. PAMI, 17,* 72–76

Cheriet, M., Said, J.N., & Suen, C.Y. (1998). A recursive thresholding technique for image segmentation. *IEEE Trans. Image Process., 7,* 918–920

Christopoulou, E.B., Skodras, A., Georgakilas, A.A., & Koutchmy, S. (2003). Wavelet analysis of umbral oscillations. *Astrophys. J., 591,* 416–431

Chui, C.K. (1992). *Wavelets: A tutorial in theory and applications*. San Diego, CA: Academic

Cochran, W.O., Hart, J.C., & Flynn, P.J. (1996). Fractak volume compression. *IEEE Trans. Vis. Comput. Graph., 2,* 3113–3332

Coifman, R.R., Meyer, Y., & Wickerhauser, M.V. (1992). *Wavelet analysis and signal processing, wavelets and their applications,* Boston, MA: Jones and Bartlett, pp. 153–178

Datta, S. (2003). Fractal structure of the Horsehead nebula (B 33). *Astron. Astrophys, 401,* 193–196

Daubechies, I. (1988). Orthonormal bases for compactly supported wavelets. *Commun. Pure Appl. Math., 41,* 909–996

Daubechies, I. (1992). *Ten lectures on wavelet.* Philadelphia: SIAM

Elmegreen, B.G., & Elmegreen, D.M. (2001). Fractal structure in galactic star fields. *Astron. J., 121,* 1507–1511

Falconer, K. (1992). Fractal geometry, mathematical foundations and applications. New York: Wiley

Falconer, K.J. (1989). Dimensions and measures of quasi-similar sets. *Proc. Am. Mathl. Soc., 106,* 543–554

Feder, J. (1998). *Fractals.* New York: Plenum

Frick, P., Beck, R., Berkhuijsen, E.M., & Patrickeyev, I. (2001). Scaling and correlation analysis of galactic images, *MNRAS, 327,* 1145–1157

Gabor, D., (1946). Theory of communications. *J. Inst. Elect. Eng. (London), 93,* 429–457

Gaite, J., Manrubia, S. (2002). Scaling of voids and fractality in the galaxy distribution. *Mon. Not. R. Astron. Soc., 335,* 977–983

Gallagher, P.T., Phillips, K.J.H., Harra-Murinion, L.K., & Keenan, F.P. (1998). Properties of quiet Sun EUV network. *Astron. Astrophys., 335,* 733–745

Grape, A. (1995). Introduction to wavelets. *IEEE Comput. Sci. Eng., 2,* 2

Gross, D., Liu, Z., Tsukada, K., & Hanasaki, K. (2000). Experimenting with pixel-level NDT data fusion techniques. *IEEE Trans. Inst. Measure., 49,* 5

Han, M. (1995). Luminosity structure and objective classification of galaxies. *Astrophys. J., 442,* 504–522

Heck, A., & Perdang, J.M. (1991). Applying fractals in astronomy, Berlin Heidelberg New York: Springer

Jacquin, A.E. (1993). Fractal image coding – a review. *Proc. IEEE, 81,* 1451–1465

Jain, A.K. (1997). Fundamentals of digital image processing. Englewood Cliffs, NJ: Prentice Hall

Jain, A.K, & Dubes, R.C. (1988). Algorithms for clustering data. Englewood Cliffs, NJ: Prentice Hall

Kaiser, G. (1994). *Friendly guide to wavelets.* Boston: Birkhaeuser

Kasparis, T., Charalampidis, D., Georgiopoulos, M., & Rolland, J. (2001). Segmentation of textured images based on fractals and image filtering. *Pattern Recognit., 34,* 1963–1973

Klonowski, W. (2000). Signal and image analysis using chaos theory and fractal geometry. *Mach. Graph. Vis., 9*(1/2), 403–431

Lawrence, J.K., Ruzmaikin, A.A., & Cadavid, A.C. (1993). Multifractal measure of the solar magnetic field. *Astrophys. J., 417,* 805

Lee, C.K., & Lee, W.K. (1998). Fast fractal image block coding based on local variances. *IEEE Trans. Image Process., 7,* 888–891

Lekshmi, S., Revathy, K., & Prabhakaran Nayar, S.R. (2003). Galaxy classification using fractal signature. *Astron. Astrophys., 405,* 1163–1167

Lekshmi, S. (2003). *Fractal techniques in image compression, analysis and classification of images and remote sensed data,* Ph.D. Thesis, Kerala University

Li, H., Liu, K.J.R., & Lo, S.C.B. (1997). Fractal modeling and segmentation for the enhancement of micro calcifications in digital mammograms. *IEEE Trans. Med. Imaging, 16,* 785–798

Li, H., Manjunath, B.S., & Mitra, S.K. (1995). Multi sensor image fusion using the wavelet transform, *Graph. Models Image Process., 57,* 235–245

Luo, D. (1998). *Pattern recognition and image processing,* Chichester, UK: Horwood

Mallat, G. (1999). *A wavelet tour of signal processing,* 2nd edn., London: Academic

Mallat, S.G. (1989). A theory for multiresolution signal decomposition: The wavelet representation. *IEEE Trans. Pattern Anal. Mach. Intell., 11,* 674–693

Mandal, M.K., Panchanathan, S., & Aboulnasr, T. (1994). *Wavelets for image compression.* Proceedings of the IEEE international symposium on time-frequency and time-scale analysis

Mandelbrot, B.B. (1977). *Fractals – form, chance and dimension.* San Francisco, CA: Freeman

Mandelbrot, B.B. (1983). *The fractal geometry of nature,* Oxford: Freeman

Meyer, Y., & Ryan, R.D. (1993). *Wavelets, algorithms and applications,* Philadelphia, PA: SIAM

Meyer, Y. (1991). *Wavelets and applications.* Springer: Berlin Heidelberg New York

Milovanov, A.V. (1994). Large-scale structure of clusters of galaxies and spontaneous polymerization on fractal geometry. *Astron. Rep., 38,* 314–320

Molinari, E., & Smareglia, R. (1998). Neural network method for galaxy classification. *Astron. Astrophys., 330,* 447–452

Murtagh F., Starck J.L., & Bijoui A. (1995). Image restoration with noise suppression using a multiresolution support. *Astron. Astrophys., 112,* 179–189

Naim, A., Lahav, O., Sodre, L., Jr., & Storrie-Lombardi, M.C. (1995). Automated morphological classification of APM galaxies by supervised artificial neural networks. *MNRAS, 275,* 567–590

Nayar, S.R.P., Radhika, V.N., Ramadas, P., & Revathy, K. (2002). Wavelet analysis of solar and solar wind parameters. *Solar Phys., 212,* 207–211

O'Neil, T.C. (1997). The multi-fractal spectrum of quasi self-similar measures. *J.Mathl. Anal. Appl., 211,* 233–257

Odewahn, S.C. (1995). Automated classification of astronomical images. *Pub. Astron. Soc. Pacific, 107,* 770–773

Oritz, A., Solanki, S.K., Domingo, V., Fligge, M., & Snahuja, B. (2002). On the intensity contrast of solar photospheric faculae and network elements. *Astron. Astrophys., 388,* 1036–1047

Peleg, S., Naor, J., Hartley, R., & Avnir, D. (1984). *Multiple resolution texture analysis and classification,* IEEE Transactions on PAMI, PAMI-6, No. 4, July 1984

Pentland, A.P. (1984). Fractal based description of natural scenes. *IEEE Trans. PAMI, 6,* 661–674

Perdang, J. (1990). Astrophysical fractals: An overview and prospects. *Vis. Astron., 33,* 249–294

Petgen, H., Jurgens, H., Saupe, D. (1991). *Fractals for the class room - Part I, Introduction to fractals and chaos.* Berlin Heidelberg New York: Springer

Philip, N.S., Wadadekar, Y., Kembhavi, A., & Joseph, K.B. (2002). A difference boosting neural network for automated star-galaxy classification. *Astron. Astrophys.,* 1119–1126

Press, W.H,, Teukolsky, S.A., Vellering, W.T., & Flannery, B.P. (1996). *Numerical recipes.* Cambridge: Cambridge University Press

Raju, G. (2002). *Image processing in astrophysics.* Ph.D. Thesis, University of Kerala

Revathy, K., Raju, G., & Prabhakaran Nayar, S.R. (1998). *Wavelet based image compression and zooming, IT for the new generation.* New Delhi: Tata McGraw Hill, pp. 191–196

Revathy K., Raju, G., & Prabhakaran Nayar, S.R. (2000). Image zooming by wavelets. *Fractals, 8,* 247–253

Revathy, K., & Lekshmi, S. (2003). Neuro-fuzzy classifier for astronomical images. *Fractals, 11,* 289–294

Revathy, K., Lekshmi, S., & Prabhakaran Nayar, S.R. (2004). *Fractal based fuzzy technique for detection of active regions from solar images.* Proceedings of the SIPWorkII

Revathy, K., Lekshmi, S., & Prabhakaran Nayar, S.R. (2005). Fractal based fuzzy technique for detection of active regions from solar images. *Solar Phys., 228,* 43–53

Rockinger, O., & Benz, A.G. (1996). *Pixel level fusion of image sequence using wavelet frames,* Proceedings of the 16th Leeds applied shape research workshop, Leeds University Press

Said, A., & Perlman, W.A. (1996). A new, fast and efficient image code based on set partitioning in hierarchical trees. *IEEE Trans. CSVT, 6,* 243–250

Sanz, J.L., Cayon, L., Matinez-Gonzalez, E., Ruiz, G.A., Diaz, F.J., Argueso, F., Silk, J., & Toffolati, L. (1999). Analysis of CMB maps with 2D wavelets. *Astron. Astrophys. Suppl. Ser., 140,* 99–105

Sato, T., Matsuoka, M., & Takayasu, H. (1996). Fractal image analysis of natural scenes and medical images. *Fractals, 4,* 463–468

Shapiro J.M. (1993). Embedded image coding using zero trees of wavelet coefficients. *IEEE Trans. Signal Process., 41*(12), 3445–3462

Slezak, E., Bijaoui, A., & Mars, G. (1990). Identification of structures from galaxy counts, use of the wavelet transforms. *Astron. Astrophys., 227,* 301–316

Slezak, E., De Laparent, V., & Bijaoui, A. (1993). Objective detection of voids and high density structures in the first CfA redshift survey slice. *Astrophys. J., 409,* 517–529

Starck, J.L., & Murtagh, F. (1998). Automatic noise estimation from the multiresolution support. *Pub. Astro. Soc. Pacific, 110,* 193–199

Starck, J.L., & Murtagh, F. (2001). Astronomical image and signal processing. *IEEE Signal Process., 18,* 30–40

Starck, J.L., Murtagh, F., & Bijaoui, A. (1995a). Multi resolution support applied to image filtering and deconvolution. *Graph. Model. Image Process., 57,* 420–431

Starck, J.L., Murtagh, F., & Bijaoui, A. (1995b). *Multiresolution and astronomical image processing.* Astronomical data analysis software and systems IV, ASP Conference Series, Vol. 77

Starck, J.L., Murtagh, F., & Bijaoui, A. (1998). *Image processing and data analysis: The multi scale approach.* Cambridge: Cambridge University Press

Stern, I. (1997). On fractal modeling in astrophysics: The effect of lacunarity on the convergence of algorithms for scaling exponents. In G. Hunt, & H.E. Payne (Eds.), *Astronomical data analysis software and systems VI*, ASP Conference Series, Vol. 125. Astronomical Society of the Pacific

Stevens, R.T. (1995). *Understanding self similar fractals.* New York: Prentice Hall

Tang, Y.Y., Tao, Y., & Lam, E.C.M. (2002). New method for feature extraction based on fractal behaviour. *Pattern Recognit., 35,* 1071–1081

Tao, L., Du, Y., Rosner, R., & Cattaneo, F. (1995). On the spatial distribution of magnetic fields on the solar surface. *Astrophys. J., 443,* 434–443

Tarakanov, P.A. (2004). Formation of a fractal structure of giant molecular clouds in the galaxy. *Astrophys., 47,* 343–351

Turner, M.J., Blackedge, J.M., & Andrews, P.R. (1998). *Fractal geometry in digital imaging.* London: Academic

Vehel, J., & Mignot, P. (1994). Multifractal segmentation of images. *Fractals 2*(3), 187–205

Voss, R. (1985). *Random fractals: Characterisation and measurement, scaling phenomena in disordered systems,* In R. Pynn, & A. Skjeltorps (Eds.). New York: Plenum

Vuduc, R. (1997). *Image segmentation using fractal dimension GEOL 634.* New York: Cornell University

Watson, S.R., & Preminger, D.G. (2000). *Solar feature identification using contrasts and contiguity.* American Astronomical Society, SPD Meeting-32-01.24

Zaritsky, D. (1995). Spectral classification of galaxies along the hubble sequence. *Astron. J., 110,* 1602–1612

Printing: Krips bv, Meppel
Binding: Stürtz, Würzburg